T0221443

EVOLUTIONARY DEVELOPMENTAL BIOLOGY OF CRUSTACEA

CRUSTACEAN ISSUES

15

General editor
RONALD VONK
*Institute for Biodiversity and Ecosystem Dynamics,
University of Amsterdam*

EVOLUTIONARY DEVELOPMENTAL BIOLOGY OF CRUSTACEA

Edited by
GERHARD SCHOLTZ
Institut für Biologie/Vergleichende Zoologie, Humboldt-Universität zu Berlin, Germany

A.A. BALKEMA PUBLISHERS / LISSE / ABINGDON / EXTON (PA) / TOKYO

Library of Congress Cataloging-in-Publication Data

A Catalogue record for this book is available from the Library of Congress

Printed in The Netherlands by Krips, Meppel

Published by: A.A. Balkema Publishers, a member of Swets & Zeitlinger Publishers
www.balkema.nl and www.szp.swets.nl

ISSN 0168-6356
ISBN 90 5809 637 8

Why should there not be a natural law connecting a starting and a finishing state of a system but not covering the intermediary state? (Only one must not think of causal efficacy)

Ludwig Wittgenstein

Contents

III *Morphology and phylogeny*

Preface

The idea for this book dates back to the last century, more precisely, to the year 1993. At that time I spent a sabbatical at the University of New South Wales studying some aspects of the embryonic development of the Australian freshwater crayfish *Cherax destructor* in the laboratory of David Sandeman. Analyzing cell lineage and gene expression patterns in the crayfish embryos and inspired by the fact that Sydney was the place where Donald T. Anderson worked and wrote his highly influential book on the "Embryology and Phylogeny in Annelids and Arthropods" I thought it was time to review the more recent literature on crustacean development because Anderson's book was at that time already 20 years old. Therefore, I contacted Frederick Schram who was then the general editor of the *Crustacean Issues*, and he agreed that a volume on embryonic development might be a good idea. For several reasons I was not able to complete the project for about a decade, which, in the end, turned out to be highly advantageous. In the meantime, the world of crustacean embryology and development had dramatically changed. The rise of the molecular approach towards the development of non-model organisms led to numerous important new insights and perspectives of crustacean development and the development of arthropods in general. Several long-standing morphological questions such as head-segmentation and tagmatization can now be interpreted against a new factual background. This period has also seen the first cell lineage studies of crustacean early cleavage using injected fluorescent dyes as lineage markers providing new insights into cell fate, germ layer formation, and morphogenesis. On the other hand, views on the phylogenetic relationships of major arthropod groups have changed. Crustacea are now almost universally believed to be the sister group of insects rather than myriapods, the traditional insect sister group, and the polyphyly of arthropods is now a completely outdated concept in the face of overwhelming evidence for arthropod monophyly. Among other developmental features, several aspects of neurogenesis proved to be key characters for this fresh approach to arthropod phylogeny.

I think this recent development of crustacean phylogeny, evolution, and development in the context of arthropods is well reflected in this present volume of *Crustacean Issues*. Moreover, I hope that together all the chapters show that the new developments have to be associated with more traditional approaches in order to make a complete story. Since arthropod development and evolution are such expansive disciplines with numerous papers appearing every month, a book like this is by its very nature outdated from the time of publication. Nevertheless, I think that the reviews compiled in this volume touch so many general aspects of crustacean development, morphology, and evolution that this *Crustacean Issue* will be a standard reference for many years to come.

In an introductory chapter, I discuss the implications of the typological *Bauplan* and *phylum* concepts versus historical concepts such as *ground pattern* and *monophylum* for the formulation of conceptual questions in evolutionary developmental biology. The importance of a phylogenetic systematic approach to comparative developmental biology is stressed. The following contributions to the book are arranged into three main sections comprising themes that are important in crustacean evo-devo. The first section "Genes and body organization" includes papers discussing the results of *Hox* gene expression in various crustacean taxa and their implications for our views on tagmatization in Crustacea and arthropods in general. The section starts with the chapter of Deutsch, Mouchel-Vielh, Quéinnec & Gibert, which deals with the highly derived body organization of cirripedes and, in particular, the parasitic rhizocephalans from the perspective of gene expression patterns mainly of the *Hox* complex. The authors address the problem of whether cirripedes possess an abdomen and they provide evidence for a vestigial larval abdomen. A similar approach can be found in the contribution of Abzhanov & Kaufman. These authors describe the expression patterns of the *Hox* gene complex in two malacostracan crustaceans. They are able to show that there is a similar pattern in a crayfish and a woodlouse with distinct differences that can be correlated with the morphological differences between these two animals. They furthermore discuss the convergent refinement of *Hox* gene expression between insects and malacostracans, which in both cases led to a functional but convergent subdivision of the trunk. The paper of Schram & Koenemann discusses the data on *Hox* gene expression in the various crustaceans in relation to tagmatization. They define the various terms used for describing the subdivision of crustacean trunks, in particular, the abdomen. Based on their conclusions the authors develop a model for the evolution of tagmatization and segmentation patters and discuss the possibility that crustaceans are not monophyletic.

The section on "Cells and segments" comprises three articles dealing with aspects of segment formation at the cellular and genetic levels and the formation of segmental structures such as neurons, ganglia, and limbs. Dohle, Gerberding, Hejnol & Scholtz address the segmentation complex of crustaceans. In their paper they describe the stereotypic cell division pattern during germ band formation and segmentation of various malacostracans. These patterns are unique amongst the arthropods and allow a cell-by-cell interpretation of segmental gene expression and morphogenesis of segmental structures such as limbs, ganglia etc. The comparison with non-malacostracans reveals profound differences and clearly indicates malacostracan monophyly. Whitington presents a detailed overview of our present knowledge of crustacean neurogenesis. In contrast to the situation in insects, little is known about the genetic basis of crustacean neurogenesis. However, the shared similarities of crustaceans and insects found in the patterns of pioneer neurons and the neuroblasts form an important point in the discussion of arthropod phylogenetic relationships. The paper of Williams addresses the problem of the great diversity of limbs in crustaceans. Is there any similar basal differentiation pattern in the highly diverse types of limbs found in crustaceans? She discusses approaches to homologizing limb parts across crustaceans and other arthropods based on shared ontogenetic patterns. The chapter ends with a general discussion of the relationships between gene expression and homologization of morphological structures.

The two articles of the third section "Morphology and phylogeny" beautifully demonstrate that morphological ontogenetic characters are very powerful in resolving phylogenetic relationships. Høeg, Lagersson & Glenner use characters of the cypris larva, in particular the so-called lattice organs, in a comprehensive way to establish a cladogram of thecostracan crustaceans which clarifies in many aspects the view of cirripede evolution, including the highly derived parasitic forms, by confirming the monophyly of the Cirripedia. In his very comprehensive and detailed study on the larval development of branchiopod crustaceans including fossil taxa, Olesen pursues a similar approach. He describes numerous larval characters with high information content for the phylogeny of Branchiopoda. The results corroborate in many instances the phylogeny of this interesting crustacean group gained from molecular data sets. Surprisingly, he is able to show quite convincingly that a fossil group might not be as closely associated with branchiopods as was previously thought.

Last, but not least, I want to thank all the people involved in making the book. First of all I thank the authors for their valuable contributions. The reviewers did a great job in making this book a fully referenced, scientific source. I am especially grateful to the general editor of the *Crustacean Issues* series, Ronald Vonk, for his help and the continuous interest in the progress of the book. Special thanks and deep gratitude must be extended to Stefan Koenemann and Frederick Schram who did all the copy editing, type setting, CRC preparation, color layout, and spelling and grammar reconciliations. This book would simply not have appeared on time without their work. The publisher's staff, in particular Beatrice Huisman, showed great patience with me as editor and always extended the deadline for final submission of the manuscripts.

Gerhard Scholtz, Berlin, 2003

INTRODUCTION

Baupläne *versus* ground patterns, phyla *versus* monophyla: aspects of patterns and processes in evolutionary developmental biology

GERHARD SCHOLTZ

Humboldt-Universität zu Berlin, Institut für Biologie/Vergleichende Zoologie, Berlin, Germany

1 INTRODUCTION

The great diversity of their body organization, segmentation patterns, tagmatization, limb types, larval forms, cleavage, and gastrulation modes unmatched by any other arthropod group makes crustaceans desirable and well-suited objects for studies dealing with questions at the interface of evolution and development. In fact, this has a long tradition in biology and even Charles Darwin (1859) took cirripede crustaceans as examples to show interdependencies between ontogeny and evolution. Shortly after Darwin, Fritz Müller (1864) used crustacean larval development to formulate the theory of recapitulation that was popularized by Haeckel (1866) as the biogenetic law (Scholtz 2000; Breidbach 2002; Richardson & Keuck 2002).

The modern interest in evolutionary developmental biology (evo-devo) results from the powerful molecular genetic approach. Genes and their products that underlie body architecture, segment formation, limb differentiation etc. can now be made visible in expression patterns related to morphological structures. Furthermore, the function of these genes can be tested in an increasing number of animal species by RNA injection and similar techniques. However, research of this kind in crustaceans is still relatively rare – despite the obvious advantages crustaceans have to offer.

The new techniques of evo-devo led to the hope or claim that the study of *Hox* gene expression and function in a series of animals can "explain" the evolutionary transition of one body organization into another one, often referred to as "*Baupläne*".

In the following, I discuss some general and controversial aspects of evo-devo by using mainly examples from crustaceans and their allies.

2 THE *BAUPLAN*/PHYLUM CONCEPT

We have to consider the concepts of evo-devo that are still not settled (Arthur 2002). To many people, the concepts and questions involved in this field are as follows (see Gould 1989; Hall 1999; Carroll et al. 2001; Collins & Valentine 2001):
-   the animal kingdom is subdivided into phyla
-   each phylum is characterized by a *Bauplan*

- a *Bauplan* is a set of conservative characters that are typical for one group but distinctively different from a *Bauplan* of another group
- *Baupläne* arose in a relatively short period during the early Cambrian (Cambrian explosion)
- in the Cambrian, there were many more *Baupläne* than there are today
- no new phyla have appeared since the Cambrian
- *Baupläne* between phyla are so different that they have to be bridged by mechanisms not known to former evolutionary biologists
- in particular, the *Hox* genes, which are responsible for the establishment of the body structures along the body axis of metazoans, present the clues for comprehending the possibility of a sudden dramatic change of body organization
- the latter point serves as the solution for understanding and bridging the variety of *Baupläne* we had in the Cambrian and today

However, one has to ask whether these are the problems and the questions that are suited to characterize a new field of research such as evolutionary developmental biology (Budd 1998, 1999; Breidbach 2002), because the points listed above are strongly characterized by typological thinking (see Richardson et al. 1999; Breidbach 2002). I want to discuss some problematical aspects of these concepts in the following.

3 PROBLEMS WITH THE *BAUPLAN*/PHYLUM DUE TO ITS TYPOLOGICAL BASIS

Both the *Bauplan* and the phylum are not sharply defined concepts and are both burdened with typological thinking. What a *Bauplan* is lies more or less in the eye of the beholder (Valentine & Hamilton 1997; Wray & Strathman 2003). Is there a vertebrate *Bauplan*, an arthropod *Bauplan*, a crustacean *Bauplan*? Or within crustaceans can we speak of a malacostracan *Bauplan*, or a branchiopod *Bauplan*? How are these different levels related to each other? Is there a hierarchical scheme for *Baupläne*? All this is neither clearly defined, nor based on a clear method. Generally, a *Bauplan* is seen as a more or less conservative set of morphological structures or characters "typical" for a large (sometimes not so large, see below) group of animals often referred to as a phylum. Hence, the *Bauplan* is an abstraction of commonality of characters among a phylum and therefore not really based on a historical approach. The *Bauplan* concept is asymmetrical insofar as it stresses similarities within the phyla, but emphasizes the differences between these phyla. In other words, each phylum has its own (*Bauplan*) theme, which shows some variation, but among the phyla we find entirely different themes that show no connections or transitions. Accordingly, the *Bauplan* concept reveals its typological, unhistorical heritage from pre-evolutionary concepts such as the "embranchements" of Cuvier (1817).

We encounter corresponding problems with the phylum concept (Budd & Jensen 2000). There is no reproducible definition of what a phylum is. For the most part, any large taxonomic group with a common *Bauplan* is treated as a phylum, but sometimes also small taxa such as Lorcifera or Cycliophora are considered as phyla because of their unique *Bauplan*. Hence, how many animal phyla are recognized is completely arbitrary (Ax 1995). Accordingly, the numbers of metazoan phyla differ considerably between authors (5: Siewing 1985; 21: Wehner & Gehring 1990; 31: Nielsen 2001; 35: Carroll et al. 2000; 37: Hall 1999). How are these phyla related to each other? Do they show a hierarchical relationship, or are they placed beside each other at the same level? As is true for the

Bauplan, the phylum concept is based on a largely unhistorical viewpoint. Even more problematic, the *Bauplan* and the phylum concepts appear as examples of circular reasoning: a *Bauplan* makes up a phylum, which is characterized by its *Bauplan*.

Similar problems are involved with the concept of the establishment of the *Bauplan* during ontogeny (here often called "body plan"). Again we encounter typological concepts, the so-called "phylotypic stage" (Sander 1983) or "zootype" (Slack et al. 1993), which are thought to represent the *Bauplan* of a phylum in its clearest form during ontogeny. Both these concepts are founded on the assumed stability of this stage in development either morphogenetically (phylotypic stage) or genetically (zootype). In principle, these views resemble the law formulated by von Baer (1828) that embryos exhibit most morphological similarities in early stages and diverge with advanced development. This implies that early stages show no or little freedom for evolutionary change because any change would have dramatic (negative) effects on subsequent development. In contrast to von Baer's laws, the approaches of the phylotypic stage and zootype take into account that there is variation in early development that sometimes is not reflected in the appearance of later stages including the adult (see below). However, these approaches adhere to typological views in that they establish an evolutionarily conserved invariant stage throughout the members of a phylum, which is considered to play an especially important role for body organization and which, in contrast to the traditional *Bauplan*, is shifted to early ontogenetic levels. As with the adult *Bauplan* we face the same problems concerning definition, circularity, and hierarchical relationships of the various phylotypic stages. Moreover, the relationships between the ontogenetic and the adult aspects of *Bauplan* are entirely unresolved (if both are real one aspect must be more important, but which one: the ontogenetic or the adult aspect?). Last but not least, the claim for stability of the phylotypic stage has been criticized and rejected (Richardson et al. 1997; Scholtz 1997).

The general problem of the *Bauplan*/phylum concept can be summarized as follows: there is a methodological switch involved, on the one hand, *similarity* is emphasized within the phylum (*Bauplan*), and, on the other hand, *differences* are stressed between phyla. This makes phyla and *Baupläne* seemingly distinct entities but causes a problem with regard to the origin of and the relationships between the different *Baupläne* or phyla. However, there is no objective and clear methodological borderline between the use of similarity or difference. Phyla and *Baupläne* can thus not form the basis for evolutionary considerations.

4 TOWARDS A PHYLOGENETIC CONCEPT TO EVO-DEVO

The only way to avoid these inconsistencies is the entirely different non-typological but historical approach of phylogenetic systematics, or cladistics, developed by Hennig (1966) and later refined (see Wiley 1981; Ax 1984; Wägele 2000). According to phylogenetic systematics the principle for grouping organisms is not overall similarity but common descent or genealogy. Here only monophyletic groups are accepted, which are defined as comprising a stem species and *all* descendant species. Monophyletic groups are established on the basis of shared evolutionary novelties (apomorphies). There is no *a priori* need that apomorphies occur throughout all representatives of a monophyletic group. In general, the focus of this approach lies on shared similarities and not on differences. The organization of the body of the parasitic rhizocephalans dramatically differs from any other crustacean, even other cirripedes, but their affinities with cirripedes are established by the shared apomorphic characters of the nauplius and cypris larvae not found in other crustaceans (Høeg et al. 2003) (see below). Monophyletic groups are part of larger monophyletic groups forming an encaptic hierarchical system with different degrees of phylogenetic relationships based on the sequence of branching events. Discussions about the question of whether the Cycliophora, to give an example, are a phylum or not are meaningless. The important aspects are, are the Cycliophora monophyletic and which other monophyletic taxon is the sister group of the Cycliophora?

The set of characters for the stem species of each monophyletic group can be reconstructed to yield the so-called ground plan or ground pattern (Ax 1984). This is done based on the distribution of characters in a given tree using the parsimony approach. Accordingly, the ground pattern forms a blend of older (plesiomorphic) characters, inherited from the ancestors, and the apomorphic characters, which have newly evolved in the lineage leading to the stem species of a monophylum (Sudhaus & Rehfeld 1992; Jenner & Schram 1999). Of course, the reconstruction of the ground pattern is not only restricted to characters of the adult stem species, but also should include the reconstruction of the development of the stem species. Accordingly, the ground pattern is entirely different from a *Bauplan*. Of course, there is room for interpretations and misinterpretations, but the ground pattern is a definable and clear concept since it is based on the monophyletic entities that are reconstructed with a transparent, repeatable method. There is no need for the conservation of the ground pattern in a given monophyletic group – members can be highly modified. As is true for monophyletic groups, the ground patterns form an encaptic hierarchy of ground patterns at different levels. The crustacean ground pattern (if crustaceans are monophyletic) evolved earlier than the malacostracan or branchiopod ground patterns, the decapod ground pattern existed before the ground pattern of brachyuran crabs because the stem species of hierarchical higher (more inclusive) groups must have originated before the stem species of its subgroups. This hierarchy allows the step-wise evolutionary transition from one ground pattern to another with clearly defined levels. And this happened continuously since the Cambrian – evolution of dramatically new animal forms did not stop and nobody who knows only the Cambrian arthropod fauna would be able to predict or expect the morphological and developmental patterns of, e.g., the stem species of mites, winged insects, and cirripede crustaceans.

It becomes clear that there are no unbridgeable gaps between the so-called *Baupläne* of so-called phyla, but rather a stepwise relationship of monophyletic taxa and their corresponding ground patterns with a stepwise transformation of characters. All this is organized in a hierarchical pattern. And even the most comprehensive large monophyletic groups of Metazoa or Bilateria originated by speciation with only relatively few differences between the two daughter species. The differences we see among recent large monophyletic groups occurred subsequently in the history of their lineages. This evolutionary process is sometimes obscured by extinction of the organisms showing the stepwise alteration of form. In these cases, the fossil record is of great help. However, a non-typological approach is again needed to recognize and interpret fossils as stem lineage (*sensu lato*) representatives of the crown groups of monophyletic taxa (Budd 1999).

All this leads to different questions for evo-devo. There is no need to search for mechanisms bridging, for instance, the arthropod and the vertebrate *Bauplan*. The ground patterns of arthropods and vertebrates are both the product of a series of transformations starting with the last common ancestor of both groups that most likely is the stem species of the entire Bilateria. This stem species was neither an arthropod nor a vertebrate, but it possessed the bilaterian features shared by insects and vertebrates such as a bilateral symmetry, an anterior pole with a "brain" (nerve plexus), a mesoderm etc. At the genetic level the anterior pole was characterized by expression of the *orthodenticle* gene group, and the *Hox* gene cluster shaped the body along its longitudinal axis (Carroll et al. 2000).

From the phylogenetic systematic perspective the concept of the so-called Cambrian explosion is also problematic (Budd & Jensen 2000; Fitch & Sudhaus 2002). Given that all the major large metazoan monophyletic groups existed during the early Cambrian, we have to date their origins back to Precambrian times (Fortey 2001). Moreover, there is an increasing number of fossils from the Precambrian that suggest an earlier occurrence of Metazoa (see Chen et al. 2000). The concept of the Cambrian explosion is again partly based on an unhistorical *Bauplan*/phylum approach. In a strict phylogenetic perspective of animal evolution, there is no need for a *Bauplan* and consequently no need for a special period of quick evolution of *Baupläne* (Budd & Jensen 2000; Fitch & Sudhaus 2002). This becomes evident if we accept that not the crown groups of the large taxa appeared but in most cases just stem lineage representatives are found in the Cambrian fossil record which do not exhibit the entire suite of apomorphies of the large "modern" monophyla (Budd & Jensen 2000; Fitch & Sudhaus 2002).

No matter what kind of use of developmental data we make in an evo-devo framework, it is important to follow the logic of phylogenetic systematics or cladistics (see also Bang et al. 2000), i.e., one has to think in terms of homology, apomorphic versus plesiomorphic characters, monophyletic taxa, sister groups, parsimony, and ground patterns.

5   WHAT IS THE USE OF ONTOGENY IN PHYLOGENY AND EVOLUTIONARY BIOLOGY?

The roles of the inclusion of developmental data into phylogenetic and evolutionary analyses are manifold and several accounts have been presented in the literature (e.g. Kluge & Strauss 1985; Kluge 1988; Mabee 2000; Arthur 2002). In respect to phylogenetic reconstruction, which is a key issue for any discussion at the evolution/development interface, ontogenetic data can provide an *a priori* contribution (additional characters, homology, ontogenetic criterion), or there is an *a posteriori* use of developmental data for evolutionary considerations based on a phylogenetic framework (transformations, recapitulation, character reversals). In any case, it is crucial to apply the concept of homology. Only homologous characters contain information for phylogenetic inferences, and the reconstruction of character transformation is only meaningful if homologous characters are considered.

5.1   *The inclusion of ontogenetic characters in phylogenetic analyses increases the number of phylogenetically informative characters*

Ontogeny contributes much to the reconstruction of phylogeny basically because it enlarges the set of characters that can be used for phylogenetic analyses. Thereby, two different aspects are important. One aspect is the sharing of early ontogenetic characters such as gene expression patterns, cleavage patterns or larval structures. Examples for crustaceans are the distinct *Hox* gene expression pattern of malacostracans (Abzhanov & Kaufman 2003), *Distal-less* expression patterns unifying Mandibulata (Scholtz 2001), the special cleavage pattern of amphipods, which is an apomorphy for this group (Scholtz & Wolff 2002), and the frontal horns of the nauplii, which are found only in cirripede crustaceans (Høeg et al. 2003). The other aspect deals with characters whose occurrence is restricted to early ontogenetic stages in some taxa whereas they are also found in adults of other groups. The phylogenetic affinity of barnacles and in particular parasitic rhizocephalans with arthropods can be only convincingly revealed with the larval occurrence of articulated limbs, compound eyes, and other arthropod characteristics (Deutsch et al. 2003).

5.2   *Ontogeny can be used to polarize character states*

There is a debate as to whether or not an ontogenetic method can be generally used in systematics to polarize character stages for phylogenetic reconstruction. Some authors provide evidence in favor of the so-called ontogenetic criterion (Nelson 1978; Meier 1997), others have a critical view concerning the general applicability of an ontogenetic criterion (Kluge 1988; Mabee 2000). In any case, there are clear examples where ontogeny can be used to polarize characters. The megalopa larvae of asymmetrical hermit crabs show a symmetrical pleon similar to that of lobsters and crayfishes indicating that the curved asymmetric abdomen of adult paguroids is apomorphic (see Richter & Scholtz 1994). Another example is the seventh pleon segment in malacostracans. This is found in adult leptostracans, whereas in eumalacostracans it occurs transiently during embryogenesis and fuses with the sixth pleon segment (Scholtz 1995). The adult condition without a seventh pleon segment is apomorphic for the Eumalacostraca (Richter & Scholtz 2001).

5.3   *Ontogenetic data strengthen the plausibility of homology assumptions*

Similar development can strongly contribute to the inference on, and the claim for homology of characters. One example is the homology of particular pioneer neurons between various insects that is substantiated by similar cell lineages leading to these neurons and by corresponding expression of genes involved in neurogenesis (Whitington & Bacon 1997). Another example is the alignment of segments of arthropods using the anterior boundary of *Hox* gene expression patterns that led to a new insight into the homology of the segments of the chelicerae in Chelicerata and of the (first) antennae of mandibulates (Damen et al. 1998; Telford & Thomas 1998). However, it must be noted that differences in development do not necessarily lead to the rejection of homology. There are numerous examples of homologous characters that are formed via different developmental pathways, i.e., early developmental stages have been evolutionarily altered while the later stages remained conserved. This can relate to altered gene expression that leads to homologous downstream products, e.g., homologous arthropod limbs with different early gene expression patterns (Abzhanov & Kaufman 1999; Jokusch et al. 2000), or to differences in cell division patterns that lead to homologous arrangements and characteristics of cells in later stages (Dohle & Scholtz 1988; Scholtz & Dohle 1996).

5.4   *Analysis of evolutionary character transformation*

Developmental data can be and are often used in a phylogenetic context by mapping ontogenetic patterns onto a given phylogeny in order to draw conclusions not only about evolutionary transformations of ontogeny and development itself, but also about the transformation of the resulting morphological structures (see below for the problems involved in this). The change of expression patterns and function of *Hox* genes in arthropods can serve as a good example for this, e.g., the gene *fushi tarazu*, which evolutionarily starts with a *Hox* function and ends as a pair-rule segmentation gene (Hughes & Kaufman 2002; Deutsch et al. 2003). However, it is clear that the study of development cannot bridge the gap between the patterns we observe in living beings (extinct or extant) and the evolutionary process that is only inferred from the comparison of these patterns. All we can achieve is a more refined view of transformation, but this view always remains an interpolation or extrapolation since evolutionary processes cannot be directly observed. The study of development under the evolutionary perspective allows a better understanding of character transformations because we can now clearly define the point where structures with a similar ontogenetic beginning start to diverge at the levels of gene expression, gene regulation and morphogenesis. For example, the stenopodous limbs of the raptorial waterflea *Leptodora kindtii* originated clearly from phyllopodous limbs (Olesen et al. 2001). This can be deduced from the distribution pattern of limb types among branchiopod crustaceans based on phylogenetic analyses (Braband et al. 2002; Olesen 2003). Including only adult structures in our analysis just allows the conclusion that one limb type (phyllopodous) was evolutionarily transformed into or replaced by another limb type (stenopodous). To gain a more detailed picture, however, it is necessary to include ontogenetic data. These reveal that the early *Anlage* of both limb types is very similar up to a certain point, and this holds true for gene expression and morphogenesis. We now have a much more detailed perspective of when and how the alteration took place ontogenetically. We can say a change like this does not require the rebuilding of limb ontogeny right from the early *Anlage*. However, this is of course no causal explanation for the change.

5.5 *Explanation for character reversals and recapitulation*

The inclusion of ontogenetic data in our perspective of phylogeny and evolution allows explanations for character reversals. It has been suggested that the free-living nauplius of malacostracans is a secondary achievement that was only possible because the nauplius stage was conserved as a so-called egg nauplius in the taxa with yolky eggs and an abbreviated development (Scholtz 2000). The evolution of secondary apposition eyes from superposition eyes within crabs (Brachyura) and hermit crabs (Anomala) can be explained with the fact that all decapod larvae, even those of brachyurans and anomalans, possess apposition eyes. The re-evolution of adult apposition eyes would thus be a case of neoteny (Richter 2002). Both examples can also serve as evidence for the existence of recapitulation of ancestral characters. Recapitulation has even been found at the levels of gene expression. In the malacostracan *Cherax destructor,* there are more stripes of the segment polarity gene *engrailed* than morphological segments, indicating an origin of malacostracan crustaceans from ancestors with a higher number of segments (Scholtz 1995). However, one has to be careful. It is not possible to infer the existence or quality of a morphological adult structure from early gene expression data alone since there is no necessary relation. In particular, this is when non-homologous structures develop via similar (homologous?) gene expression patterns. An example for this is the homologization of crustacean epipodites and hexapod wings based on similar early expression patterns of the *apterous* and *nubbin* genes (Averof & Cohen 1997). However, neither the position, nor the specific structural quality of the adult morphology of crustacean epipodites and insect wings indicate their homology. Furthermore, the plesiomorphic absence of wings (and epipodites) in basal hexapods, such as endognathans, archaeognathans, and zygentomans, and the lack of epipodites in maxillopodan crustaceans dispute the presence of corresponding structures in a common ancestor of crustaceans and hexapods (see also Bang et al. 2000). Other examples for this are the different, non-homologous eye types, e.g., vertebrate lens eye and arthropod compound eye, and appendage types, e.g., tube feet of echinoderms, vertebrate and arthropod limbs, of Metazoa, which rely on homologous genes *Pax-6* (eyes) and *Distal-less* (appendages) and their expression (see below and Nielsen & Martinez (2003) for discussion who coined the term "homocracy" for this phenomenon).

5.6   Ontogenetic transformation processes serve as paradigms for evolutionary transformation processes

The study of directly observable developmental processes allows analogous inferences on evolutionary processes (Patterson 1983). Both processes, evolutionary and ontogenetic, represent series of patterns (stages) in time that are principally independent from each other and that are free to become modified (see below). A principle difference between ontogenetic and phylogenetic processes, however, is the linearity of ontogenetic transformation, whereas phylogenetic transformation requires the reconstruction of ancestors as the starting point for transformation analysis.

6   PATTERN AND PROCESS - THE COMPARATIVE APPROACH

6.1   *Ontogeny and evolutionary patterns and processes*

Ontogeny appears as a transformation process that is directly observable. In contrast, evolutionary transformation processes are only indirectly inferred from comparative analyses of observable patterns. As in all historical sciences the gap between these observable patterns and historical (evolutionary) processes can principally not be bridged. We have no time machine for the direct observation of historical events and processes, and the theory of evolution is the prerequisite for interpreting the present patterns as the product of history. Hence, a causal explanation of evolutionary transformational change based on comparative developmental studies seems impossible. Even if we knew the entire regulatory genetic network, the cell division patterns and morphogenesis, and we note distinct genetic differences that can be addressed via experimental manipulations, we only have access to recent species which themselves underwent evolutionary changes. A true experiment simulating evolution is only possible if we could use a direct ancestor of the species under investigation. Therefore, I agree with the conclusion of Wagner (2001) that even experiments do not provide a causal explanation of evolutionary innovations.

6.2   *Developmental processes as series of patterns*

Kluge & Strauss (1985) emphasize that the gradual nature of ontogeny does not allow it to be cut in a non-arbitrary manner and that therefore a different approach to the use of ontogenetic data in phylogenetic analyses is necessary. In contrast to this, I think the analysis of development requires the subdivision of developmental processes into a sequence of discrete elements with each being characterized by a distinct pattern (see Scholtz & Dohle 1996). In other words, processes are treated as a series of patterns (and perhaps a process is nothing else as neurophysiology suggests, Hoffman 1998). There is no fundamental difference to the treatment of characters in general for which spatial, or in the case of ontogeny spatial and temporal, delimitation implies in any case an unavoidable arbitrary element. Sometimes ontogeny might be even advantageous with respect to discrete entities since at the level of cell division patterns we find distinct elements that allow the recognition of series of discrete states. In developmental biology, the discrete patterns are traditionally called developmental or ontogenetic stages. Stages can relate to the entire developing organism or to parts of it. Each of these stage-patterns of a sequence can be compared horizontally between the species/taxa under investigation with respect to similarity and evolutionary transformation. This can be done at any level from genetics to morphogenesis and at any stage. The result is a mosaic of similarities and differences of pattern sequences. In this way, comparisons contribute to resolving the general question of developmental biology: what kinds of relationships exist between ontogenetic stages? This question is similar to the general problem of developmental biology: what is the nature of the relation between the genotype and the phenotype, and what role does development play with respect to the morphology of the adult organism? Is ontogeny the stepwise realization of the phenotype based on the genotype? Is one stage the cause for the subsequent ones? For instance, the recognition of differences between species found with a pattern at a particular stage, with conserved previous stages, indicates evolutionary freedom of this relatively late stage in regard to previous stages, and falsifies a causal nexus to previous stages. The recognition of differences between species found with a pattern at a particular stage with conservation of subsequent stages indicates evolutionary freedom of this early stage and falsifies a causal nexus to subsequent stages.

6.3   *The independence of developmental stages*

There is growing evidence at all developmental levels from gene expression, cell division patterns to morphogenesis that virtually every single stage can be evolutionarily altered without necessarily affecting the following stages (Dohle 1989; Wagner & Misof 1993; Raff 1996; Scholtz & Dohle 1996; Abzhanov & Kaufman 1999; Jokusch et al 2000; Davis & Patel 2002). This clearly proves the principal independence of ontogenetic stages and disputes the discrimination between causal and temporal ontogenetic sequences (see Alberch 1985). One can further ask whether the adult morphology is the result of the preceding development or just one stage among others? Again, there is evidence that the terminal stage plays no role different from earlier stages in terms of independence, as is revealed by cases of neoteny where adult stages are lost and thus cannot be the goal or final cause of development. If this is true, one has to treat development as a series of independent patterns that all can be individually altered evolutionarily, including the terminal stage. Often the question is: how have developmental processes changed during evolution generating the innovations of morphology that are reflected in the diversity we see today? This view perhaps emphasizes too much the terminal stage as the most important part of development. Rather, we have to state that evolutionary change can occur at any level and at any stage of development, and this change of the level, or stage, *is* the innovation itself. This contradicts the view of a deterministic development with an invariant causal relation between all stages and levels. Moreover, this clearly stands in contrast to any typological concept putting stress on the greater importance of early stages for setting up any kind of *Bauplan*. In other words, the evolutionary freedom of all developmental stages including early ones falsifies laws or rules such as those of von Baer about the increasing difference between embryos of later stages or the concepts of phylotypic or zootypic stages. Even if we assume that von Baerian rules or the existence of a phylotypic stage holds true for a number of cases, we can never make predictions of whether the next case studied obeys this "law". However, one has to discriminate between normal individual development in one species and the comparison of development of different species under an evolutionary perspective. During normal development it might not be possible to alter or eliminate a stage experimentally because the ability to regulate is not sufficiently present in this particular species. In contrast, the experiments done by nature in the course of evolution reveal the principal possibility to alter any developmental aspect.

6.4   *Later stages cannot be inferred from earlier stages*

If this principal independence of developmental stages holds true, it is impossible to infer the nature of advanced stages from the pattern of earlier stages in ontogeny or *vice versa*. For instance, gene expression alone does not allow us to infer the existence of a specific adult morphology. In other words, the homology of (morphologically non-similar) adult stage characters cannot be deduced from the similar (even homologous) early developmental stages.

This seems to be an important step for analyzing the evolution of developmental processes. As a consequence of this independence and the resulting impossibility to make any inference between stages it appears necessary that the ground pattern for every single stage or level has to be independently reconstructed, and only subsequently the developmental process of the stem species can be reconstructed in order to gain a picture of the plesiomorphic development of a monophyletic group as the starting point for evolutionary considerations. The ground pattern has to be reconstructed with respect to each individual ontogenetic feature be it gene expression, cell division or morphogenetic stages. We have to treat every comparable stage individually, i.e. we have to compare the expression of a gene with the expression of the corresponding gene we must compare cellular levels etc. The proof of an evolutionary change of the role of genes is a good example for this. Just the existence of a *fushi tarazu* gene in an arthropod does not imply whether it plays a role during segmentation, neurogenesis, or the specification of a body region (Hughes & Kaufman 2002) (see above).

Another instance for this is eye evolution in bilaterians. It is not possible to conclude from the existence and the early expression of the *Pax-6/eyeless* gene that the resulting morphological character is an arthropod compound eye or a vertebrate or molluscan lens eye – all is possible, as has been shown by experimental and comparative studies (Gehring 2001). If we reconstruct the ground pattern of the bilaterian stem species with respect to eye development and adult eye morphology, we can safely conclude that the *Pax-6/eyeless* gene was present and expressed based on the distribution among a variety of distantly related bilaterian taxa. However, it is also safe to postulate that the adult eye of the bilaterian stem species was neither an arthropod compound eye nor a vertebrate lens eye; both solutions imply too many convergent losses. In this case it is difficult to reconstruct the original bilaterian eye because the distribution of eye types provides us with no clear picture – if there was an eye (as a compact organ) at all present. This view leads directly to the homology aspect since it is possible to homologize the *Pax-6* genes between the various bilaterians, but evidently it is not possible to claim homology between the various eye types based on the fact that the same gene is involved in their formation because the gene and its expression are much older than the specific morphological eye types such as compound eye or lens eye.

In summary I think we are just at the beginning of an emerging non-typological approach to evolutionary developmental biology. Due to its manifold forms and developmental patterns the fascinating group of Crustacea can play an important part for the achievement of this goal.

ACKNOWLEDGEMENTS

My thoughts benefited from many discussions with Wolfgang Dohle, Stefan Richter, and the students of our seminars on comparative zoology and phylogenetics. Greg Edgecombe, Graham Budd, and Tino Pabst improved the manuscript. To all I express my sincere thanks.

REFERENCES

Abzhanov, A. & Kaufman, T.C. 1999. Homeotic genes and the arthropod head: expression patterns of the *labial, proboscipedia*, and *Deformed* genes in crustaceans and insects. *Proc. Natl. Acad. Sci. USA* 96: 10224-10229.

Abzhanov, A. & Kaufman, T.C. 2003. HOX genes and tagmatization of the higher Crustacea (Malacostraca). In: Scholtz, G. (ed), *Crustacean Issues 15, Evolutionary Developmental Biology of Crustacea*: 43-74. Lisse: Balkema.

Alberch, P. 1985. Problems with the interpretation of developmental sequences. *Syst. Zool.* 34: 46-58.

Arthur, W. 2002. The emerging conceptual framework of evolutionary developmental biology. *Nature* 415: 757-764.

Averof, M. & Cohen, S.M. 1997. Evolutionary origin of insect wings from ancestral gills. *Nature* 385: 627-630.

Ax, P. 1984. *Das Phylogenetische System*. Stuttgart, Gustav Fischer.

Ax, P. 1995. *Das System der Metazoa I*. Stuttgart: Gustav Fischer.

Bang, R., DeSalle, R. & Wheeler, W. 2000. Transformationalism, taxism, and developmental biology in systematics. *Syst. Biol.* 49: 19-27.

Braband, A., Richter, S., Hiesel, R. & Scholtz, G. 2002. Phylogenetic relationships within the Phyllopoda (Crustacea, Branchiopoda) based on mitochondrial and nuclear markers. *Mol. Phyl. Evol.* 25: 229-244.

Breidbach, O. 2002. The former synthesis – some remarks on the typological background of Haeckel's ideas about evolution. *Theory Biosci.* 121: 280-296.

Budd, G.E. 1998. Arthropod body-plan evolution in the Cambrian with an example from anomalocarid muscle. *Lethaia* 31: 197-210.

Budd, G.E. 1999. Does evolution in body patterning genes drive morphological change – or vice versa? *BioEssays* 21: 326-332.

Budd, G.E. & Jensen, S. 2000. A critical reappraisal of the fossil record of the bilaterian phyla. *Biol. Rev.* 75: 253-295.

Carroll, S.B., Grenier, J.K. & Weatherbee, S.D. 2001. *From DNA to Diversity*. Malden: Blackwell Science.

Chen, J.-Y., Oliveri, P., Li, C.-W., Zhou, G.-Q., Gao, F., Hagadorn, J.W., Peterson, K.J. & Davidson, E.H. 2000. Precambrian animal diversity: Putative phosphatized embryos from the Doushantuo Formation of China. *Proc. Natl. Acad. Sci. USA* 97: 4457-4462.

Collins, A.G. & Valentine, J.W. 2001. Defining phyla: evolutionary pathways to metazoan body plans. *Evol. Dev.* 3: 432-442.

Cuvier, G. 1817. *Le règne animal.* Paris: Déterville.

Damen, W.D., Hausdorf, M., Seyfarth, E.-A. & Tautz, D. 1998. A conserved mode of head segmentation in arthropods revealed by the expression pattern of *Hox* genes in spider. *Proc. Nat. Acad. Sci. USA* 95: 10665-10670.

Darwin, C. 1859. *On the Origin of Species by Means of Natural Selection, or the Preservation of Favoured Races in the Struggle for Life.* London: John Murray.

Davis, G. K. & Patel, N. H. 2002. Short, long, and beyond: molecular and embryological approaches to insect segmentation. *Annu. Rev. Entomol.* 47: 669-699.

Deutsch, J.S., Mouchel-Vielh, E., Quéinnec, É. & Gibert, J.-M. 2003. Genes, segments and tagmata in cirripedes. In: Scholtz, G. (ed), *Crustacean Issues 15, Evolutionary Developmental Biology of Crustacea*: 19-42. Lisse: Balkema.

Dohle, W. 1989. Differences in cell pattern formation in early embryology and their bearing on evolutionary changes in morphology. *Geobios, mémoire spécial* 12: 145-155.

Dohle, W. & Scholtz, G. 1988. Clonal analysis of the crustacean segment: the discordance between genealogical and segmental borders. *Development* 104, *Suppl.*: 147-160.

Fitch, D.H.A. & Sudhaus, W. 2002. One small step for worms, one giant leap for "*Bauplan?*" *Evol. Dev.* 4: 243-246.

Fortey, R. 2001. The Cambrian explosion exploded? *Science* 293: 438-439.

Gehring, W.J. 2001. *Wie Gene die Entwicklung steuern. Die Geschichte der Homeobox.* Basel: Birkhäuser.

Gould, S.J. Wonderful Life: *The Burgess Shale and the Nature of History.* New York: Norton.

Haeckel, E. 1866. *Generelle Morphologie der Organismen.* Berlin: Georg Reimer.

Hall, B.K. 1999. *Evolutionary Developmental Biology.* Dordrecht: Kluwer Academic Publishers.

Hennig, W. 1966. *Phylogenetic Systematics.* Urbana: University of Illinois Press.

Høeg, J.T. Lagersson, N.C. & Glenner, H. 2003. The complete cypris larva and its significance in thecostracan phylogeny. In: Scholtz, G. (ed), *Crustacean Issues 15, Evolutionary Developmental Biology of Crustacea* : 197-215. Lisse: Balkema.

Hoffman, D.D. 1998. Visual Intelligence. How we create what we see. New York: Norton & Co.

Hughes, C.L. & Kaufman, T.C. 2002. Hox genes and the evolution of the arthropod body plan. *Evol. Dev.* 4: 459-499.

Jenner, R.A. & Schram, F.R. 1999. The grand game of metazoan phylogeny: rules and strategies. *Biol. Rev.* 74: 121-142.

Jokusch, E.L., Nulsen, C., Newfeld, S.J. & Nagy, L.M. 2000. Leg development in flies versus grasshoppers: differences in *dpp* expression do not lead to differences in the expression of downstream components of the leg patterning pathway. *Development* 127: 1617-1626.

Kluge, A.G. 1988. The characterisation of ontogeny. In: Humphries, C.J. (ed), *Ontogeny and Systematics*: 57-81. New York: Columbia University Press.

Kluge, A.G. & Strauss, R.E. 1985. Ontogeny and systematics. *Ann. Rev. Ecol. Syst.* 16: 247-268.

Mabee, P.M. 2000. The usefulness of ontogeny in interpreting morphological characters. In: Wiens, J.J. (ed), *Phylogenetic Analysis of Morphological Data*: 84-114. Washington: Smithsonian Institution Press.

Meier, R. 1997. A test and review of the empirical performance of the ontogenetic criterion. *Syst. Biol.* 46: 699-721.

Müller, F. 1864. *Für Darwin.* Leipzig: Engelmann.

Nelson, G.J. 1978. Ontogeny, phylogeny, paleontology, and the Biogenetic Law. *Syst. Zool.* 27: 324-345.

Nielsen, C. 2001. *Animal Evolution.* 2nd Edition. Oxford: Oxford University Press.

Nielsen, C. & Martinez, P. 2003. Patterns of gene expression: homology or homocracy? *Dev. Genes Evol.* 213: 149-154.

Olesen, J. 2003. On the ontogeny of the Branchiopoda (Crustacea): contribution of development to phylogeny and classification. In: Scholtz, G. (ed), *Crustacean Issues 15, Evolutionary Developmental Biology of Crustacea*: 217-269. Lisse: Balkema.

Olesen, J., Richter, S. & Scholtz, G. 2001. The evolutionary transformation of phyllopodous to stenopodous limbs in the Branchiopoda (Crustacea) - Is there a common mechanism for early limb development in arthropods? *Int. J. Dev. Biol.* 45: 869-876.

Patterson, C. 1983. How does phylogeny differ from ontogeny? In: Goodwin, B. C., Holder, N. & Wylie, C. G. (eds), *Development and Evolution*: 1-31. Cambridge: Cambridge University Press.

Raff, R.A. 1996. *The shape of life*. Chicago: University of Chicago Press.

Richardson, M.K. & Keuck, G. 2002. Haeckel's ABC of evolution and development. *Biol. Rev.* 77: 495-528.

Richardson, M.K., Hanken, J., Gooneratne, M.L., Pieau, C., Raynaud, A., Selwood, L. & Wright, G.M. 1997. There is no highly conserved embryonic stage in the vertebrates, implications for current theories of evolution and development. *Anat. Embryol.* 196: 91-106.

Richardson, M.K. Minelli, A. & Coates, M.I. 1999. Some problems with typological thinking in evolution and development. *Evol. Dev.* 1: 5-7.

Richter, S. 2002. Evolution of optical design in Crustacea. In: Wiese, K. (ed), *The Crustacean Nervous System*: 512-524. Berlin: Springer Verlag.

Richter, S. & Scholtz, G. 1994. Morphological evidence for a hermit crab ancestry of lithodids (Crustacea, Decapoda, Anomala, Paguroidea). *Zool. Anz.* 233: 187-210.

Richter, S. & Scholtz, G. 2001. Phylogenetic analysis of the Malacostraca (Crustacea). *J. Zool. Syst. Evol. Research* 39: 113-136.

Sander, K. 1983. The evolution of patterning mechanisms: gleanings from insect embryogenesis and spermatogenesis. In: Goodwin, B. C., Holder, N. & Wylie, C. G. (eds), *Development and Evolution*: 137-158. Cambridge: Cambridge University Press.

Scholtz, G. 1995. Expression of the *engrailed* gene reveals nine putative segment-anlagen in the embryonic pleon of the freshwater crayfish *Cherax destructor* (Crustacea, Malacostraca, Decapoda). *Biol. Bull.* 188: 157-165.

Scholtz, G. 1997. Cleavage, germ band formation and head segmentation: the ground pattern of the Euarthropoda. In: Fortey, R.A. & Thomas, R.H. (eds), *Arthropod Relationships*: 317-332. London: Chapman and Hall.

Scholtz, G. 2000. Evolution of the nauplius stage in malacostracan crustaceans. *J. Zool. Syst. Evol. Research* 38: 175-187.

Scholtz, G. 2001. Evolution of developmental patterns in arthropods – the analysis of gene expression and its bearing on morphology and phylogenetics. *Zoology* 103: 99-111.

Scholtz, G. & Dohle, W. 1996. Cell lineage and cell fate in crustacean embryos - a comparative approach. *Int. J. Dev. Biol.* 40: 211-220.

Scholtz, G. & Wolff, C. 2002. Cleavage, gastrulation, and germ band formation of the amphipod *Orchestia cavimana* (Crustacea, Malacostraca, Peracarida). *Contrib. Zool.* 71: 9-28.

Siewing, R. (ed.) 1985. *Lehrbuch der Zoologie, Bd. 2, Systematik*. Stuttgart: Gustav Fischer Verlag.

Slack, J.M.W., Holland, P.W.H. & Graham, C.F. 1993. The zootype and the phylotypic stage. *Nature* 361: 490-492.

Sudhaus, W. & Rehfeld, K. 1992. *Einführung in die Phylogenetik und Systematik*. Stuttgart: Gustav Fischer Verlag.

Telford, M.J. & Thomas, R.H. 1998. Expression of homeobox genes shows chelicerate arthropods retain their deutocerebral segment. *Proc. Nat. Acad. Sci. USA* 95: 10671-10675.

Valentine, J.W. & Hamilton, H. 1997. Body plans, phyla and arthropods. In: Fortey, R.A. & Thomas, R.H. (eds.), Arthropod Relationships: 1-9. London: Chapman & Hall.

Wägele, J.-W. 2000. *Grundlagen der Phylogenetischen Systematik*. München: Verlag Dr. Friedrich Pfeil.

von Baer, K.E. 1828. *Entwickelungsgeschichte der Thiere*. Königsberg: Bornträger.

Wagner, G.P. 2001. What is the promise of developmental evolution? Part II: a causal explanation of evolutionary innovations may be impossible. *J. Exp. Zool. (Mol. Dev. Evol.)* 291: 305-309.

Wagner, G.P. & Misof, B.Y. 1993. How can a character be developmentally constrained despite variation in developmental pathways? *J. Evol. Biol.* 6: 449-455.

Wehner, R. & Gehring, W. 1990. *Zoologie*. 22.Aufl. Stuttgart: Thieme.

Wiley, E.O. 1981. Phylogenetics: *The Theory and Practice of Phylogenetic Systematics*. New York: Wiley Interscience.

Wray, G.A. & Strathmann, R.R. 2002. Stasis, change, and functional constraint in the evolution of body plans, whatever they may be. *Vie Milieu* 52: 189-199.

I GENES AND BODY ORGANIZATION

Genes, segments, and tagmata in cirripedes

JEAN S. DEUTSCH, EMMANUELE MOUCHEL-VIELH, ÉRIC QUÉINNEC AND JEAN-MICHEL GIBERT [1]

Université P. et M. Curie, Paris, France

[1] *present address: Department of Zoology, Cambridge, UK*

1 INTRODUCTION

1.1 *What are the cirripedes? Are they molluscs or crustaceans?*

Darwin spent no less than eight years studying and writing about Cirripedia. He published four monographs, on living and on fossil cirripedes in 1851 and in 1854 [1]. *"Almost every one who has walked over a rocky shore knows that a barnacle or acorn shell is an irregular cone"*, says Darwin (1854: 33), that is found settling on any solid substratum, like rocks or shells or crustacean carapaces. *"Within this shell,"* continues Darwin, *"the animal's body is lodged; and through a slit in the lid, it has the power of protruding six pairs of articulated cirri or legs."* Hence the name "cirripedes".

Here comes the first interesting point: the passage from anatomy to taxonomy. Taxonomy is closely related to comparative anatomy, body plans, and therefore, to development and evolution. On the basis of their external shelly appearance, cirripedes were initially classified as molluscs, but their body is articulated. Cuvier, nevertheless, continued to place the Cirripedia among the Mollusca, while recognizing the inconsistency of their internal morphology, more similar to that of annelids and arthropods, segmented animals that he had grouped together in the "Embranchement" of Articulata. In 1830, J. Vaughan Thompson was the first to show that the larvae of the typical sessile barnacle *Balanus* were nauplii. As nauplii are typical crustacean larvae, this established without doubt that the Cirripedia belong to the Crustacea [2]. Cuvier died in 1832, and zoologist H. Milne-Edwards, to whom Darwin dedicated his 1854 monograph, after some hesitation, included them in the Crustacea (1852).

---

[1] "The Origin of Species" was published in 1859.

[2] *"Straus was the first who, in 1819, maintained that Cirripedes were most closely allied to Crustacea. But this view was disregarded, until J. Vaughan Thompson's capital discovery, in 1830, of their metamorphoses, since which time, Cirripedes have been almost universally admitted amongst the Crustaceans"* (Darwin 1854: 9). *"J. Vaughan Thompson, a British army surgeon based in Ireland, demonstrated unequivocally for the first time that the settlement of Balanus as a shelled animal was preceded by swimming larval stages, the first of which was obviously a nauplius"* (Anderson 1994: 2).

1.2   *What are the cirripedes? Some very peculiar cirripedes*

1.2.1   *Thoracica*

Following this initial confusion, there was an inevitable second taxonomic dispute: which animals are and which are not to be included within the Cirripedia? This debate lasted until very recently.

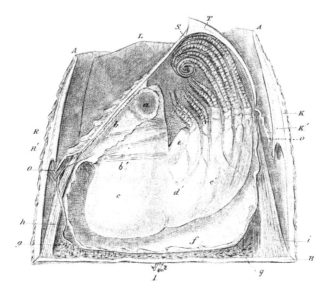

Figure 1. Internal anatomy of the adult thoracican barnacle *Balanus*. From Darwin 1854 (plate XXX, fig. 1). The head is highly modified. No abdominal part is present behind the sixth thoracic segment bearing cirri. In this drawing, there is no penis. In certain thoracic barnacle species, the penis is a transient structure, which can be lost and gained.

In his books, Darwin described a number of new cirripede species, in extremely precise detail. He defined the terms, still in use, to designate the various parts of the shell. He studied their life history, including sexual life. For instance, he discovered "complemental" males in some species that are otherwise hermaphrodite. He devoted his monographs almost exclusively to those animals that are the most familiar and commonly known barnacles: pedunculate barnacles such as *Lepas anatifera* (first monograph, 1851) or sessile ones such as the acorn barnacle *Balanus* (second monograph on living barnacles, 1854) (Fig. 1). All these cirripedes belong to the order (or super-order [3]) Thoracica. This term, which was coined by Darwin, is highly significant for our present purpose. These cirri-pedes are called "Thoracica" because they are composed virtually entirely of a thorax [4].

---

[3] As we consider that all taxonomic ranks above the species level have no biological meaning, being merely useful for classification, we will use the terms and ranks given in the "Traité de Zoologie", vol. VII-1, p. 7 (Forest 1994), throughout the present text.

[4] Darwin gave this definition of the Thoracica: "*Cirripedia having a carapace, consisting either of a capitulum on a peduncle, or of an operculated shell with a basis. Body formed of six thoracic segments, generally furnished with six pairs of cirri; abdomen rudimentary, but often bearing caudal appendages)*", (1854: 30).

The head is deeply transformed in the adult, and the abdomen is highly reduced and present only during the larval stages.

1.2.2 *Apoda and Acrothoracica*

In addition to the Thoracica, Darwin created two new cirripede orders, the "Apoda" and the "Abdominalia" on the basis of differences in body plans [5]. He devotes only a few pages to the Apoda and the Abdominalia at the end of his 1854 book. The "Apoda" comprised a single species, *Proteolepas*, based on a single specimen. However, *"that the little barnacle parasite* Proteolepas bivincta *Darwin was not itself a cirripede was confirmed by Bocquet-Védrine (1979) who demonstrated that it was an epicaridan isopod"* (Newman 1987: 4). In 1854, Darwin included only one species, *Cryptophialus minutus*, in the second order, the Abdominalia. He hesitated about including another cirripede, *Alcippe lampas* (now renamed *Trypetesa lampas*) [6]. They were called "Abdominalia" because he thought that the cirri are carried by abdominal segments [7].

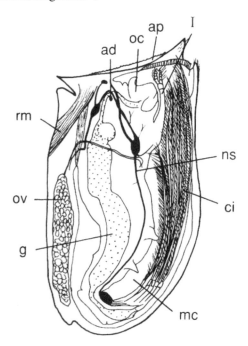

Figure 2. Internal anatomy of the acrothoracican cirripede *Berndtia*. From Anderson 1994 (fig. 2.16, p. 65, after Utinomi 1950). Note the cleft separating the first cirrus (I), transformed into a maxilliped, from the five other cirri (ci). This species possesses the full complement of six pairs of thoracic cirri.

---

[5] *"I divide the Cirripedia into three orders, - the Thoracica, Abdominalia, and Apoda, between which the fundamental difference consists in the limbs or cirri being thoracic in the first, abdominal in the second, and entirely absent in the third"* (1854: 20).

[6] In fact, in the 1851 volume, he considered that this super-order *"contains the remarkable burrowing genus Alcippe, and a second genus, or rather family, obtained by me on the coast of South America"* (i.e. *Cryptophialus*) (p. 2).

[7] Darwin defined the "Abdominalia" as: *"Cirripedia, having a flask-shaped carapace; body consisting of one cephalic, seven thoracic, and three abdominal segments; the latter bearing three pairs of cirri; the thoracic segments without limbs; (...) larva (...) without thoracic legs, but with abdominal appendages"* (1854: 563).

However, Darwin himself wrote: "*I have several times tried to persuade myself, with no success, into the belief that I have somehow misunderstood the homologies of the thoracic segments and cirri of* Alcippe *and* Cryptophialus" (1854, footnote: 565). In fact, the cirri of *Cryptophialus* are thoracic, not abdominal, as shown by Gruvel (1905), who coined the term "Acrothoracica" in place of "Abdominalia", including *Alcippe* (= *Trypetesa*) together with *Cryptophialus* within the Acrothoracica. The differences in body plans between the two orders are not as large as Darwin thought. Indeed, the Acrothoracica differ from the Thoracica in the first three thoracic segments. When present, the first pair is transformed into maxillipeds. In some species it is reduced or even absent, i.e., *Trypetesa*. Behind the first segment, the thorax is elongated and bent in such a way that the following segments are separated from the first (Fig. 2). In many acrothoracican species the 2nd and 3rd pairs of cirri are absent. Thus, in acrothoracican cirripedes, contrary to Darwin's view, only a part of the thorax is devoid of limbs, not its entire length, and all cirri are carried by thoracic segments.

1.2.3 *Rhizocephala and Ascothoracida*

Darwin did not mention another group, the Rhizocephala. All rhizocephalans are parasites, mostly of other crustaceans, including barnacles themselves. As parasites, their adult morphology is highly modified. In fact, it is modified to such a point that no trace of metamery can be seen any longer. The largest part of the adult (the "interna") is composed of filaments (roots) that invade the host, pumping nutrients from its hemolymph. The external part ("externa") is nothing but a sack containing the ovaries and the dwarf males, themselves parasites of the female. The embryos develop within this sack. When they are mature, they simultaneously burst out as hundreds of nauplius larvae. In 1836, J. Vaughan Thomson described the typical rhizocephalan *Sacculina carcini* (Fig. 3) and recognized its larvae as akin to those of the more common barnacles. Later on, Yves Delage described the life cycle of *Sacculina* (1884). Recently, J. Høeg and his collaborators have added the final touch to the strange and wonderful life story of "root-heads" and the way they infest their host (Høeg 1987; Glenner & Høeg 1995; Glenner et al. 2000).

Figure 3. *Sacculina carcini.* The externa of the rhizocephalan parasite can be seen on this green crab (*Carcinus maenas*) taken from the seashore of Roscoff (Brittany, France).

A similar set of debates took place around another group of crustaceans, the Ascothoracida. They were first discovered and described by Lacaze-Duthiers at the end of the 19th century. "Ascothoracida" can be translated as "thorax in a sack". Indeed, their thorax is enclosed within a carapace, from which the abdomen springs out. This abdomen is composed of "*four or five segments*" (Schram 1986: 479) and is followed by a telson with well-developed caudal rami. Most ascothoracids are parasites, but some species are free-swimming.

The question whether the Ascothoracida and the Rhizocephala belong to the Cirripedia was debated for years and was only recently solved. All possible taxonomies (and phylogenies) have been proposed, sometimes successively by the same author(s), with Ascothoracida in and Rhizocephala out, both in (Borradaile et al. 1958: 342; Barnes et al. 1988: 255; Ruppert & Barnes, 1994: 789), both out (Meglitsch & Schram 1991, p. 492; Newman et al. 1969) or Ascothoracida out and Rhizocephala in. The latter situation is now generally accepted (Schram & Høeg 1995): the Rhizocephala are true cirripedes, even closer to the Thoracica than the Acrothoracica, and the Ascothoracida are the sister group of the Cirripedia. This phylogeny is supported both by morphological arguments (Høeg 1992; Anderson 1994; Schram & Høeg 1995; Glenner et al. 1995; Høeg & Kolbasov 2002; Høeg et al. 2003) and molecular data (Spears et al. 1994; Billoud et al. 2000; Gibert 2000).

Due to their adaptation to parasitism, there are no morphological characters left in adults of the Rhizocephala that can be compared to the Thoracica and the Acrothoracica. However, several larval characters support the inclusion of the Rhizocephala within the Cirripedia, and the exclusion of the Ascothoracida [8]. For instance, the nauplii of cirripedes, including the Rhizocephala, possess a pair of frontal horns, which are absent in nauplii of any other crustacean species. Ascothoracid nauplii do not have frontal horns. In all cirripedes, including the Rhizocephala, the naupliar stages are followed by a typical new larval stage, the cypris (Fig. 4), which precedes metamorphosis.

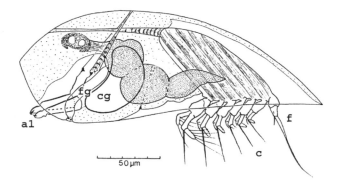

Figure 4. The cypris larva of the rhizocephalan *Sacculina carcini*. Modified from Mourlan et al. (1985). al = antennule; c = cirri; cg = cement gland; f = furca; fg = frontal gland. In grey: central nervous system.

---

[8] The importance of developmental characters in establishing phylogenetic relations has been underscored by Darwin in "*The Origin of Species*": chap. XIII, "*Mutual affinities of organic beings*": "*Embryonic characters are the most important of any in the classification of animals*" (p. 403); "*In most cases, the larvae though active, still obey more or less closely the law of embryonic resemblance. Cirripedes afford a good instance of this.*" (p. 420).

The cypris larvae of all cirripedes are astonishingly similar, even across the orders. The last larval stage of the Ascothoracida was at one time also called a cypris, although it looks somewhat different, and is no longer considered as a true homologue of the cirripede cypris (see Høeg et al. 2003). In addition, the three cirripede orders share a very peculiar and obviously apomorphic type of sperm (Healy & Anderson 1990; Høeg 1992).

1.3   *From taxonomy to evo-devo: the body plan of cirripedes*

1.3.1   *Segments*
Evolutionary interpretations of developmental genetics must be founded on a firm phylogeny of the species in question. This phylogenetic background is now established for the Cirripedia. Based on phylogeny and on comparative anatomy, we shall now try to define the ground plan of the Cirripedia. As in all crustaceans where a naupliar stage is present, four segments are formed in the embryo at hatching: the ocular-protocerebral segment, sometimes referred to as the "acron" (segment 0) [9], the antennular (a1) segment (segment 1), the antennal (a2) segment (segment 2) and the mandibular (mn) segment (segment 3) (Scholtz 1995b, 1997; Quéinnec 2001). Two other cephalic segments, maxillulary (mx1) and maxillary (mx2) and six thoracic segments develop during later naupliar stages. The thoracic appendages are not present before the cypris stage. As in all crustaceans, the posterior extremity of the body is formed by a telson, already present at hatching, often bearing a furca.

1.3.2   *Tagmata*
From this segment formation, tagmata can be derived. Tagmata are groups of segments, which are often viewed as functional units. For instance, the head would be a feeding and sensory unit, the thorax would be devoted to locomotion, and so on. Attention must be paid that too often the same word is used to designate a tagma in two different groups of animals, although the homologies are not established even among the Crustacea [10]. Moreover, the division between segments belonging to adjacent tagmata is sometimes arbitrary, or driven by a preconceived "*Bauplan*" or "archetype" in the mind of the observer.

1.3.3   *The naupliar head*
The first four segments, which are completed on hatching, form a well-defined entity, which can be called "anterior head" or "naupliar head" (Quéinnec 2001). The segments of this part of the head are difficult to define on the basis of the brain parts. Indeed, in the adult cirripedes, the supra-esophageal brain is often condensed into a single mass. Nonetheless, a clear distinction between proto- deuto- and tritocerebrum can be made in larval stages (Mourlan et al. 1985). In all cirripedes, a median nauplius eye, already present in the first nauplius stage, is maintained at all larval stages, including the cypris. Except in rhizocephalans, it persists in adults. In all cirripedes except in some rhizocephalans,

---

[9] Not all authors consider this more anterior part of the body to be a segment (see Scholtz 1997, 2001). The number of segments in the head of arthropods has long been a matter of debate. See for example the discussion in Rogers & Kaufman (1997).

[10] "*In nearly all cases, the post-cephalic somites can be further grouped into regions or tagmata distinguished by the shape of the somites or the character of their appendages. In descriptive carcinology two such regions are commonly distinguished as* thorax *and* abdomen, *but it must be pointed out that there is no morphological equivalence between the tagmata so named in different groups.*" (Calman 1909: 6).

compound eyes, homologous to those of other crustaceans and insects, form at the cypris stage. They are usually lost in the juvenile. The eyes belong to the first ("ocular-protocerebral") segment. The following three segments bear appendages, the uniramous antennule, the biramous antenna and the mandible. These appendages are present in all cirripede nauplii, even those, like rhizocephalans, which do not feed, being lecithotrophic and even devoid of any digestive system. Indeed, these cephalic appendages are used to swim rather than to feed. In all cirripedes, these appendages regress at the cypris stage [11], with the exception of the antennules, which undergo a great transformation. The antennules are used as a settlement device when the cypris finds its place or host. Like the cypris itself, this transformation and function of the antennule is a synapomorphy of the Cirripedia.

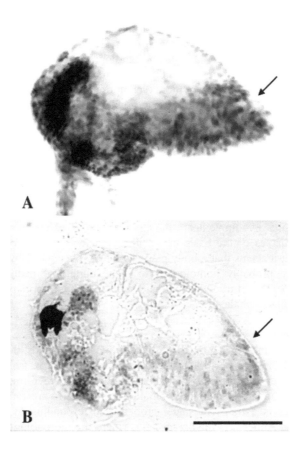

Figure 5. Anatomy of a *Sacculina carcini* nauplius. Nauplius stage II, nuclear staining by haemalaun; upper view: *in toto*; lower view: sagittal section. Arrow: furrow between thorax and abdomen. (by Éric Quéinnec and Monique Guimonneau).

---

[11] "*All naupliar swimming, feeding and flotatory features are lost. The dorsal shield extends as a bivalve carapace enclosing the body (...). Frontolateral horns and shield spines are lost. (...) The mandibles are reduced to gnathobasal rudiments associated with exposed but equally rudimentary and non-functional maxillules and maxillae.* " (Anderson 1994: 22).

1.3.4   *The gnathal head*
The following two segments bear the maxillules and the maxillae. They develop during the naupliar stages, like the thoracic segments. The maxillules and maxillae of the rhizocephalans, which do not feed as larvae, do not develop, to the extent as the mere presence of one or both of their gnathal segments has been questioned [12]. These two segments form the "gnathal" or "posterior" head (Quéinnec 2001). They are related to the rest of the trunk (thorax plus "abdomen") by their mode of development. However, they are distinct from the thorax by two criteria: i) their appendages develop before the thoracic ones; and ii) during larval development, a furrow, corresponding to a curve in the A/P axis of the body, clearly distinguishes it from the thorax (Fig. 5). This bend may be the reason why Darwin misunderstood the homologies between the tagmata of the Thoracica and of the so-called "Abdominalia" (= Acrothoracica). Indeed, in acrothoracicans, a secondary curve separates the anterior thorax from the posterior one, bearing cirri. In addition, their first pair of cirri is transformed into maxillipeds. These maxillipeds are actually used for feeding, together with the other mouthparts. In his attempt to homologize the body plan of cirripedes with that of malacostracans, Darwin described this segment bearing the maxillipeds as part of a "cephalothorax" region, leaving a bare thorax, composed of only two segments, and a limb-bearing abdomen of three segments. After Gruvel (1905), these acrothoracican features have been considered as derived, and thus not included within the ground plan of cirripedes.

1.3.5   *The head (cephalon)*
What we call here the naupliar (= anterior) head and the gnathal (= posterior) head, is usually considered as a single tagma, the head or cephalon [13]. In all cirripedes, the head is highly modified in the adult. The compound eyes of the cypris are shed, the three ocelli composing the naupliar eye move apart, the anterior head expands and forms the peduncle in pedunculate barnacles or the basis in sessile barnacles. Rhizocephalan adults are so highly modified that it seems that a body plan cannot be revealed. However, the roots that invade the host are probably homologous to the peduncle of pedunculate thoracicans (Bresciani & Høeg 2001). If true, the peduncle being a cephalic formation, this would fully justify the name "Rhizocephala".

1.3.6   *The thorax*
There is less difficulty in defining the thorax. As in most arthropods, thoracic segments are formed progressively from a posterior growth zone, located just in front of the telson. At the last naupliar instar (nauplius IV) all thoracic segments are formed and the buds of thoracic appendages appear. Segmented biramous appendages are formed just before the cypris stage. There is a transient stage, probably corresponding to what Darwin calls "*larval stage 2*" and Anderson "*the pre-cypris*", where the morphology of a cypris can be seen within the metanaupliar cuticle, just before molting. The cypris is indeed the stage

---

[12] In many cirripedes, the maxillules and maxillae are highly reduced appendages. In rhizocephalans they are not formed at all. However, *engrailed* staining shows that the maxillulary and maxillary segments are formed (Gibert et al. 2000).

[13] See Calman (1909), p. 5: "*The three anterior pairs of limbs (antennules, antennae and mandibles), which alone are present in the nauplius larva, show peculiarities of structure and development which seem to place them in a different category from the succeeding limbs, and there is some ground for regarding the three corresponding somites as belonging to a 'primary head' region. For descriptive purpose, however, it is convenient to treat the two following somites also as cephalic.*"

where the thoracic limbs become externally visible (Fig. 4). The external appearance of the cypris is quite similar in all cirripedes: a large anterior cephalic region followed by a six-segmented thorax. The cypris of the Acrothoracica possess the same six pairs of thoracic appendages as the other cirripedes, indicating that the loss of the 2nd and 3rd pairs of cirri is a secondary event (Turquier, 1972).

In thoracicans and acrothoracicans, the thoracic biramous appendages are modified into plumose cirri during the metamorphosis that transforms the cypris into a juvenile barnacle. The acrobatic movements accompanying this metamorphosis have been described in detail by Walley (1969) and Glenner & Høeg (1993) for the Thoracica and by Turquier (1970) for the Acrothoracica (see review in Anderson 1994). In the Thoracica, the result is that the A/P axis of the animal is no longer a straight line, and the dorsoventral axis is no longer related to "up" and "down". The head forms the base of the animal. The cirri, which are "ventral" thoracic appendages, are oriented laterally in pedunculate and upward in sessile thoracicans. In the Acrothoracica, the body of the adult is U-shaped, with a deep furrow separating the posterior-most thoracic segments from the rest of the body.

1.3.7 *A genital tagma?*

Before going into the ticklish question of the abdomen, it is worth saying a word about the genitalia. In all crustaceans, the male genitalia are located at the border between thorax and abdomen. The female apertures are often located one segment more anteriorly. Do the genitalia belong to the thorax or to the abdomen? Both views have been proposed, although the thorax seems to have more supporters. In fact, the argument maybe a circular one, since the trunk is sometimes divided into "thorax" and "abdomen" on the mere basis of the location of the genitalia [14]. In addition, it has been proposed that the segments bearing the genital appendages constitute a tagma *per se*, different from both the thorax and the abdomen. This has been supported by Averof & Akam (1995), on the basis of the expression pattern of the Hox gene *Abdominal-B* (*AbdB*) in the branchiopod crustacean *Artemia*, where it is expressed only in the segments bearing the male or female genital apparatus. In *Artemia*, these genital segments precede the limb-less abdomen [15].

Acrothoracicans have separate sexes, the male being reduced to a dwarf (Høeg 1995). Contrary to Delage's initial thinking, the rhizocephalans also have separate sexes. One or two dwarf males parasitize the female, herself a parasite of a crustacean. Most thoracican cirripedes are hermaphrodites, while some species also possess the "complemental" dwarf males discovered by Darwin. In the Thoracica, the penis, very large relative to the animal's size, is in effect located in or just behind the last thoracic segment. On the other hand, the female aperture is located in the first thoracic segment. This anterior location is found in all the Thecostraca, a clade comprising the Cirripedia, the Ascothoracida and certain other minor crustacean groups. With the "genital tagma" hypothesis, we are thus faced with a problem: in these crustaceans either the female genital segment does not belong to the so-called "genital" tagma, or the genital "tagma" is not continuous.

---

[14] *"In the Branchiopoda, they may denote respectively the regions in front and behind the genital apertures"* (Calman 1909: 6). *"The precise boundary between thorax and abdomen is sometimes difficult to fix. The names (…) are in this respect inconsistently applied, denoting in some groups limb-bearing and limb-less regions, in others the sections of the trunk which lie before and behind the genital openings."* (Borradaile et al. 1958: 349).

[15] *"In the majority of the Anostraca, the first two apodous somites, namely, the 12th and 13th of the trunk, are more or less completely fused, forming a 'genital segment' on which the genital ducts open."* (Calman 1909: 47).

1.3.8   *Last but not least: the abdomen controversy*

In 1994, Anderson published a summary of present knowledge on cirripedes, which was lacking since the monographs of Darwin and Gruvel. He wrote: "*As maxillipodan crustaceans, cirripedes have an organization based on six head segments, six thoracic segments and five abdominal segments, although the abdomen of all cirripedes is greatly reduced*". This is a typically circular argument based on the prejudice of an archetypal *Bauplan*. Cirripedes *must* have an abdomen of five segments "*because*" they are maxillopodans. And why should they belong to the "Maxillopoda" (if the "Maxillopoda" were monophyletic)? *Because* of their body plan! Let us look at the facts, based on observations.

Adults. Textbooks generally agree that there is no abdomen in adult cirripedes [16] (Borradaile et al. 1958; Schram 1986; Brusca & Brusca 1990; Ruppert & Barnes 1994, Newman 1997). Adult rhizocephalans do not present any segmentation at all. Although the interna could possibly be seen as deriving from a cephalic part of the body, the externa is not much more than a sack (hence the name "*Sacculina*") containing the ovaries and the hyper-parasitic males. In most adult acrothoracicans, the caudal furca is lost and there is nothing at the rear of the body following the sixth thoracic segment, so there is nothing that can be viewed as an abdominal segment (Turquier 1970 and Fig. 2). In adult thoracicans, the only structure behind the sixth pair of cirri is the penis and, in some pedunculate thoracicans, the furca. As we saw above, the mere presence of a penis is a weak argument in favor of the existence of an abdomen. Darwin, when he defined the "*Names given to the different parts of Cirripedes*" did not mention an abdomen in adult thoracicans [17]. Fig. 6 shows in detail the posterior extremity of the body of the pedunculate barnacle *Lepas anatifera*, as drawn in Darwin's book (1851).

Hence, if there is an abdomen in cirripedes, it is only a transient structure, restricted to the larval stages. This is an additional argument to separate the Ascothoracida from the Cirripedia.

Nauplius stages. When the nauplius develops, it can be seen that the hind part of the trunk does not form a single lobe but surprisingly forms two. Depending on the species and on the naupliar stage, these two lobes are more or less prominent, although always present. In schematic drawings of a sagittal section of the body of a late nauplius, the anus opens *in between* these two hind lobes (Fig. 7; Walley 1969, p. 258) [18]. Each lobe ends with spines or "processes". In addition, the dorsal shield of the nauplius also possesses cuticle ornaments, forming spines, such that up to three such "processes" can be observed at the bottom of the advanced nauplius [19]. The presence of several lobes and "processes" or "appendages" has generated a great confusion in the literature. Which is the abdomen, or abdomen

---

[16] Schram 1986, p. 539, among the characters used to derive the first cladistic analysis of Crustacea "*The Cirripedia* sensu stricto *(...) have lost the abdomen*"; Borradaile et al. 1958, p. 341-2, recognize "*no abdominal somites*" in Thoracica and Acrothoracica, but include the "Apoda" and the Ascothoracida among the Cirripedia.

[17] "*The body consists of the* thorax *supporting the cirri, and of an especial enlargement, or downward prolongation of the thorax, which includes the stomach, and which I have called the* prosoma" (1851, p. 6). However, he wrote in 1854, p. 65-66, about the sessile barnacles: "*The prosciformed penis lies folded under the thorax; and I believe (from what is seen in the anomalous genus* Proteolepas*) that it normally arise from the ventral surface of the terminal point of the rudimentary abdomen.*" But we have seen that *Proteolepas*, the unique member of the "Apoda", is not in fact a cirripede.

[18] This can be observed only in the Thoracica. Indeed, in the Acrothoracica, the digestive apparatus is reduced, and there is no anus (Turquier 1970) and in the Rhizocephala, there is no digestive tube at all.

[19] These spines or "processes" can reach a huge size, up to several times the size of the nauplius' body, as in the pedunculate barnacle *Lepas anatifera*.

Anlage? The lobes and their processes are given different names: what one author calls "abdominal", another calls "thoracic" while a third uses "thoraco-abdominal" [20].

The dorsal-most lobe possesses an unpaired process, while the ventral-most lobe has a forked one. To our eyes, this designates the latter as the furca. The lobe presenting this process must be the telson. We suggest that the other more dorsal lobe is the abdominal *Anlagen*. It could seem strange that this abdominal region so often bears a spiny process. However, these ornaments, like those of the dorsal shield, are mere cuticular formations that cannot be used as evidence for a true appendage.

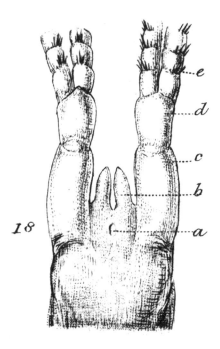

Figure 6. The posterior extremity of the body of an adult thoracican. From Darwin (1851) plate X, fig. 18. Legend by Darwin: *Caudal appendages, and basal segment of the sixth pair of cirri, of* Lepas anatifera*; (a) anus; (b) caudal appendages; (c) lower segment of pedicel of sixth cirrus; (d) upper segment of ditto; (e) basal segments of the two rami.* No abdomen is present between the sixth thoracic segment and the furca.

---

[20] Runnström (1924-25) describes two hind spines in a sketch of the nauplius I of *Balanus balanoides*. One is dorsal, called "*caudalstachel*", and is probably the shield spine. The other, linked to a lobe, is unpaired but terminates by a forked end. It is called "*ventraler abdominalanhang*" (p. 7). On the sketch of the nauplius II stage (p. 9), a new lobe appears more ventrally, bearing a "process" made of two pieces, which makes three hind spines, whereas the primary lobe, which is now more dorsal in location, is still called "*ventraler abdominalanhang*".

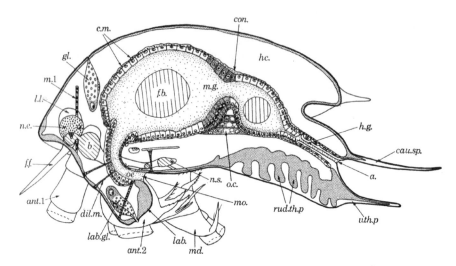

Figure 7. Internal anatomy of a *Balanus* nauplius. A schematic drawing of the internal anatomy of a stage V nauplius, from Walley (1969). Note the two hind lobes and that the anus opens in between the superior and the inferior lobe. *a.* = anus; *cau. sp.* = caudal spine; *v. th. p.* = ventral thoracic process.

Cypris stage. Under the term "*Metamorphoses*" Darwin (1851) gave a general description of the development of pedunculate barnacles. He described an abdomen in the cypris, which he called "*larva, last stage*" or "*pupa*"(p. 19): "*The abdomen is small, and its structure might easily be overlooked without careful dissection of the different parts: it consists of three segments; the first can be seen to be distinct from the last thoracic segment, bearing the sixth pair of limbs, only from the fold of the epimeral (= pleural) element, and from its difference in shape; the second segment is very short, but quite distinct; the third is four or five times as long as the second, and bears at the end two little appendages, each consisting of two segments, the lower one with a single spine, and the upper one with three, very long, plumose spines, like those of the rami of the thoracic limbs. The abdomen contains only the rectum and two delicate muscles running into the two appendages, between the bases of which the anus is seated.*" Two drawings of the cypris of *Lepas australis* illustrate Darwin's words (1854, plate XXX). This species, which he collected in the South American seas, was easier to observe since its size is about 3 mm, 10 times as large as the cypris of the common barnacles from European seas. Reading Darwin today, it is clear that the extremity of the abdomen would now be considered as the non-segmented posterior part of the body: *i* in Darwin's drawing would be the forked appendage of the telson and *h* would be the telson proper (Fig. 8). In all crustaceans, the anus opens in the telson [21].

---

[21] Calman (1909: 4): "*At the posterior end of the body is a terminal segment known as the telson, upon which the anus opens*"; and p. 7: "*The telson (...) has not the value of a true somite*"; see also the discussion by Schram (1986: 6-8).

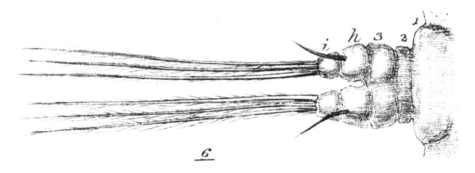

Figure 8. Abdomen of the cypris of Lepas australis by Darwin. Darwin (1854), plate XXX, fig. 6; legend: (1) first abdominal segment, attached to the posterior thoracic segment; (2) second abdominal segment; (3) third or last abdominal segment; (h) lower segment; (i) upper segment of caudal appendage.

In his "Monographie des Cirrhipèdes", Gruvel (1905: 448) shows a drawing of the cypris of the same *L. australis*, showing a three-segmented structure in front of the furca-bearing telson. Although it is designated as an "*appendice caudal*" in the figure legend, it seems to be the abdomen. However, this picture might simply be a re-interpretation of Darwin's writings. Since then, apart from copies of Gruvel's drawing, a number of drawings and photographs from light or scanning electron microscopes (SEM) of cypris larvae from various species have been published. Nobody showed a segmented structure in addition to the telson [22] until…1999! (Kolbasov et al. 1999). In this work, the authors describe the tiny, but segmented, abdomen of an acrothoracican cypris, documented by SEM. The form and location of this abdomen is very similar to that drawn on Gruvel's picture. Not only it is very small compared to the thorax, but it is also situated dorsal to it. This dorsal situation, surprising at first glance, can easily be explained by the difference in proportions between the two tagmata: the great development of the thorax could have "pushed" the reduced abdomen dorsally.

2 GENES IN DEVELOPMENT

We have undertaken an "evo-devo" approach to the body plan of the Cirripedia. The first step of such a study is to clone genes homologous to genes involved in the development of *Drosophila*. The second step is to draw the expression patterns of the cloned genes and to compare them to the patterns of homologous genes in other species. Comparative developmental genetics thus provides new characters for comparative anatomy and embryology. A third step is to assay the function of the genes under study. Up to now, most "evo-devo" studies have been limited to the first two steps, due to the fact that functional genetic assays were only available in a handful of model organisms. But now, thanks to the use of new vectors for transgenesis and to the new method of RNA interference (RNAi), functional genetic assays have become feasible in an increasing number of animal species.

---

[22] In fact, a single "segment" bearing the furca, which is obviously the telson, has been called the "abdomen" by many authors.

In our own "evo-devo" studies on cirripedes, we first accumulated molecular data in a sample of species representative of the three orders of the Cirripedia and in some other crustaceans, including the ascothoracid *Ulophysema* as a sister group. For expression studies, we chose the rhizocephalan *Sacculina carcini* as a model organism. Indeed, all cirripede larvae are very much alike. A single *Sacculina* sack produces several hundreds of nauplii in a single burst. The larvae are easy to raise: being lecithotrophic, they do not feed throughout their four naupliar stages and their cypris life. The drawback is that *Sacculina* larvae present some derived characters, such as loss of a digestive tube.

2.1 *Segmentation genes*

2.1.1 engrailed
In *Drosophila, engrailed* has been shown to play a pivotal role in segment determination. It is expressed in the posterior compartment of every segment and of every appendage, where it determines the parasegmental border. This segmental function appears to be preserved in all arthropods where it has been studied. We found two *engrailed* genes, which we named *engrailed.a (en.a)* and *engrailed.b (en.b)*, in all cirripede species studied, but only one in the sister taxon *Ulophysema oeresundense* (Gibert et al. 1997; Gibert 2000; Fig. 9). This provides an additional support to the phylogeny of Cirripedia, including the Rhizocephala but not the Ascothoracida.

The expression pattern of both *Sacculina engrailed* genes has been studied in detail, by immuno-staining and *in situ* hybridization for *en.a*, by *in situ* hybridization alone for *en.b* (Quéinnec et al. 1999; Gibert et al. 2000) (Fig. 10A). As expected, both genes are expressed in parasegmental stripes, in every segment of the trunk. The metameric expression of *en.b* is more intense and lasts longer than that of *en.a*. In contrast, *en.a* presents a late phase of expression in the central nervous system (CNS) that has not been detected for *en.b*. For both *engrailed* genes, no expression has been detected in the naupliar (anterior) head. On the other hand, expression was found in the two maxillary segments, mx1 and mx2, indicating that at least a trace of these segments is present in *Sacculina* larvae, even though the corresponding appendages do not develop. Behind these two segments, six stripes corresponding to the six thoracic segments appear in an anterior to posterior progression. And, more importantly for our purpose, up to five very tiny stripes are present in a hind and dorsal position. The domain where these tiny stripes appear is physically separated by a furrow from the thoracic region (Fig. 5). We interpret this body part as a "vestigial abdomen". We believe that it corresponds to the dorsal hind lobe described in cirripede nauplii by several authors (see above).

2.1.2 ScaPouIII
In *Sacculina carcini*, we have cloned and studied a gene belonging to the third family of POU transcription factors, which we named *ScaPouIII* (Gibert et al. 2002). This gene is the ortholog of the *Drosophila* gene *ventral-vein-less* (*vvl*). In situ hybridization revealed that it is expressed in stripes in the thorax. Moreover, the *ScaPouIII* stripes are located in the anterior compartment of segments and of forming appendages, a pattern complementary to the *engrailed* one. Given its pattern and timing of expression, we hypothesize that *ScaPouIII* is involved in the maintenance of *engrailed* expression in *Sacculina*. Interestingly enough, *ScaPouIII* is not expressed in the vestigial abdomen.

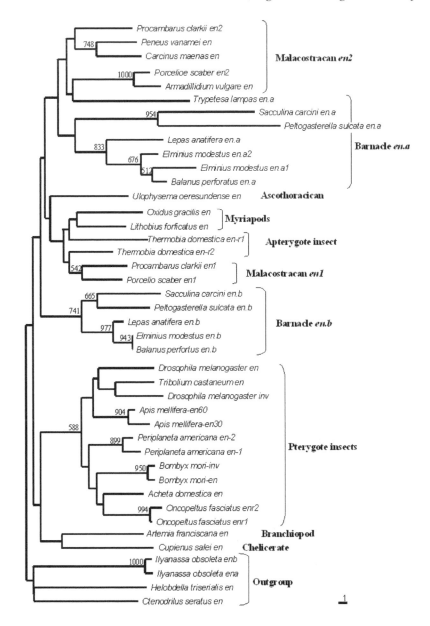

Figure 9. Phylogenetic analysis of arthropod *engrailed* genes. Because only partial sequences are known in many instances, a short fragment of the Engrailed proteins could be aligned (alignment available on request). This fragment comprises the COOH part of the homeodomain and the beginning of the following conserved domain. The tree presented here is a distance tree by Neighbor Joining (NJ). Figures at nodes are Bootstrap Proportions (BP) on 1000 replicas. Data from cirripedes, from the ascothoracid *Ulophysema*, from the crab *Carcinus*, from the isopod *Armadillidium* and from the myriapod *Lithobius* are from Gibert et al. 1997 and Gibert 2000. All other sequences are available on GenBank.

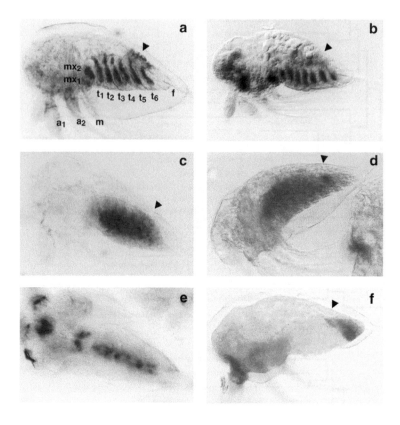

Figure 10. [See Plate 1 in color insert.] Expression pattern of some developmental genes in *Sacculina* larvae. In situ hybridization on Nauplius III larvae. (a) *engrailed.b*; (b) *ScaPouIII*; (c) *Antp*; (d) *Ubx*; (e) *Diva*; (f) *caudal*. Lateral views, except in e, ventral view. Arrowhead: vestigial abdomen.

A striped expression pattern of *ScaPouIII* was completely unexpected, since *vvl* is not a segmentation gene in *Drosophila*. In other crustaceans, *vvl* homologues have been cloned from the branchiopod *Artemia franciscana* and from the malacostracans *Porcellio scaber* (woodlouse) and *Procambarus clarkii* (crayfish). Expression data have been reported in *Porcellio* only. In this isopod crustacean, *vvl* is expressed in the proximal end of each appendage in formation (Abzhanov & Kaufman 2000). It is therefore possible that a repeated expression pattern of *vvl* might be ancestral in crustaceans.

2.2   Hox *and* para-Hox *genes*

2.2.1   *The* Hox *repertoire in Cirripedia*
The role of the *Hox* genes in designing the body plan of the bilaterian animals is so conspicuous that they have been called "the architect genes" (McGinnis & Kuziora 1994). Hence, *Hox* genes were obvious candidates to study in cirripedes.   We described the

repertoire of the *Hox* genes in the Cirripedia (Mouchel-Vielh et al. 1998, 1999). We studied three cirripede species representative of the three orders, i.e., the acorn barnacle *Elminius modestus* as a representative of the Thoracica, *Trypetesa lampas* as a representative of the Acrothoracica and *Sacculina carcini* as a representative of the Rhizocephala. Using the same methods we studied the ascothoracid *Ulophysema oeresundense* as an out-group. Using PCR on genomic DNA or cDNA, we amplified short fragments of the genes of interest. The amplified fragments made it possible to identify the orthology relations of the obtained sequences with the corresponding *Drosophila* genes.

The results are summarized in Fig. 10. Out of the 10 *Hox* genes composing the basal repertoire in the Pan-Arthropoda (Grenier et al. 1997) and more largely in Ecdysozoa (de Rosa et al. 1999), we found only eight in every cirripede species studied. The absence of an ortholog of the *zen/Hox3* class was not a surprise. These genes diverged in sequence in the insect lineage (Falciani et al. 1996), and also in a number of crustacean species (Balavoine, pers. comm.). A member of this group in cirripedes might well have escaped our screening.

More surprising was the absence of the *abdominal-A* (*abdA*) representative. Our search for *abdA* was successful in the ascothoracid *Ulophysema*. In addition to PCR, we looked for *abdA* in cirripedes by screening cDNA and genomic DNA libraries, with no positive result. The *abdA* gene is closely related to the *Ultrabithorax* (*Ubx*) gene (Akam at al. 1988; Balavoine 1997). It is specific to the Ecdysozoa (Grenier et al. 1997; de Rosa et al. 1999). In all arthropods where it is known, the sequence of *abdA* is very much conserved in the homeobox and flanking regions. In cirripedes, its sequence could have diverged to such a point that it escaped our screening. Alternatively, this gene may be absent from the genome of the Cirripedia.

2.2.2 *The expression of* Hox *genes in* Sacculina

In the rhizocephalan *Sacculina carcini*, the *Hox* genes *Antennapedia* (*Antp*) and *Ubx* are expressed in largely overlapping domains comprising the whole thorax and the maxillulary and maxillary segments (Mouchel-Vielh et al. 2002). Neither *Antp* nor *Ubx* is expressed in the abdomen (Fig. 10). This agrees with the situation in other crustaceans (Averof & Akam 1995; Abzhanov & Kaufman 2000a, b, 2003).

2.2.3 *The* Diva *gene of* Sacculina

During the course of our *Hox* screening, we retrieved a gene that we named *Diva*, for "*Divergent Antennapedia*" because, although somewhat divergent, its sequence is close to that of the *Antennapedia*-like sub-class of Hox genes. We demonstrated that it is a homologue of the divergent *Hox* gene *fushi tarazu* (*ftz*), involved in segmentation in *Drosophila*. In the chelicerate *Archegozetes longisetosus*, the expression pattern of a *ftz* ortholog is homeotic-like (Telford 2000). This kind of pattern has not been found in *Sacculina*, where *Diva* is expressed only in the central nervous system of the head and thorax (Mouchel-Vielh et al. 2002).

2.2.4 *The* caudal *gene of* Sacculina

A family of genes, not linked to the *Hox* complex(es), but phylogenetically related to the *Hox* genes, has been called the "*para-Hox*" genes by Peter Holland (Brooke et al. 1998). The most well-known member of this family is the *caudal* (*cad*) gene. First discovered in *Drosophila* (Mlodzik et al. 1985; Macdonald & Struhl 1986), like so many developmental genes, *caudal* has been found preserved in sequence from the origin of the Eumetazoa

(Gauchat et al. 2000). As its name indicates, it is expressed in the posterior terminal end of the body of bilaterian animals. We were therefore interested in studying *caudal* in cirripedes (Rabet et al. 2001). It is first expressed in the so-called "caudal papilla" of the embryo, which is assumed to be the *Anlage* of the posterior growth zone that generates the segments in larvae. Indeed, *caudal* is largely expressed in the trunk segments in early nauplius stages, but its expression domain later regresses posteriorly. In fact, it is expressed in the telson, like the *caudal* homologue of the decapod crustacean *Procambarus* (Abzhanov & Kaufman 2000b). However, *caudal* is never expressed in the vestigial abdomen of *Sacculina* at any point of larval development.

3 DISCUSSION: THE ABDOMEN OF CIRRIPEDES

Without a doubt, the crustacean ancestral way of living was a freely swimming mode. All members of the Cirripedia have abandoned this ancestral behavior for a fixed life. As a correlate, they present a strikingly modified body plan. In the adult, the head is profoundly modified, the cephalic appendages being lost. They simply sit on their head, which generates the device used to fix them on their substratum or host. In non-parasitic species their thorax turns ventral side up, bearing modified legs, the cirri. Their abdomen is transient, being reduced to a small vestige present only in larval stages.

We looked for the genetic developmental mechanisms that underlie these morphological changes by studying developmental genes in the cirripede *Sacculina carcini* and comparing the results with the expression pattern of homologous genes in *Drosophila melanogaster* and in other crustaceans. We focused on the much-debated question of the abdomen of cirripedes.

First, the expression of *engrailed* genes revealed unexpected tiny stripes in a dorsal posterior region of the larvae. We interpret this pattern as the relics of a genetic mechanism leading to the formation of an abdomen. In fact, this region corresponds to a dorsal lobe that has been thoroughly observed in cirripede larvae by numerous authors. The expression of *engrailed* has been shown in all studied arthropods to correlate with the parasegmental border. Although *engrailed* is a key regulator of segmental *determination*, generation of *engrailed* stripes is not sufficient for the eventual *formation* of segments. Indeed, Scholtz has shown that posterior-most abdominal *engrailed* stripes in the crayfish *Cherax destructor* correspond to segments that are never morphologically formed (Scholtz et al. 1994; Scholtz 1995a). In the present case, the dorsal striped pattern that we observed in *Sacculina* larvae correlates with morphological observations (Kolbasov et al. 1999; Blin et al. 2003).

Thus, *engrailed* expression data answer the first set of questions: Is there an abdomen in cirripedes? And what is the abdomen? In addition, it shows that at least a part of the genetic mechanism leading to the formation of a segmented abdomen is active, and, being based on *engrailed*, conserved by comparison to what is known in other arthropods. This leads to another question: why does the cirripede abdomen remain rudimentary?

We noticed an unexpected feature concerning the timing of appearance of these abdominal *engrailed* stripes. The thoracic *engrailed* stripes are progressively set up from anterior to posterior during naupliar development. This is particularly clear in the case of *en.b*, which is more intensively expressed than *en.a*. However, for both *engrailed* genes, the abdominal stripes are formed simultaneously, long before the sixth thoracic stripe appears. In fact, they can be seen as early as *engrailed* expression can be detected (Gibert et al. 2000). This shows that the regulatory mechanisms that promote *engrailed* expression differ between the thorax and the abdomen. Moreover, given that we can reasonably infer that in the ancestral arthropod *engrailed* was expressed earlier in the fore part of the trunk than in the hind part, a heterochronic change probably occurred during the evolution of the lineage leading to the Cirripedia.

We have also shown that *ScaPouIII*, the homologue of the *vvl* gene of *Drosophila* is expressed in thoracic stripes in *Sacculina*. From this pattern, we can speculate that it plays a role in segment determination. As its expression pattern is complementary to the *engrailed* stripes in the thorax, *ScaPouIII* could be a negative regulator of one or both *Sacculina engrailed* genes. Other lines of indirect evidence support this hypothesis. Gel retardation assays showed that the VVL protein can bind to putative *cis*-regulatory sequences of the *Sacculina engrailed.a* gene, and the *vvl* gene regulates the expression of a transgene driven by these sequences in *Drosophila* (Gibert et al. 2002). *ScaPouIII* is not expressed in the vestigial abdomen. This can be taken as another indication that the mechanisms leading to the determination and/or the formation of segments in the abdomen differ from those in the thorax. An analogous situation is known in *Drosophila*, where the timing and the regulation of the cephalic *engrailed* stripes differ from those of the trunk stripes (Schmidt-Ott & Technau 1992; Rogers & Kaufman 1996).

In order to gain more insights into the process of abdominal development, we looked for *Hox* and *Hox*-related genes. In an extensive survey of *Hox* genes, we failed to find the homologue of the *abdA* gene in cirripedes (see above). We concluded from our negative result that *abdA* was either lost or deeply derived in sequence in the studied cirripedes (Fig. 11). In both cases, this genetic event, linked to the divergence of the cirripede lineage from the other thecostracans, could be related to the reduction of the abdomen. In crustaceans, the expression of an *abdA* gene has been studied in the branchiopod *Artemia franciscana* (Averof & Akam 1995) and in two malacostracans, the isopod *Porcellio scaber* and the decapod *Procambarus clarkii* (Abzhanov & Kaufman 2000a, b). In the former, its expression pattern largely overlaps that of *Ubx*, being expressed only in the thorax, but in the malacostracans, it is specifically expressed in the abdomen (= pleon). It is the only *Hox* gene currently known to be expressed in this tagma in crustaceans. On the other hand, it should be pointed that the homology between the pleon of the malacostracans and the "abdomen" of non-malacostracan crustaceans is discussed (Walossek & Müller 1997; Gruner 1997; Deutsch 2001).

The *Hox* genes *Antp* and *Ubx* are not expressed in the vestigial abdomen of *Sacculina* larvae. This agrees with the situation in other crustaceans (Averof & Akam 1995; Abzhanov & Kaufman 2000a, b). We also studied the expression of a related gene, which we called *Diva*. The homeobox sequence of *Diva* is closely related to the homeoboxes of the median group of *Hox* genes, comprising *Sex combs reduced (Scr) Antp, Ubx* and *abdA*. *Diva* is in fact the ortholog of the *Drosophila* gene *ftz*. In the mite *Archegozetes*, the *ftz* homolog is expressed in the opisthosoma, at least in the early stages (Telford 2000). *Diva* is not expressed in the abdomen either.

On the other hand, the *Hox* gene *AbdB* is expressed in the vestigial abdomen. This expression pattern in *Sacculina* contrasts with that in *Artemia*, where *AbdB* is expressed exclusively in genital segments, and not in the post-genital "abdomen". This fits the proposal of Damen & Tautz (1999) that the roles of *AbdB* in the specification of the hind part of the body may be independent from its role in the specification of the genitalia. It would be interesting to study the expression pattern of *AbdB* in other non-malacostracan crustaceans, such as copepods. In *Drosophila,* the *para-Hox* gene, *caudal,* is critical for specifying the hind abdominal body segments (Schulz & Tautz 1995; Moreno & Morata 1999). The *caudal* homologue of *Sacculina* is not expressed in the vestigial abdomen. Again, this lack of expression may be related to the lack of complete development of this tagma. Alternatively, in *Sacculina*, the role of *caudal* could have been taken by the related *Hox* gene *AbdB*.

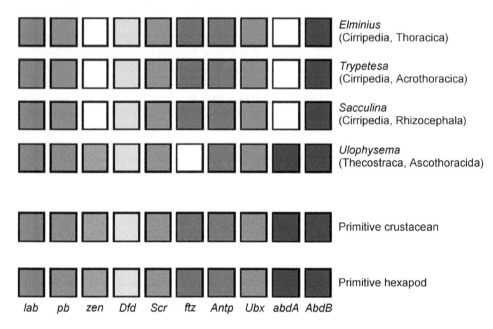

Figure 11. [See Plate 2 in color insert.] Cirripede *Hox* genes. Cirripede and ascothoracid *Hox* genes from Mouchel-Vielh et al. 1998. The Hox repertoire in a primitive crustacean is inferred from current knowledge on branchiopod and malacostracan genes: *Artemia franciscana* (Averof & Akam 1993, 1995); *Carcinus maenas* (Mouchel-Vielh et al. 1998 and our unpublished results); *Procambarus clarkii* and *Porcellio scaber* (Abzhanov & Kaufman 2000a, b). Hexapod primitive repertoire from various sources. Empty squares mean no information.

4 CONCLUSION

To summary, given our current knowledge, the evolution of the Cirripedia toward a fixed life and adaptive modification of their body plan seems to have been accompanied by multiple changes in the genetic basis of their development.

Important general conclusions can be drawn from our studies on these animals:
1) Gain by duplication (*en.a* and *en.b*), and putative loss or high divergence (*abdA*), of developmental genes add strong evidence to phylogenetic analyses.
2) Comparing developmental genes in phylogenetically related species helps infer what could be the ancestral state. In particular, in our study of the Cirripedia putative genetic changes from the ancestor to the extant species occurred at different levels: i) genomic levels (see point 1); ii) mutations in coding sequences, although the neutral or adaptive

nature of these mutations remains to be demonstrated; and iii) mutations at the level of *cis*-acting regulatory elements and/or *trans*-acting factors, as indirectly shown by changes in gene expression.

3) Changes in body plans can be correlated with changes in the genetic bases of development. However, even what appears to be a single morphological character, such as the strong reduction of the abdomen, cannot be correlated to a single change in developmental genes. This may simply reflect the interconnection of multiple developmental genetic pathways in the generation of any morphological structure.

REFERENCES

Abzhanov, A. & Kaufman, T.C. 2000a. Crustacean (malacostracan) Hox genes and the evolution of the arthropod trunk. *Development* 127: 2239-49.

Abzhanov, A. & Kaufman, T.C. 2000b. Embryonic expression of the Hox genes of the crayfish *Procambarus clarkii* (Crustacea, Decapoda). *Evol. Dev.* 2: 271-283.

Abzhanov, A. & Kaufman, T.C. 2003. HOX genes and tagmatization of the higher Crustacea (Malacostraca). In: Scholtz, G. (ed), *Crustacean Issues 15, Evolutionary Developmental Biology of Crustacea*: 43-74. Lisse: Balkema.

Akam, M., Dawson, I. & Tear, G. 1988. Homeotic genes and the control of segment diversity. *Development* 104: 123-133.

Anderson, D.T. 1994. *Barnacles. Structure, function, development and evolution.* London: Chapman & Hall.

Averof, M. & Akam, M. 1995a. *Hox* genes and the diversification of insect and crustacean body plans. *Nature* 376: 420-423.

Averof, M. & Akam, M. 1995b. Insect-crustacean relationships: insights from comparative developmental and molecular studies. *Phil. Trans. R. Soc. Lond. B* 347: 293-303.

Balavoine, G. 1997. The early emergence of platyhelminths is contradicted by the agreement between 18S rRNA and Hox genes data. *C. R. Acad. Sci. Paris [III] (Life Science)* 320: 83-94.

Barnes, R.S.K., Calow, P. & Olive, P.J.W. 1988. *The Invertebrates: a new synthesis.* Oxford: Blackwell Pub.

Billoud, B., Guerrucci, M.A., Masselot, M. & Deutsch, J.S. 2000. Cirripede phylogeny using a novel approach: molecular morphometrics. *Mol. Biol. Evol.* 17: 1435-45.

Blin, M., Rabet, N., Deutsch, J. & Mouchel-Vielh, E. 2003. Possible implication of Hox genes *Abdominal-B* and *abdominal-A* in the specification of genital and abdominal segments in cirripedes. *Dev Genes Evol.* 213: 90-96.

Borradaile, L.A., Potts, F.A. & Kerkut, G.A. 1958. *The Invertebrata. A manual for the use of students.* Cambridge: Cambridge University Press.

Bresciani, J. & Høeg, J.T. 2001. Comparative ultrastructure of the root system in rhizocephalan barnacles (Crustacea: Cirripedia: Rhizocephala). *J. Morph.* 249: 9-42.

Brooke, N.M., Garcia-Fernandez, J. & Holland, P.W.H. 1998. The ParaHox gene cluster is an evolutionary sister of the Hox gene cluster. *Nature* 392: 920-922.

Brusca, R.C. & Brusca, G.J. 1990. *Invertebrates.* Sunderland: Sinauer.

Calman, W.T. 1909. *Crustacea,* London: Adam & Charles Black.

Damen, W.G. & Tautz, D. 1999. *Abdominal-B* expression in a spider suggests a general role for *Abdominal- B* specifying the genital structure. *J. Exp. Zool.* 285: 85-91.

Darwin, C. 1851. *A monograph on the subclass Cirripedia, with figures of all the species. The Lepadidae or pedunculated cirripedes.* London: Ray Society.

Darwin, C. 1854. *A monograph on the subclass Cirripedia, with figures of all the species. The Balanidae, the Verrucidae, etc.* London: Ray Society.

de Rosa, R., Grenier, J., Andreeva, T., Cook, C., Adoutte, A., Akam, M., Carroll, S. & Balavoine, G. 1999. Hox genes in brachiopods and priapulids and protostome evolution. *Nature* 399: 772-776.

Delage, Y. 1884. Évolution de la Sacculine (*Sacculina carcini* Thomps.) crustacé endoparasite de l'ordre nouveau des Kentrogonides. *Arch. Zool. Exp. Gén.* II: 417-736.

Deutsch, J.S. 2001. Are Hexapoda members of the Crustacea? Evidence from comparative developmental genetics. *Ann. Soc. Entomol. Fr.* 37: 41-49.

Falciani, F., Hausdorf, B., Schröder, R., Akam, M., Tautz, D., Denell, R. & Brown, S. 1996. Class 3 *Hox* genes in insects and the origin of *zen*. *Proc. Natl. Acad. Sci. USA* 93: 8479-8484.

Forest, J. 1994 Les crustacés. Définition, formes primitives et classification. In: Grassé, P. P. (ed), *Traité de Zoologie. Crustacés*: 1-8. Paris: Masson.

Gauchat, D., Mazet, F., Berney, C., Schummer, M., Kreger, S., Pawlowski, J., & Galliot, B. 2000. Evolution of Antp-class genes and differential expression of Hydra Hox/paraHox genes in anterior patterning. *Proc. Nat.l Acad. Sci. USA* 97: 4493-8.

Gibert, J.-M. 2000. Les gènes *engrailed* et le plan d'organisation des crustacés cirripèdes. PhD thesis, Université P et M Curie, Paris.

Gibert, J.M., Joannin, N., Blin, M., Rigolot, C., Mouchel-Vielh, E., Queinnec, É. & Deutsch, J.S. 2002. Heterospecific transgenesis in *Drosophila* suggests that *engrailed.a* is regulated by POU proteins in the crustacean *Sacculina carcini*. *Dev. Genes Evol.* 212: 19-29.

Gibert, J.M., Mouchel-Vielh, E. & Deutsch, J.S. 1997. *engrailed* duplication events during the evolution of barnacles. *J. Mol. Evol.* 44: 585-594.

Gibert, J.M., Mouchel-Vielh, E., Quéinnec, É., Deutsch, J.S. 2000. Barnacle duplicate *engrailed* genes: divergent expression patterns and evidence for a vestigial abdomen. *Evol. Dev.* 2: 194-202.

Glenner, H., Grygier, M.J., Høeg, J.T., Jensen, P.J. & Schram, F.R. 1995. Cladistic analysis of the Cirripedia Thoracica. *Zool. J. Linn. Soc.* 114: 365-404.

Glenner, H., Høeg, J.T. 1993. Scanning electron microscopy of metamorphosis in four species of barnacles (Cirripedia, Thoracica, Balanomorpha). *Mar. Biol.* 117: 431-439.

Glenner H, Høeg JT 1995. A new motile, multicellular stage involved in host invasion by parasitic barnacles (Rhizocephala). *Nature* 377: 147-150.

Glenner, H., Høeg, J.T., O'Brien, J.J. & Sherman, T.D. 2000. Invasive vermigon stage in the parasitic barnacles *Loxothylacus texanus* and *L. panopei* (Sacculinidae): closing of the rhizocephalan life cycle. *Mar. Biol.* 136: 249-256.

Grenier, J.K., Garber, T.L., Warren, R., Whitington, P.M. & Carroll, S. 1997. Evolution of the entire arthropod Hox gene set predated the origin and radiation of the onychophoran/arthropod clade. *Curr. Biol.* 7: 547-553.

Gruner, H.-E. 1997. Les Crustacés. Segmentation, tagmes et appendices. In: Grassé, P.P. (ed) *Traité de Zoologie, Vol VII-I*: 9-47. Paris: Masson.

Gruvel, A. 1905. *Monographie des cirrhipèdes ou thécostracés*. Paris: Masson.

Healy, J.M. & Anderson, D.T. 1990. Sperm ultrastructure in the Cirripedia and its phylogenetic significance. *Rec. Aust. Mus.* 42: 1-26.

Høeg, J.T. 1987. Male cyprid metamorphosis and a new larval form, the trichogon, in the parasitic barnacle *Sacculina carcini* (Crustacea: Cirripedia: Rhizocephala). *Phil. Trans. R. Soc. Lond. B* 317: 47-63.

Høeg, J.T. 1992a. The phylogenetic position of the Rhizocephala: are they truly barnacles? *Acta Zool.* 73: 323-326.

Høeg, J.T. 1992b. Rhizocephala. In: Harrison, F. W. & Humes, A. G. (eds) *Microscopic anatomy of invertebrates. Crustacea*: 313-345. New-York: Wiley-Liss.

Høeg, J.T. 1995. Sex and the single cirripede: A phylogenetic perspective. In: Schram, F.R. & Høeg, J.T. (eds), *New frontiers in Barnacle evolution:* 195-207. Rotterdam: Balkema.

Høeg, J.T. & Kolbasov, G.A. 2002. Lattice organs in y-cyprids of the Facetotecta and their significance in the phylogeny of the Crustacea Thecostraca. *Acta Zool.* 83: 67-79.

Høeg, J.T. Lagersson, N.C. & Glenner, H. 2003. The complete cypris larva and its significance in thecostracan phylogeny. In: Scholtz, G. (ed), *Crustacean Issues 15, Evolutionary Developmental Biology of Crustacea*: 197-215. Lisse: Balkema.

Kolbasov, G.A., Høeg, J.T. & Elfimov, A.S. 1999. Scanning electron microscopy of acrothoracican cypris larvae (Crustacea, Cirripedia, Acrothoracica, Lithoglyptidae). *Contrib. Zool.* 68: 65.

Macdonald, P.M. & Struhl, G. 1986. A molecular gradient in early *Drosophila* embryos and its role in specifying the body pattern. *Nature* 324: 537-45.

McGinnis, W. & Kuziora, M. 1994. The molecular architects of body design. *Sci. Am.* 270: 58-61.

Meglitsch, P.A. & Schram, F.R. 1991. *Invertebrate Zoology*. Oxford: Oxford University Press.

Mlodzik, M., Fjose, A. & Gehring, W.J. 1985. Isolation of *caudal*, a *Drosophila* homeobox-containing gene with maternal expression, whose transcripts form a concentration gradient at the pre-blastoderm stage. *EMBO J.* 4: 2961-2969.

Moreno, E. & Morata, G. 1999. *Caudal* is the Hox gene that specifies the most posterior *Drosophila* segment. *Nature* 400: 873-7.

Mouchel-Vielh, E., Blin, M., Rigolot, C. & Deutsch, J.S. 2002. Expression of a *fushi-tarazu* homologue in a cirripede crustacean. *Evol. Dev.* 4: 76-85.

Mouchel-Vielh, E., Rigolot, C. & Deutsch, J.S. 1999. The complement of Hox genes and their timing of expression in barnacles. In: Schram, F. R. &. v. Vaupel Klein, J. C. (eds), *Crustaceans and the biodiversity crisis*: 115-124. Leiden: Brill.

Mouchel-Vielh, E., Rigolot, C., Gibert, J.-M. & Deutsch, J.S. 1998. Molecules and the body plan: the *Hox* genes of Cirripedes (Crustacea). *Mol. Phyl. Evol.* 9: 382-389.

Mourlan, A., Turquier, Y. & Baucher, M.-F. 1985. Recherches sur l'ontogenèse des Rhizocéphales. II. Organisation anatomique de la cypris libre de *Sacculina carcini*. *Cah. Biol. Mar.* 26: 281-300.

Newman, W.A. 1987. Evolution of Cirripedes and their major groups. *Crustacean Issues* 5: 3-42.

Newman, W.A. 1997. Sous-classe des Cirripèdes (*Cirripedia* Burmeister, 1834) Super-ordre des Thoraciques et des Acrothoraciques (*Thoracica* Darwin, 1854 - *Acrothoracica* Gruvel, 1905). In: Grassé, P.P. (ed), *Traité de Zoologie, Vol. VII-II*: 453-540. Paris: Masson.

Newman, W.A., Zullo, V.A. & Withers, T.H. 1969. Cirripedia. In: Moore, R. C. (ed), *Treatise on invertebrate paleontology*: R206-R292. Kansas City: The Geological Society of America & the University of Kansas.

Quéinnec, É. 2001. Insights into arthropod head evolution. Two heads in one: the end of the "endless dispute"? *Ann. Soc. Entomol. Fr.* 37: 51-69.

Quéinnec, É., Mouchel-Vielh, E., Guimonneau, M., Gibert, J.M., Turquier, Y. & Deutsch, J.S. 1999. Cloning and expression of the *engrailed.a* gene of the barnacle *Sacculina carcini*. *Dev. Genes Evol.* 209: 180-185.

Rabet, N., Gibert, J.M., Quéinnec, É., Deutsch, J.S. & Mouchel-Vielh, E. 2001. The *caudal* gene of the barnacle *Sacculina carcini* is not expressed in its vestigial abdomen. *Dev. Genes Evol.* 211: 172-178.

Rogers, B.T. & Kaufman, T.C. 1996. Structure of the insect head as revealed by the EN protein pattern in developing embryos. *Development* 122: 3419-3432.

Rogers, B.T. & Kaufman, T.C. 1997. Structure of the insect head in ontogeny and phylogeny: a view from Drosophila. *Int. Rev. Cytol.* 174: 1-84.

Runnström, S. 1924-25. *Zur Biologie und Entwicklung von* Balanus balanoides *(Linné)*. Bergen.

Ruppert, E.E. & Barnes, R.D. 1994. *Invertebrate Zoology*. Fort Worth: Saunders.

Schmidt-Ott, U. & Technau, G.M. 1992. Expression of *en* and *wg* in the embryonic head and brain of Drosophila indicates a refolded band of seven segment remnants. *Development* 116: 111-125.

Scholtz, G. 1995a. Expression of the *engrailed* gene reveals nine putative segment-anlagen in the embryonic pleon of the freshwater crayfish *Cherax destructor* (Crustacea, Malacostraca, Decapoda). *Biol. Bull.* 188: 157-165.

Scholtz, G. 1995b. Head segmentation in Crustacea - an immunocytochemical study. *Zoology* 98: 104-114.

Scholtz, G. 1997. Cleavage, germ band formation and head segmentation: the ground pattern of the Euarthropoda. In: Fortey, R.A. & Thomas, R.H. (eds), *Arthropod relationships*: 317-332. London: Chapman & Hall.

Scholtz, G. 2001. Evolution of developmental patterns in arthropods - the analysis of gene expression and its bearing on morphology and phylogenetics. *Zoology* 103: 99-111.

Scholtz, G., Patel, N.H. & Dohle, W. 1994. Serially homologous *engrailed* stripes are generated via different cell lineages in the germ band of amphipod crustaceans (Malacostraca, Percarida). *Int. J. Dev. Biol.* 38: 471-478.

Schram, F.R. 1986. *Crustacea*. Oxford: Oxford University Press.

Schram, F.R. & Høeg, J.T. 1995. New frontiers in barnacle evolution. In: Schram, F.R. & Høeg, J.T. (eds), *Crustacean issues*: 297-312. Rotterdam: Balkema.

Schulz, C. & Tautz, D. 1995. Zygotic caudal regulation by hunchback and its role in abdominal segment formation of the *Drosophila* embryo. *Development* 121: 1023-8.

Spears, T. & Abele, L.G. & Applegate, M.A. 1994. Phylogenetic study of Cirripedes and selected relatives (Thecostraca) based on 18S rDNA sequence analysis. *J. Crust. Biol.* 14: 641-656.

Telford, M.J. 2000. Evidence for the derivation of the *Drosophila fushi tarazu* gene from a Hox gene orthologous to lophotrochozoan *Lox5*. *Curr. Biol.* 10: 349-352.

Turquier, Y. 1970. Recherches sur la biologie des Cirripèdes Acrothoraciques. III. La métamorphose des cypris femelles de *Trypetesa nassaroides* et de *T. lampas. Arch. Zool. Exp. Gen.* 111: 573-628.

Turquier, Y. 1972. Contribution à l'étude des Cirripèdes Acrothoraciques. *Arch. Zool. Exp. Gen.* 113: 499-551.

Walley, L.J. 1969. Studies on the larval structure and metamorphosis of *Balanus balanoides* (L.). *Phil. Trans. R. Soc. Lond. B* 256: 237-280.

Walossek, D. & Müller, K.J. 1997. Cambrian 'Orsten' type arthropods and the phylogeny of Crustacea. In: Fortey, R.A. & Thomas, R.H. (eds), *Arthropod relationships*: 139-153. London: Chapman & Hall.

Hox genes and tagmatization of the higher Crustacea (Malacostraca)†

ARHAT ABZHANOV[1] AND THOMAS C. KAUFMAN
Howard Hughes Medical Institute, Department of Biology, Indiana University, Bloomington, USA

[1] *present address: Harvard Medical School, Department of Genetics, Boston, USA*

†The authors would like to dedicate this paper to the memory of Steven J. Gould and Andre Adoutte – we will long remember their numerous and significant contributions to the field of evolutionary biology

1 INTRODUCTION

1.1 *Phylogenetic relations within Arthropoda and Mandibulata*

The Phylum Arthropoda is the most species-rich and morphologically diverse animal group on the planet. Since their appearance in the Early Cambrian and subsequent radiation, arthropods have come to inhabit and dominate the vast majority of ecological habitats. One key to this success is postulated to have been the ability of arthropods to evolve increasingly complex body plans and specialized appendages (Brusca & Brusca 1990) and the origin of this diversity has been a matter of scientific debate for many years. From the arthropod groups that existed in the early Cambrian, four have persisted to the Present: the Chelicerata, Myriapoda, Crustacea and Hexapoda. For the purposes of this discussion we favor the hypothesis that the Crustacea represent a monophyletic group. None of the data presented here are inconsistent with this idea; however, admittedly none of them prove the point either. The crustaceans are an extremely interesting group of arthropods as they are represented by many species that have distinct and often-disparate body plans. The Higher Crustacea (Malacostraca) is the most derived group of crustaceans and is characterized by a high degree of specialization of the body into several distinct functional and morphological units, called tagmata. In contrast, some of the more basal crustaceans have less tagmatized body plans, a morphology thought to be similar to the pattern that existed in the common mandibulate ancestor. Therefore, in order to understand the evolution of the Crustacea it is important to understand the underlying developmental mechanisms that produce large-scale morphological features, such as the patterns of tagmosis, within this group. Moreover, there is an increasing body of evidence that comparable developmental processes operate in diverse arthropod lineages, suggesting fundamentally similar principles driving morphological variation. Thus information obtained in the study of the Crustacea can be extrapolated to provide further insights into the ontogeny and phylogeny of the Hexapoda, Myriapoda and Chelicerata. In this work we will focus on the recent advances taken toward understanding the role that the *Hox* cluster, a grouping of important developmental genes, may have played during evolution of the malacostracan body plan.

Although arthropod phylogeny remains a highly active and much-debated field of investigation, contemporary morphological, molecular and developmental studies point to specific relationships among extant arthropods. The Hexapoda, Crustacea and Myriapoda, the three classes that share a six-segmented head [1] with antennae, jaws and other mouthparts, are grouped together into the Mandibulata (Whitington et al. 1993; Scholtz 1995a; Rogers & Kaufman 1997). Additionally, several recent phylogenetic analyses point to the conclusion that the Crustacea are closely related to the Hexapoda (Boore et al. 1995, 1998; Cook et al. 2001; Dohle 2001; Giribet et al. 2001). The two groups share a number of important morphological features and the apparent phylogenetic proximity of the Crustacea and Hexapoda allows for a more meaningful comparison of the developmental mechanisms that may play a role in setting up the adult body plan. This in turn can lead to a better understanding of the possible roles that changes in developmental networks play in evolution.

1.2   *Tagmatization in higher crustaceans*

The most obvious morphological characteristic of an arthropod body is the number and nature of the tagmata, groups of segments united by similar morphology and/or function. Members of the Myriapoda (thought to be basal to both the Crustacea and Hexapoda) and some orders of crustaceans possess only two major body tagmata: the head and trunk (Beklemishev 1964; Schram 1986; Brusca & Brusca 1990; Walossek & Müller 1997). The primitive mandibulate trunk is a long, multi-segmented section of the body often with locomotory and/or respiratory appendages on each segment. The appendages can be used for walking (Myriapoda) or swimming (Crustacea), but, importantly, all of the appendages are generally similar to each other within a single tagma (Brusca & Brusca 1990). The best-studied example of a group with a subdivided or tagmatized trunk is the Hexapoda. In this group the trunk is partitioned into a three-segmented thorax bearing legs and in some cases wings and a limbless eleven-segmented abdomen. Such an arrangement is thought to be advantageous in terrestrial environments and may have been key to the success of the Hexapoda on land. Another example of a tagmatized mandibulate trunk[2] is found in the Malacostraca. Similar to the Hexapoda, the Malacostraca have divided their trunk into functionally and morphologically distinct tagmata called the pereion and pleon (Siewing 1963; Beklemishev 1964; Lauterbach 1975; Schram 1986; Brusca & Brusca 1990). For example, in the adult isopod *Porcellio scaber* the pereion bears seven pairs of uniramous walking legs whereas the pleon has five pairs of small biramous limbs used for breathing and one pair of uropods (Fig. 1; Schram 1986; Abzhanov & Kaufman 1999a). Moreover, as in the case of the derivation of the hexapod trunk tagmata, the combined pereion and pleon of malacostracans have been proposed to be homologous to the less derived "thorax" of cephalocarid, branchiopod and maxillopod crustaceans. Additionally, some current models

---

[1] We realize that the number of segments in the head is a somewhat contentious issue (for a recent review see Scholtz 2001); however, for the purposes of this discussion we adopt the view that there are six segments in the ancestral mandibulate head.

[2] The term "trunk" used here and after refers to the segmented region of the body posterior to the head extending to the genital openings. The post-genital region of, for example the branchiopod *Artemia* while considered a tagma by some does not accumulate *Hox* gene products (Averof & Akam 1995) and thus falls out of the purvue of this discussion (see also Schram & Koenemann 2003).

suggest that the malacostracans are the only crustaceans with a truly tagmatized trunk (Siewing 1963; Lauterbach 1975; Briggs et al. 1992; Walossek & Müller 1997; Scholtz 2001).

1.3   Hox *genes and development*

Progress over the past 20 years in developmental genetics of the fruit fly *Drosophila melanogaster* has provided important insights into the natural mechanics and genetic regulation of ontogeny. Most importantly, several genes involved in specific developmental processes, such as segmentation and patterning, have been identified. The group of homeotic or *Hox* genes provides an excellent example of this type of gene. These loci are physically linked in clusters in *Drosophila* and in vertebrates and have been found to be required for the proper specification of unique cell fates along the main body axis (Lewis 1978; 1981; Morata & Kerridge 1981; Lawrence & Morata 1994; Roch & Akam 2000). It has also been shown that genes closer to the 3′ end of the *Hox* cluster are expressed more anteriorly in the embryo, with *labial* (*lab*), the proximal-most gene, having the most anterior expression domain boundary and *Abdominal-B* (*Abd-B*) the distal-most gene being expressed most posteriorly (McGinnis & Krumlauf 1992). This phenomenon, called co-linearity, between a gene's anterior boundary of expression and its position in the cluster appears to be highly conserved (Beeman et al. 1993; Averof & Akam 1995, Denell et al. 1996; Kourakis et al. 1997). Co-linearity often results in staggered, i.e., stair-like, anterior expression boundaries. The *Hox* genes specify the body plan and basic morphology of *Drosophila* and do so through their encoded homeodomain containing transcription factor protein products (Kaufman et al. 1990; Carroll 1995). The evolution of the *Hox* cluster is thought to predate the divergence of arthropods, as a complete set of *Hox* genes has been found in several closely related groups of the Arthropoda (Grenier et al. 1997; Falciani et al. 1996). It is evident therefore that most arthropod body plans, both simple and complex, likely evolved in the context of a single conserved *Hox* cluster. Thus, based on the important role these genes play in the specification of segmental identity in hexapods, it seems likely that changes in expression and function of the homeotic genes are important for the development and evolution of distinct arthropod forms. Moreover, since the *Hox* genes control the identities of complex structures, like appendages or segments, and since mutations in these genes frequently result in considerably altered morphology, they are often viewed as tools for providing an understanding of the nature of large morphological differences. Accordingly, information on the *Hox* genes from various species has been used to understand evolutionary relationships such as divergence, homology and convergence in arthropod body plans. Nevertheless, we need a more complete understanding of the evolution of *Hox* genes themselves, i.e., their appearance, function, expression patterns and regulation, to see exactly how useful these genes are in solving long-standing problems of arthropod phylogeny and morphological evolution.

1.4   Hox *genes in evo-devo studies*

In order to understand which processes are specific to *Drosophila* and which are more relevant to all hexapods and other arthropods, our laboratory and others have embarked a series of studies aimed at comparing the expression patterns and amino acid sequence conservation of developmentally important genes, particularly the *Hox* genes. Toward this end, representatives of several phylogenetically more basal hexapod orders have been sampled (Beeman et al. 1993; Warren & Carroll 1995; Rogers & Kaufman 1997; Brown et al. 1999; Peterson et al. 1999). Most hexapods were found to have conserved *Hox* expression patterns, which parallels their invariable body plan of three tagmata. As previously mentioned, the three tagmata of the Hexapoda include a six-segmented head, a thorax of three and an abdomen of eight to eleven segments. The observed level of conservation in the hexapods suggest that *Hox* gene expression patterns are essential for establishing the basic body plan and argues for similar roles for these genes in all of the Hexapoda (Beeman et al. 1993; Warren & Carroll 1995; Brown et al. 1999; Peterson et al. 1999). Importantly, the deployment of the *Hox* genes has also been analyzed in the brine shrimp *Artemia franciscana* (Crustacea, Branchiopoda), which is considered to have a primitive level of organization of the metameres within the Mandibulata (Averof & Akam 1995). This primitive mandibulate *Bauplan* consists of a head of six segments and a homonomous trunk, i.e., an undifferentiated trunk bearing similar appendages on all segments. Averof and Akam (1995) found that *Antennapedia* (*Antp*), *Ultrabithorax* (*Ubx*) and *abdominal-A* (*abd-A*), three genes required for development of the thorax and abdomen in hexapods, are co-expressed in the homonomous, undifferentiated "thorax/trunk" of *Artemia*. These authors suggested that the 13-segmented limb bearing *Artemia* "thorax" (+ genital segments) corresponds to both the thorax and abdomen of hexapods, while the limbless "abdomen" which does not accumulate *Antp*, *Ubx* or *abd-A* gene products corresponds to hexapod post-genital segment(s) (Averof & Akam 1995). Moreover, the differences seen between the *Drosophila* and *Artemia* patterns of Hox deployment were hypothesized to have arisen concomitant with the tagmatization of the hexapod trunk (ibid.). We have expanded the studies of crustacean *Hox* gene expression into representatives of another key group, the Malacostraca, which, as noted, is characterized by a more derived and tagmatized body plan. Here, we will summarize the results from these and other recent comparatives analyses of the expression patterns of *Hox* genes and conclude that these genes have been essential in the diversification of the malacostracan and hexapod body plans. We argue that *Hox* genes have modified their expression patterns and roles and have thus provided novel genetic and developmental environments. This has been accomplished by the acquisition of more spatially and temporally defined expression domains that define new tagmatic patterns from an otherwise identical series of segments. These new conditions could then be exploited during evolution to produce morphologically distinct segments and appendages within these novel domains. We conclude that the same processes that led to the emergence of the basic malacostracan *Bauplan* likely have also been utilized in setting up the observed morphological differences amongst the orders within the Malacostraca.

2 THE CRUSTACEANS *Porcellio scaber* (ISOPODA) AND *Procambarus clarkii* (DECAPODA) AS MALACOSTRACAN MODELS

2.1 Porcellio scaber

The common woodlouse, the isopod *Porcellio scaber* belongs to the Malacostraca, which is the largest group of living crustaceans. The development of *P. scaber* is epimorphic and terminates in the hatching of a larva that has a majority of the morphological attributes of the adult isopod. The principal morphological difference between the (manca) larva and the adult is the absence of the posterior eighth pair of thoracic appendages (Beklemishev 1964; Kaestner 1970). The ancestral adult malacostracan crustacean has three tagmata: a head of six segments with jaws, a thorax with eight pairs of appendages and an abdomen (pleon) with six pairs of small appendages used for locomotion and/or respiration (for a more complete discussion of this body plan see: Richter & Scholtz 2001 and Scholtz 2001). In modern isopods, the first pair of thoracic limbs has been transformed into mouthpart maxillipeds (Schram 1986; Fig. 1A-C). Frequently, in adult isopods these appendages have flat proximal articles (basipodites) that enclose the mandibles and the two pairs of maxillae and serve as a "lower lip", a function analogous to that of the hexapod labium (Brusca & Brusca 1990). Electron micrographs of developing *Porcellio* embryos at different stages have revealed some interesting changes during morphogenesis (Fig. 1; Abzhanov & Kaufman 1999). The main tagmatic borders are established very shortly after germ band elongation and at this point the three tagmata are clearly distinct. At this stage of development (30-50%, staging is as in Whitington et al. 1993), the thorax has eight post-gnathal segments with the first pair of thoracic appendages (T1) developing as uniramous legs, a condition reminiscent of the above mentioned malacostracan ancestral state (Fig. 1A; Abzhanov & Kaufman 1999a). This pair of appendages develops as legs until the 60-70% developmental stage. At about the 75% stage, the T1 legs "transform" into mouthpart maxillipeds via a series of morphological changes and fusion of the T1 segment with the head capsule (Fig. 1A-C). This transformation is accompanied by radical changes in the expression patterns of several appendage genes (Abzhanov & Kaufman 2000d and below). The resulting adult morphologies of legs and maxillipeds are quite dissimilar. Averof and Patel (1997) report that in many crustacean species, including some malacostracans, Ubx and Abd-A "trunk" Hox encoded proteins are not detected in the anterior trunk segments and appendages if they have been co-opted as maxillipeds. In good agreement with their observation, *Ubx* and *abd-A* are not detected in the maxillipeds of *Porcellio* at any point in development (Fig. 1N, O; Abzhanov & Kaufman 1999a). Since this first trunk segment appendage develops initially as a "leg" with the same morphology as the legs on the more posterior thoracic segments but does not accumulate *Ubx* or *abd-A*, we conclude that neither the Ubx nor Abd-A products play an instructive role in early thoracic segment and leg morphogenesis.

As a result of the T1 leg to maxilliped transformation as seen in *Porcellio*, the anterior boundary of the adult isopod pereion is shifted posteriad relative both to the early embryonic and ancestral position. Consequently, after the 50-60% stage, the pereion contains only six pairs of walking appendages all sharing similar morphology. As noted however, the adult *Porcellio* body plan includes a pereion of seven, not six, segments bearing pereiopods. As it turns out, the most posterior pereionic segment remains limbless until after hatching and only develops its appendage after molting, therefore once again changing the developing body plan (in terms of appendage composition). The pleon (abdomen) is six segments long in both embryos and adults and is functionally subdivided into two parts. The anterior five segments bear short flat biramous appendages and the last, sixth segment a pair of uropods. The anterior appendages are modified for gas exchange (breathing) and protection against evaporation. The uropods are often used for swimming in aquatic species. In *Porcellio*, however, the uropods are a pair of long spiky biramous appendages.

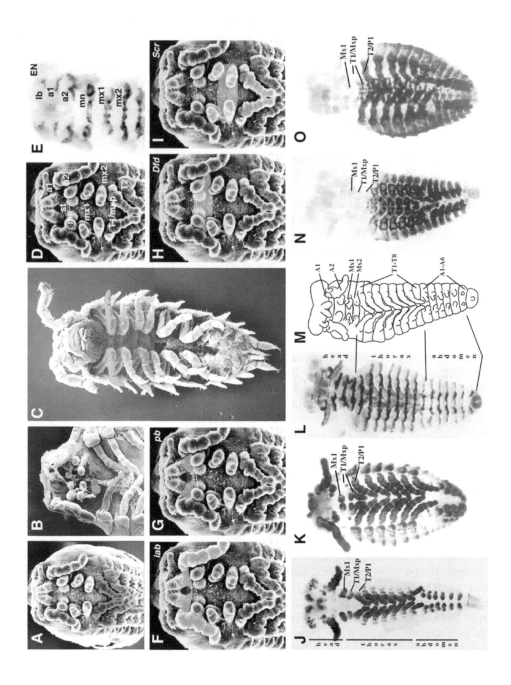

2.2 Procambarus clarkii

The red swamp crayfish *Procambarus clarkii* belongs to the Decapoda. This group has, independently from other malacostracans, modified up to three pairs of their anterior legs into maxillipeds (Schram 1986; Richter & Scholtz 2001). All three pairs of maxillipeds in *P. clarkii* have different morphologies (Fig. 2). The first pair (T1/mxp1) is the most similar to the second maxillae (mx2) while the third pair (T3/mxp3) has many leg-like features (Fig. 2E-H; Brusca & Brusca 1990). In several ways the segments bearing the maxillipeds constitute an additional tagma especially in terms of their novel morphology and associated functional capacities. Overall, almost every decapod appendage and segment develops a set of fundamentally unique features. In the crayfish in particular, this variety of appendages is used for sensing, mastication, food-handling and collecting, breathing, walking and swimming and ensures that its body plan is one of the most sophisticated within the Mala-costraca and, arguably, in all of the Arthropoda. All adult segments bearing gnathal and post-gnathal masticatory appendages are fused together into an extended cephalon (Fig. 2; Brusca & Brusca 1990). The cephalization process has proceeded in advanced decapods to an extent not observed in any other group of Mandibulata. The pereionic segments are also fused and attached to the cephalon thus forming an extensive "cephalothorax" covered with a carapace. Moreover, decapod pereionic appendages are often morphologically diverse and multi-functional. For example, in the crayfish and related taxa, the first pair of limbs (T4/P1) possesses large claws (chelae) that are used for food handling and in defense (Fig. 2I). The second and third pairs of pereionic limbs (T5/P2 and T6/P3) are used for both food gathering and locomotion, while the fourth and fifth pairs of pereiopods (T7/P4 and T8/P5) are used primarily for walking and grooming (Fig. 2J, K).

Figure 1. [See Plate 3 in color insert.] The isopod body plan and development as represented by the woodlouse *Porcellio scaber*. (A-C) Scanning electron micrographs (SEMs) of the 45-50% stage (A), 75-80% stage (B) embryos and hatched larva (C). The maxillipeds are false colored red and the second thoracic appendages/first pereiopods are false colored green. The first thoracic limbs (A) can be seen in their transformation to maxillipeds (B, C). (D, E) Head appendages and segments of *P. scaber*, ventral view. (D) is an SEM micrograph of a 45-50% embryo, (E) Immunochemical staining with the Mab4D9 anti-En antibody of the embryo similar to (D). (F-I) Summary of the head Hox expression domains. (F) *lab* is expressed in the 2nd antennal segment and appendages, (G) *pb* is expressed in the posterior 2nd antennal segment, (H) *Dfd* expression is limited to the ventral mandibular segment, (I) mRNA accumulation pattern of *Scr* in the 1st maxillary appendages, 2nd maxillary segment and appendages and distal half of T1 maxilliped appendage. (J, K) Early (25-30%) and late (75-80%) stage embryos stained with the polyclonal antibody recognizing Dll. (L, M) Mid-stage embryo stained with antibody recognizing En to reveal the segmental boundaries and the outline of the embryo. There are twenty En stripes corresponding to the six head segments, eight embryonic thoracic segments (T1-T8) and six abdominal (pleonic) segments (A1-A6). Note that the thorax of the embryo does not directly correspond to the pereion of the adult isopod. (N, O) 45-50% and 75-80% stage embryos stained with antibody FP6.87 recognizing Ultrabithorax (Ubx) and Abdominal-A (Abd-A). *Ubx* and *abd-A* are expressed in the entirety of the developing embryonic trunk (thorax and pleon) with an anterior boundary in the posterior T1/Mxp segment. Mxp appendages do not accumulate Ubx and Abd-A. Abbreviations for segments in (B): oc/lb = ocular/labral, a1 = first antennal, a2 = second antennal, mn = mandibular, mx1 = first maxillary, mx2 = second maxillary, T1-T8 = thoracic segments, A1-A6 = pleon segments. Based on Abzhanov and Kaufman (1999a, b; 2000c).

The chimerical multifunctional nature of the appendages is reflected in their complex morphologies. This is in sharp contrast to other malacostracans including isopods where all appendages within the pereion are similar to each other (Fig. 1C; Schram 1986; Abzhanov & Kaufman 1999a).

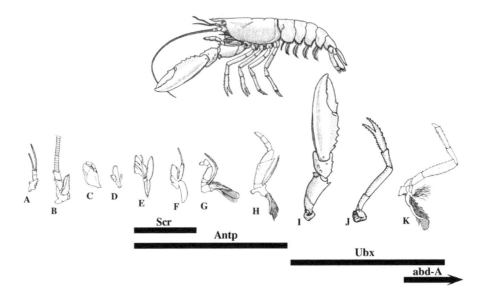

Figure 2. Drawings of the variable limb morphologies seen in the crayfish. At the top of the figure is the entire animal, below which are the various appendages from anterior (left) to posterior (right). The bars below the limbs indicate the expression domains of the *Hox* genes expressed in the posterior gnathal appendages and trunk of the mature embryo. (A) First antenna. (B) Second antenna. (C) Mandible. (D) First maxilla. (E) Second maxilla. (F) First maxilliped. (G) Second maxilliped. (H) Third maxilliped. (I) Cheliped. (J) Second leg. (K) Fourth leg. Scr = *Sex combs reduced*, Antp = *Antennapedia*, Ubx = *Ultrabithorax*, abd-A = *abdominal-A*. (Drawing modified from Hegner & Engemann 1968)

Similar to *Porcellio, Procambarus* embryos undergo direct development and hatch as miniature adults thus making this species a useful developmental model. Most of the appendages are morphologically similar to each other early in development, but their morphology changes dramatically as development proceeds. By the 60-70% stage, most appendages have unique and easily identifiable morphologies (Fig. 2; Abzhanov & Kaufman 2000b). The most interesting events take place during the development of the *Procambarus* trunk. The anterior morphological boundary of the pereion begins at the level of the T2/mxp2 segment in the 15-20% stage. However, it shifts to the T3/mxp3 segment at around the 50% stage and stays there until shortly before hatching, when the T3/mxp3 appendages become less leg-like and the T4/P1 limbs acquire their characteristic large chela (Abzhanov & Kaufman 2000b). The development of the pleon is basically similar to that of *Porcellio*. The anterior five segments bear similar small biramous swimmerets (Fig. 2) that also serve as egg-carriers in females. These are morphologically distinct from the sixth pair of pleopods that serve as a large fan-shaped tailfin used for locomotion. The morphological boundaries in the trunk established by the 80% stage are quite conspicuous and coincide with the tagmatic boundaries seen in the adult.

3 HOX GENES IN EVOLUTION AND DEVELOPMENT OF THE MANDIBULATE HEAD IN THE MALACOSTRACA

3.1 *The mandibulate head*

The six-segmented mandibulate head is postulated to be a primitive character shared by all three extant mandibulate groups: the Myriapoda, Crustacea and Hexapoda (Telford & Thomas 1995; Scholtz 1995; Scholtz 1997; Rogers & Kaufman 1997; Nielsen 2001). Furthermore, the body plan of all mandibulates, like that of the Chelicerata, is thought to have been primitively subdivided into two tagmata: the head and the homonomous, undifferentiated trunk (Brusca & Brusca 1990). It is thought that both the mandibulate head and the chelicerate prosoma are derived from a group of the anterior-most body segments, suggesting that the common ancestor of these two groups already had two distinct tagmata. However, unlike chelicerates, the mandibulate anterior tagma appears as a complex structure that consists of unique segments with specialized appendages, e.g., antennae, jaws and maxillae, that are generally not utilized for locomotion. Thus within the mandibulates, a reliable identification of homologous segments and appendages is often possible and similarities of *Hox* expression patterns can be determined. This provides a framework within which straightforward predictions of possible developmental functions of *Hox* genes can be made. The challenge is to understand how on the molecular/genetic level a relatively undifferentiated prosoma-like tagma could have evolved into the complex mandibulate head. Another important question relates to how the homologous *Hox* genes are deployed in different mandibulate heads. As alluded to above, in *Drosophila,* the cephalic *Hox* genes are modularly expressed and required for the development of specific head segments and appendages within their expression domains. Therefore, we can explore the nature of the relationship between the individual head *Hox* genes and specific segments and appendages.

Four *Hox* genes of *Drosophila* have been shown to be necessary for the correct development of the head segments and appendages - *labial* (*lab*), *proboscipedia* (*pb*), *Deformed* (*Dfd*) and *Sex combs reduced* (*Scr*). The genetic function and expression patterns of their homologs have also been examined in more basal groups of hexapods (Kelsh et al. 1994; Rogers & Kaufman 1997; Rogers et al. 1997; Peterson et al. 1999). The published expression patterns of these genes in several species of hexapods show distinct and minimally overlapping expression domains covering 1-2 segments (Fig. 3). This is in contrast to the expression patterns of their homologues in chordate and chelicerate embryos where their expression domains are broadly overlapping and several segments long (Fig. 3; Manak & Scott 1994; Damen et al. 1998; Telford & Thomas 1998; Abzhanov et al. 1999). Moreover, the "trunk" genes *Ultrabithorax* (*Ubx*), *abdominal-A* (*abd-A*) and *Abdominal-B* (*Abd-B*) are also expressed in extended overlapping domains in hexapods, chelicerates and vertebrates, this suggests that the head genes in hexapods have undergone considerable evolution to limit their expression domains. Regrettably, no data existed for cephalic *Hox* gene expression patterns for any representative of the Crustacea including the Malacostraca. To address the developmental patterning of the crustacean head we chose the woodlouse *Porcellio scaber* as a malacostracan model to study expression patterns of the

Hox genes of the head (Abzhanov & Kaufman 1999). It is important to note that our model organism is as derived as hexapods in terms of its body plan and tagmatization relative to phylogenetically more basal arthropod groups, including other crustaceans. The *Porcellio* expression patterns were interpreted under the assumptions that: 1) the Hexapoda is monophyletic, with the Zygentoma as a basal group; 2) the Mandibulata are monophyletic, with the class Crustacea the sister group of the Hexapoda; and 3) the class Chelicerata is the sister group to the Mandibulata.

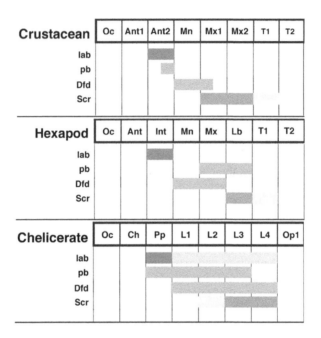

Figure 3. [See Plate 4 in color insert.] Comparison of the expression domains of the head *Hox* genes in three arthropod groups. Diagram of domains in the Crustacea (Isopoda), Hexapoda and Chelicerata. Columns represent homologous segments. Dark colored bars indicate primary domains (strong expression) of Hox expression and pastel bars indicate secondary domains (weak, transient, late or variable expression). Oc = ocular; Ant1 = first antennal (antennal, Ant in hexapods); Ant2 = second antennal (intercalary, Int in hexapods); Mn = mandibular; Mx1 = first maxillary (maxillary, Mx in hexapods); Mx2 = second maxillary (labial, Lb in hexapods); T1 = first trunk; T2 = second trunk; Ch = cheliceral; Pp = pedipalpal; L1 = first leg; L2 = second leg; L3 = third leg; L4 = fourth leg; Op1 = first opisthosomal. *lab = labial, pb = proboscipedia, Dfd = Deformed, Scr = Sex combs reduced.* Based on Kaufman et al. (1990); Rogers and Kaufman (1997); Damen et al. (1998); Telford & Thomas (1998); Abzhanov et al. (1999); Peterson et al. (1999); Abzhanov & Kaufman (2000a, b).

To provide a context for the expression domain analyses we will begin with a brief description of the segmental organization of the malacostracan head. Additionally, in order to interpret and compare the expression patterns of *lab, pb, Dfd* and *Scr* the monoclonal antibody 4D9 was used. This reagent recognizes the segment-specifying protein Engrailed (En), to reveal the segmental borders. The En antibody reveals six segments in the embryonic head of *P. scaber*: ocular (Oc), 1st antennal (Ant1), 2nd antennal (Ant2), mandibular

(Mn), 1st maxillary (Mx1) and 2nd maxillary (Mx2). This result confirms other published observations (Scholtz 1995a) (Fig. 1D, E). The labrum, the most anterior region of the head initially develops as a pair of small appendage-like structures that fuse together medially at about the 65-70% stage of development. The 1st antennal segment bears a pair of small uniramous antennae. The 2nd antennae are the largest appendages on the *Porcellio* head. The stomodeal opening resides at the level of the posterior first antennal segment and extends down to the posterior of the Ant2 segment. The mandibular En band is broad and En is expressed strongly in the developing posterior mandibular appendages. Similar expression of En is observed in the first and second maxillary segments of the posterior head (Fig. 1E). The following is a description of the cephalic *Hox* genes in *Porcellio*. The observed patterns are summarized in Fig. 3 where a comparison to the hexapod and chelicerate patterns can also be found.

3.2 labial

The most anteriorly expressed *Hox* gene in the hexapod head is *lab*. In *Drosophila, lab* expression is primarily in the intercalary segment, a small rudimentary metamere lacking appendages, located posterior to the antennal segment and anterior to the mandibular segment (Diederich et al. 1989). The exact role of *lab* in this segment is unclear as mutants do not show an obvious homeotic transformation in embryos or adults (Merrill et al. 1989). However, *lab* does appear to be important for the formation of the head in *Drosophila* as mutants do show defects in the development of certain cephalic structures (ibid). The *lab* intercalary expression domain is conserved in all of the hexapod species examined thus far (reviewed in Rogers & Kaufman 1997).

A meaningful analysis of the hexapod, crustacean and chelicerate patterns of the head *Hox* gene expression is made difficult by the uncertainty of the segmental homologies between the Chelicerata and Mandibulata. In chelicerates, the anterior-most boundary of *lab* expression is in the developing pedipalps that are thought to be homologous to the appendages of the second antennal segment. However, this expression is extended posteriorly to the fourth pair of walking limbs, which would correspond to the first pair of the thoracic appendages in hexapods and crustaceans (maxillipeds in *P. scaber*) (Damen et al. 1998; Telford & Thomas 1998). If one accepts the homology of the chelicerate pedipalpal segment with the hexapod intercalary it would appear that the anterior boundary of *lab* expression is conserved in these two groups. However, the posterior limit of the chelicerate *lab* expression domain is clearly not the same as seen in hexapods. Integumentary expression of *lab* is limited to the single (intercalary) segment while in chelicerates expression extends four full segments posteriad. Comparison of the *P. scaber lab* expression domain to that of hexapods and chelicerates reveals both conservation and change in expression (Abzhanov & Kaufman 1999b) (Fig. 1F and Fig. 3). In hexapods and malacostracans, the *lab* expression domain is restricted to the intercalary and second antennal segments, respectively. These segments are thought to be homologous based on morphological (brain anatomy and innervation pattern) and molecular (En expression pattern) data (Patel et al. 1989a, b; Scholtz 1995a). At first glance, the fact that both segments express *lab* suggests that *lab* was utilized to specify this segment prior to hexapod/malacostracan divergence and might be involved in conserved developmental processes. However, in *Porcellio, lab* is expressed in the second antennae that are large sensory and somewhat leg-like appendages, whereas as noted previously the adult hexapod intercalary segment is limbless. Interestingly, embryos of some hexapod species develop small transitory limb-buds on the intercalary segment that might be atavisms representative of a more primitive state (Beklemishev, 1964). It does not appear, however, that *lab* was directly involved in appendage loss in the hexapods, as mutations in this gene do not cause limb growth from the intercalary segment (Merrill et al. 1989). It is quite possible that *lab* contributes to the unique morphology of the second antennae as compared to the rest of the appendages found in the adult crustaceans. We conclude that the above differences in morphology in homologous segments between crustaceans and hexapods coupled with a conservation of expression are consistent with the hypothesis that *lab* has been co-opted to perform similar but clearly distinct functions in these two lineages.

3.3   proboscipedia

The *proboscipedia* (*pb*) locus is located adjacent to *lab* in the Hox cluster and is expressed in hexapods in the appendages of the maxillary and labial segments. The genetic function of *pb* in *Drosophila melanogaster* and the beetle *Tribolium castaneum* is to specify posterior mouthparts (Kaufman 1990; Shippy et al. 2000). It should be noted that the main expression domain of *pb* in these two hexapods is not co-linear with those of *lab* and *Dfd* (Cribbs et al. 1992; Randazzo et al. 1991). To explain this situation it has been proposed that the *pb* gene lost its collinear expression pattern sometime before the appearance of modern hexapods and gained a new, appendage-specific role in the maxillae and labium (Schmidt-Ott & Technau 1992). Another, much weaker, expression domain exists in the ventral portion of the intercalary segment (Rogers & Kaufman 1997). The significance of this last domain is not clear and in some groups, including *Drosophila*, it was found to be limited to mesoderm. In the apterygote hexapod *Thermobia domestica*, there is clear epidermal *pb* expression in the intercalary segment, but it is weak, late and transient (Peterson et al. 1998). It is possible that this more anterior expression may be indicative of an earlier more extensive expression domain. As is the case with *lab,* the *pb* homolog from the chelicerate *Archegozetes longisetosus* is expressed in a broad domain from the pedipalps to the third pair of walking legs where it is detected mainly in the appendages (Telford & Thomas 1998). The anterior boundary is in the pedipalpal segment that as noted above is homologous to the second antennal segment of crustaceans and the intercalary segment of hexapods (Fig. 3). Interestingly, this boundary is co-linear with both *lab* and *Dfd,* and similar to the relative expression of these genes in vertebrates and annelids (Carroll 1995). Taking into account the "out of register" expression of *pb* homologs in hexapods and the presumably vestigial domain of expression in the intercalary segment (in some hexapods), we proposed that the ancestral arthropod expression domain of *pb* and likely of all the other *Hox* genes of the head was similar to those seen in modern chelicerates (Abzhanov & Kaufman 1999b). In summary, the chelicerate and hexapod *pb* patterns have been shown to be markedly different and the hexapod expression domain is highly compartmentalized relative to chelicerates. Rather surprisingly, the *pb* expression domain of *Porcellio* was found to be different from either the chelicerate or hexapod pattern. In *Porcellio, pb* is restricted to the posterior part of the second antennal segment, and its field of expression includes neither the first nor second maxillary segments or appendages (Fig. 1G and Fig. 3). As this gene is expressed in and is required for maxillary and labial appendage development in hexapods, it clearly cannot be performing a similar function in this crustacean. Also, it would appear that hexapods and malacostracan crustaceans evolved their modern expression patterns independently from a broad chelicerate-like ancestral domain. We propose that the basic mandibulate head structures evolved prior to this event and that *pb* has been utilized to specify increasingly specialized mouthparts.

3.4 Deformed

The expression domain of the third head *Hox* gene, *Deformed* (*Dfd*), in *Drosophila melanogaster* and other hexapods includes the mandibular and maxillary segments and appendages (Kaufman et al. 1990; Rogers et al. 1997). Genetically, *Dfd* is required for the normal development and the correct morphology of sense organs of the mandibles and maxillae in *Drosophila* (Diederich et al. 1991; Kaufman et al. 1990; Merrill et al. 1987). *Dfd* expression has also been examined in three species of chelicerates, a mite *Archegozetes longisetosus* and a two spiders *Cupiennius salei* and *Steatoda triangulosa* (Damen et al. 1998; Telford & Thomas 1998; Abzhanov et al. 1999). As is the case for *lab* and *pb*, the posterior extension of *Dfd* expression relative to the hexapod pattern is reminiscent of that seen in chordates or annelids, again suggesting that it represents a more basal pattern. A comparative analysis of the *Dfd* expression domains in *P. scaber*, hexapods and chelicerates reveals that the crustacean domain is the smallest and most discrete (Fig. 3). *Dfd* expression in *Porcellio* is limited to the paragnaths (Fig. 1H and Fig. 3), mouthpart like structures associated with the developing stomodeum, whose exact embryonic origins are not known but which might be homologous to the hexapod hypopharynx (Kaestner 1970; Schram 1986; Abzhanov & Kaufman 1999b). Whatever their allegiance, paragnaths are found throughout the Crustacea in groups as diverse as remipedes, cephalocarids, copepods and malacostracans (Kaestner 1970). If we assume that the paragnaths belong to the mandibular segment then *Dfd* in *Porcellio* has a similar anterior boundary of expression to chelicerates and hexapods. However, here again the chelicerate and even the hexapod domains are clearly broader than that of *P. scaber*. Assuming that the broad chelicerate domain of expression represents a more primitive condition, then it would seem that the hexapod and crustacean patterns are derived with the Malacostracan crustaceans having the greatest reduction of expression. Importantly, *Dfd* in *P. scaber* is not expressed in the mandibles or maxillae where *Dfd* function has been shown to be required in hexapods. This suggests that *Dfd* in malacostracans has undergone an independent truncation of an ancestral domain and was subsequently utilized for aspects of mouthpart development distinct from those in the Hexapoda.

3.5 Sex combs reduced

The head-trunk border in hexapods is specified by *Sex combs reduced* (*Scr*) (Kaufman et al. 1990; Beeman et al. 1993; Rogers et al. 1996). Hexapod *Scr* homologs have also been shown to be required for the development of labial mouthparts (Pattatucci et al. 1991; Rogers et al. 1997). In *Porcellio, Scr* transcript is expressed in both pairs of maxillae (Mx1 and Mx2) and in the telopodite of the T1/Mxp1 limb (Abzhanov & Kaufman 1999a) (Fig. 1I). Expression is also observed in the posterior Mx1 segment and the T1 appendages and weakly in the outer branches of the first pair of maxillae (Mx1) (Fig. 1I). Ectodermal cells of the posterior-most part of the Mx2 segment do not accumulate detectable levels of *Scr* transcript. It is more difficult to determine the anterior boundary of the *Scr* expression domain because *Scr* expression, if any, is very weak in the Mx1 segment. The strong expression of *Scr* mRNA in the T1 appendages is not uniform and is restricted principally to the distal end of the limbs (Abzhanov & Kaufman 1999a). Although *Scr* transcript is detected in the Mx2 segment and limbs starting in early development, Scr protein does not become readily detectable in the Mx1 and T1/Mxp1 appendages until late in development. At that point, the 50-70% stage of development, its appearance coincides with transformation of the T1 leg into a mouthpart maxilliped (Abzhanov & Kaufman 1999a).

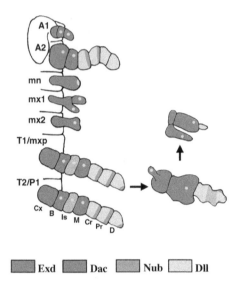

Figure 4. [See Plate 5 in color insert.] Cartoon of the expression domains of the limb specifying genes in the embryonic gnathocephalic and anterior trunk of *Porcellio scaber*. The drawing to the left shows the pattern of accumulation in early embryos when the appendage on the first trunk segment resembles those on the more posterior segments. The arrows to the right indicate the direction of transformation of this limb into the maxilliped and the changes of expression of the limb patterning genes. Segment labels: A1 = first antenna; A2 = second antenna; mn = mandibular; mx1 = first maxillary; mx2 = second maxillary; T1/mxp = first trunk/maxilliped; T2/P1 = second trunk/first pereionic. Limb segment labels: B = basis, Is = ischium, M = merus, Cr = carpus, Pr = propodus, D = dactyl. Gene names: *Exd* = *extra-denticle*, *Dac* = *dachshund*, *Nub* = *nubbin*, *Dll* = *Distal-less*. After Abzhanov & Kaufman (2000d).

To understand development of the malacostracan mouthparts in general and the molecular nature of homeotic-like leg-to-maxilliped transformation in particular, it is useful to consider the expression patterns of some of the genes involved in appendage development as it is known from *Drosophila*. The expression domains of *Distal-less* (*Dll*), *extradenticle* (*exd*), *dachshund* (*dac*) and *nubbin* (*nub*) have been analyzed in *Porcellio* (Abzhanov & Kaufman 2000d) (Fig. 4). It was found that walking legs express *exd* in the proximal podomeres (leg segments), *dac* is expressed in a band in the intermediate podomeres, Dll is accumulated in the distal segments of the limb and *nub* is expressed in several rings in the more distal segments (Fig. 4). The expression domains of *dac* and *exd* overlap extensively with that of *Dll*. While with respect to one another, *exd* and *dac* have distinct, non-overlapping domains. Assuming that the expression patterns in the walking legs represent the ground plan for all other limbs, it becomes evident that the first and second antennae, mandibles, first and second maxillae are highly modified types of appendages (Fig. 4). The mandible has a small group of *Dll*-expressing cells at the tip throughout development, has a broad band of *dac* and a proximal band of nuclear Exd. Therefore, we suggest that the mandible, although a highly reduced and no longer a clearly segmented entity, still contains representative elements of an entire limb and appears to posses remnants of the intermediate and distal parts. Both the first and second maxillae express *Exd* and *Dll* but not *dac* or *nub*, implying a basipodal nature of these limbs, i.e., they are missing all but the most proximal podomeres. This observation is important to understand the transformation of the first trunk legs into maxillipeds (Abzhanov & Kaufman 1999a). As noted, in early development the T1/Mxp leg displays an expression pattern identical to those in other pereiopods. However, as the first pair of legs ontogenetically transform into maxillipeds, the telopodite becomes more fused and reduced and the *dac* and *nub* expression domains become diffuse and then disappear (Fig. 4). In addition, during the transformation the domain of the *exd*-expressing cells in the basis podomere of the limb expands while the telopodite begins to lose its segmental nature and shrinks relative to the rest of the appendage. The originally strong band of *dac* slowly disappears as the telopodite changes to become a small palp on the maxilliped. As a result, the final segmental structure of the maxilliped appears to mimic that of the maxillae. The coincident appearance of Scr protein in the T1/Mxp limb with this morphogenetic transformation coupled with its continuous accumulation in the Mx2 appendage suggests that *Scr* is involved in the determination of maxillary identity in *Porcellio*. Moreover, based on the observed changes in limb gene expression it likely functions to suppress the loci required for walking leg identity. In less derived crustaceans, i.e., those without maxillipeds, we would anticipate that Scr accumulation would be restricted to the Mx1 and Mx2 appendages. In those groups in which posterior limbs are recruited into the gnathal apparatus, Scr might also be re-cruited. In other words, *Scr* primitively functioned in specifying the maxillae and was adopted to perform this function in the more posterior limbs. Interestingly, genetic studies on *Drosophila* have shown how the patterning of another highly specialized mouthpart, the dipteran proboscis, is controlled by the head homeotic *pb*. Notably, the absence of the *pb* locus in these flies transforms the labial palps (Mx2 in *Porcellio*) into legs and appears to regulate a similar suite of leg specifying loci (Abzhanov et al. 2001). Conversely, a change in expression pattern of the malacostracan *Scr* locus could have and apparently does have dramatic morphological consequences transforming a leg into a gnathal palp.

It is thought that all maxillipeds, including Mxp1, evolved independently in isopods and decapods (Schram 1986). Maxilliped development in *Procambarus* is, perhaps not too surprisingly, quite distinct from that of *Porcellio*. The crayfish T1/Mxp1 limbs do not grow as walking legs for any extended period of time, do not display any leg-like characteristics and already early in development become thin plate-like structures that are morphologically very similar to the Mx2 mouthparts (Schram 1986; Abzhanov & Kaufman 2000b) (Fig. 2). Similar to the situation seen in *Porcellio, Scr* transcript is found in the developing Mx2 and the T1/Mxp endopod from the early 20% to the 40-50% stages and later. Moreover, no Scr protein is detected in the T1/Mxp1 of *Procambarus* early in development ($\leq 20\%$ stage) but it does accumulate shortly after the 30% stage (Abzhanov & Kaufman 2000b). These results are consistent with the idea that the Mxp of *Porcellio* and the Mxp1 of *Procambarus* are homologous (Richter & Scholtz 2001). However, no appreciable level of *Scr* transcript or protein has been detected in the more posterior maxillipeds of *Procambarus* (T2/Mxp2 and T3/Mxp3; Fig. 2). Thus it does not appear that *Scr* is involved in development of T2/Mxp2 and T3/Mxp3 mouthparts in *Procambarus*. It should be noted, however, that the second and third maxillipeds are morphologically much more pediform and function differently from the Mx2 + T1/Mxp1 assemblage (Fig. 2). To summarize, it appears that at least in the lineages leading to *Procambarus* and *Porcellio* and perhaps in all of the Decapoda and Isopoda, *Scr* has been utilized for transformation of the first trunk appendage into a maxillae-like entity. This expansion of *Scr*'s domain of activity has occurred in the context of removal of *Ubx* expression in the first trunk segment. However, this paradigm does not seem to extend to the more posterior maxillipeds as *Scr* is not expressed during their development. Rather, evolution and development of their morphology might be linked to one of the trunk genes, e.g., *Antennapedia* as will be described in the section on the trunk *Hox* genes below.

We have proposed that the basic mandibulate head evolved prior to the establishment of the defined cephalic *Hox* gene expression domains that have been utilized to their present developmental functions independently in crustaceans and hexapods (Abzhanov & Kaufman 1999b). This hypothesis involves an intermediate hypothetical mandibulate ancestor, which did not have segment-specific expression domains and probably possessed Hox patterns of expression similar to those seen in chelicerates. The specification of individual segments and head appendages in such an animal would be dependent on the redundant and/or combinatorial functions of multiple co-expressed *Hox* genes. Subsequent evolution of more distinct expression domains could, then, facilitate evolution of the distinct morphologies of gnathal segments and appendages seen in modern hexapods and higher crustaceans.

4 HOX GENES IN EVOLUTION AND DEVELOPMENT OF THE MALACOSTRACAN TRUNK

4.1 *The trunk* Hox *genes*

The trunk *Hox* gene expression patterns have been studied most extensively in representatives of various orders of the Hexapoda (Kelsh et al. 1994; Hayward et al. 1995; Morata and Sanchez-Herrero, 1999; Peterson et al. 1999). Outside of this highly derived mandibulate group, arthropod *Hox* genes have been studied only in a branchiopod crustacean, the brine shrimp *Artemia franciscana*, and more recently, in the Chelicerata (Averof & Akam 1995; Damen et al. 1998; Telford & Thomas 1998; Abzhanov et al. 1999). A comparative analysis of the expression patterns from *Artemia* and the Hexapoda led to an intriguing hypothesis on the origin of the tagmata in the hexapod body plan (Averof & Akam 1995). As noted above it has been proposed that the appendage bearing "thorax" of *Artemia*, which displays overlapping domains of trunk *Hox* gene expression (see below), is homologous to both the thorax and abdomen of hexapods. Indeed this idea has some support among paleontologists and other researchers studying the morphological evolution of the Crustacea (Lauterbach 1975; Averof & Akam 1995, Walossek & Müller 1997). According to this model the evolution of distinct trunk tagmata is associated with and facilitated by the evolution of singular *Hox* gene expression domains and functions. Since the validity of the argument was based on only one tagmatized mandibulate trunk, that of the Hexapoda, it seemed prudent to extend the analysis to other groups that have diversified their trunks. Therefore we analyzed the Hox expression patterns in the tagmatized trunk of *Porcellio*. As noted above, this species belongs to a lineage that evolved a tagmatized trunk, producing a morphologically and functionally distinctive pereion (thorax) and pleon (abdomen), and did so independently from hexapods (Fig. 5; Lauterbach 1975; Schram 1986; Brusca & Brusca 1990; Walossek & Müller 1997).

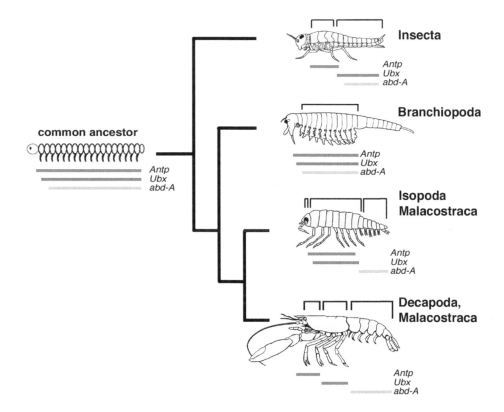

Figure 5. Evolution of trunk tagmosis in Malacostraca and related lineages shown as a model for evolution of distinct expression domains associated with changes in tagmatization. The last ancestor of hexapods and crustaceans is posited to be a mandibulate arthropod with a homonomous trunk. Hexapods and malacostracans independently have subdivided their trunks into more specialized tagmata. The tagmatic groupings are shown above the outlines of the schematic arthropods. Note, that branchiopods are thought to have an undifferentiated trunk plus several additional posterior post-genital segments. Expression domains of the posterior *Hox* genes *Antp, Ubx* and *abd A* in the trunks of malacostracans, hexapods and branchiopods are shown as bars underneath the arthropods. Based on Lauterbach (1975); Akam & Martinez-Arias (1985); Kaufman et al. (1990); Beeman et al. (1993); Kelsh et al. (1994); Boore et al. (1995); Averof & Akam (1995); Spears & Abele (1997); Wills (1997); Boore et al. (1998); Peterson et al. (1999); Abzhanov & Kaufman (2000).

Our studies have demonstrated that both the anterior and posterior boundaries and domains of all the trunk *Hox* genes investigated differ from their homologues in hexapods (Abzhanov & Kaufman 2000a). Moreover, these expression domains coincide extremely well with the tagmatic boundaries of the developing malacostracan trunk suggesting an important and direct role that these genes play in setting trunk tagmatization.

The bithorax complex (BX-C) *Hox* genes of *Drosophila melanogaster* function in the developing trunk (Duncan 1987; Kaufman et al. 1989). The different segment types in the thorax and abdomen in *Drosophila* are specified by the genes *Antennapedia* (*Antp*), *Ultrabithorax* (*Ubx*), *abdominal-A* (*abd-A*), and *Abdominal-B* (*Abd-B*) (Lewis 1981; Morata & Kerridge 1981). *Antp* is expressed in and required for the specification of the three-segmented locomotory thorax in *Drosophila* (Martinez Arias 1986; Kaufman et al. 1989) while *Ubx* and *abd-A* are involved in the development of the leg-less abdomen (Morata & Kerridge 1981; Karch et al. 1990). *Ubx* is also expressed in the posterior thorax where it is known to be involved in the development of the modified hind wings, the halteres (Lewis 1978). Some of the developmental and biochemical functions of *Ubx* and *abd-A* are similar, e.g., both are thought to suppress appendage development on the abdomen via down-regulation of *Distal-less* (*Dll*) expression (Vachon et al. 1992; Casares et al. 1996). In other hexapod orders, the expression patterns and functions of *Antp, Ubx* and *abd-A* have been shown to be very similar to those in *Drosophila* (Beeman et al. 1990; Beeman et al. 1993; Kelsh et al. 1994; Rogers & Kaufman 1997; Peterson et al. 1999; Zheng et al. 1999). We will consider the specifics of the expression patterns of hexapod trunk *Hox* genes in our comparative analysis of higher crustaceans below.

4.2   *Evolution of the trunk in Malacostraca*

The fossil record shows that the modern hexapod tagmatic pattern of thorax and abdomen evolved from ancestors with a homonomous, limb-bearing trunk (Beklemishev 1964; Kukalova-Peck 1991; 1997). This latter body plan is the proposed ancestral condition. It features an almost complete lack of trunk tagmosis and can still be found in the myriapods, a group basal to the rest of mandibulates (Brusca & Brusca 1990; Whitington & Bacon 1997; Zrzavy et al. 1997; Boore et al. 1998). In fact, a long undifferentiated trunk characterizes several orders of crustaceans and is exemplified by members of the basal order Branchiopoda such as *Artemia* (Fig. 5). As noted, in contrast to the hexapod Hox accumulation pattern, homologues of all three trunk genes, *Antp, Ubx* and *abd-A*, are co-expressed in broad domains throughout the trunk of this species (Fig. 5; Averof & Akam 1995). It is important to reiterate here that the word "trunk" in *Artemia* refers to the appendage bearing post-cephalic segments and does not include the post-genital limbless segments. *Artemia Antp* has an expression domain that extends throughout the trunk, with an anterior boundary that begins in the posterior mouthparts. *Ubx* and *abd-A* also are expressed through the trunk but have anterior boundaries that begin one and two segments posterior, respectively, to that of *Antp* (Averof & Akam 1995). This co-linear and nested pattern of *Hox* gene expression in *Artemia* is similar to what is observed in annelids, chordates and the opisthosoma of chelicerates. Thus implying that it represents the ancestral condition for the trunk *Hox* genes in all mandibulates (McGinnis & Krumlauf 1992; Averof & Akam 1995; Carroll 1995; Kourakis et al. 1997; Damen et al. 1998; Telford & Thomas 1998; Abzhanov et al. 1999).

Recent morphological and molecular phylogenies of the Crustacea demonstrate that this arthropod group comprises a monophyletic assembly. Some classes such as the Remipedia and Branchiopoda lie at a basal position while the Malacostraca are the most derived group (Brusca & Brusca 1990; Wills 1997; Whitington & Bacon 1997; Nilsson & Osorio 1997; Spears & Abele 1997). Moreover, the Crustacea are often placed in modern phylogenies as the sister group to the Hexapoda in the subphylum Mandibulata while the Myriapoda and Chelicerata are viewed as out-groups (Whitington et al. 1993; Friedrich & Tautz 1995; Dohle 1997; Wills 1997; Whitington & Bacon 1997; Zrzavy et al. 1997; Boore et al. 1998). We should mention that some studies also suggest that crustaceans may be paraphyletic with regard to the Hexapoda with the Malacostraca representing the closest sister group to hexapods (Nilsson & Osorio 1997; Whitington & Bacon 1997; Wilson et al. 2000). As noted, similar to the Hexapoda, the higher crustaceans have divided the trunk into morphologically and functionally distinct tagmata, the pereion and pleon and recent models suggest that the malacostracans are among the few if not the only crustaceans with a tagmatized trunk (Beklemishev 1964; Lauterbach 1975; Schram 1986; Brusca & Brusca 1990; Briggs et al. 1992). Furthermore, as in the case of derivation of the hexapod trunk tagma, both the pereion and pleon of malacostracans have been proposed to be homologous to and derived from the "thorax" (trunk) of branchiopods, cephalocarids and maxillopods (Lauterbach 1975; Walossek & Müller 1997; Ax 1999). This model suggests that among the many mandibulate arthropods, only the malacostracans and hexapods have evolved novel tagmata within their trunks.

All modern malacostracans share a similar body plan. Representatives of the order Leptostraca, a basal group within the Malacostraca, have eight pairs of paddle-like swimming and feeding appendages on the pereion and six pairs of appendages on the seven-segmented pleon (Brusca & Brusca 1990; Richter & Scholtz. 2001). As already mentioned, other malacostracans have modified one or more of their anterior locomotory appendages into maxillipeds and lost the seventh pleonic segment (Beklemishev 1964; Brusca & Brusca 1990; Richter & Scholtz 2001). We have reported the trunk *Hox* gene expression patterns and ideas on their possible functions in both the isopod *Porcellio scaber* and the decapod *Procambarus clarkii* (Abzhanov & Kaufman 2000). Expression patterns of *Antp, Ubx* and *abd-A* are known from *Porcellio* and patterns of *Antp, Ubx, abd-A,* and *caudal* are known from *Procambarus* . Our results show both conserved elements of the expression domains that seem to be unique to malacostracans and some unexpected changes in the expression domains that might coincide well with differences in isopod and decapod body plans.

4.3 Antennapedia

In *Artemia franciscana,* as noted, *Antp* is expressed throughout the eleven-segmented homonomous trunk (Fig. 5; Averof & Akam 1995). The anterior boundary of *Antp* expression extends into the posterior gnathal region of the head and is detected in the posterior part of Mx1 (Averof & Akam 1995). Interestingly, *Antp* expression is not uniform in the trunk but is restricted chiefly to the legs suggesting a somewhat restricted function within the developing trunk (Averof & Akam 1995). In hexapods, *Antp* has been reported to be important during development of the thoracic segments and appendages (Carroll et al. 1988). Predictably, it's principal expression domain there is in the three-segmented thorax although some weaker late expression in the abdomen is detected in several hexapod orders (Rogers & Kaufman 1997; Peterson et al. 1999).

In *Porcellio*, *Antp* is expressed in the thorax throughout development although at stages after the 60% stage some weak expression is also observed in the ventral pleon (Figs. 5, 6; Abzhanov & Kaufman 2000b). No *Antp* mRNA was detected in the cephalon, lateral pleonic ectoderm, pleonic appendages or uropods. Similar to *Artemia*, in early embryos the anterior boundary laterally coincides with the Mx1/Mx2 segmental boundary and appears to extend into the posterior ventral part of the Mx1 segment, although expression in anterior Mx2 is weaker than in the rest of the expression domain. In later stage embryos, the anterior boundary of expression is in the posterior Mx2 segment including the posterior portion of the Mx2 appendages. The posterior boundary of accumulation is in the seventh trunk/sixth pereionic (T7/P6) segment (Fig. 6). This molecular boundary corresponds to the posterior thoracic border as detected morphologically in early 30-40% stage embryos and in later mid-stage embryos. Intriguingly, the T8 segment does not accumulate detectable *Antp* transcript during embryogenesis. Moreover, it does not have any appendages pereionic or otherwise until after hatching when a pair of walking legs similar to the other pereiopods develop (Schram 1986; Abzhanov & Kaufman 2000a). It is not known whether *Antp* is expressed in T8 after hatching.

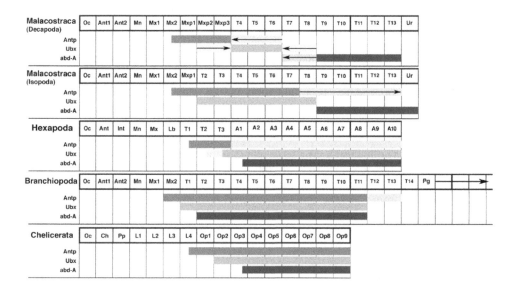

Figure 6. [See Plate 6 in color insert.] A diagram comparing the expression patterns of the trunk *Hox* genes *Antp*, *Ubx* and *abd-A* in two groups of malacostracans, hexapods, branchiopods and chelicerates. Columns represent homologous segments. The dark colors indicate strong and persistent expression domains and pastel colors indicate late, weak or transient expression domains. For the transient domains arrows indicate the expansion or contraction of the domains from early to late embryogenesis. Mxp = maxilliped; Ur = uropod; Te = telson; Pg = post-genital segments other abbreviations as in Figure 3. Based on Kaufman et al. (1990); Averof & Akam (1995); Averof & Patel (1997); Damen et al. (1998); Telford & Thomas (1998); Peterson et al. (1999); Abzhanov & Kaufman (2000).

However, if it were required for pereionic limb specification, this would certainly be our prediction. The main domain of *Antp* expression in the thorax is initiated early in development, prior to morphologically discernible tagmatization of the trunk. *Antp* transcript accumulates in the thoracic appendages in both ecto- and mesodermal tissues and there is a clear gradient along the body axis in proximodistal distribution of *Antp* transcripts within these appendages. In particular, in the coxa (coxopod segment) and the basis (basipod segment) of the more posterior pereiopods there appears to be a substantial decrease in ectodermal expression. At the same time, expression in the mesoderm appears to be at levels comparable to the more anterior pereiopods. *Antp* transcripts accumulate at detectably higher levels along the ventral midline in the thorax than in the lateral plates and this domain may mark cells of the developing central nervous system. Additionally, there is late expression in the ventral pleonic ectoderm, but its level is lower than in the ventral thorax and is likely to be associated with the developing central nervous system. The differences in the levels of *Antp* expression between thorax and pleon might be indicative of a role in specification of the different organization of the ganglia that are found in these two tagmata in adult isopods. The T8 segment also exhibits some very weak lateral expression of *Antp* early in embryogenesis. The distribution of Antp protein in *Porcellio* is currently unknown but deserves further investigation in embryos older than the 70-80% stage (Abzhanov & Kaufman 1999a). In short, it is clear that the boundaries of *Porcellio Antp* mRNA accumulation pattern coincide with the developing morphological boundaries of the embryonic tagmata through much of embryogenesis (Abzhanov & Kaufman 2000a).

The *Antp* expression pattern in *Procambarus* represents an interesting and dramatic modification of the isopod pattern (Figs. 5, 6; Abzhanov & Kaufman 2000b). The expression of *Antp* in *Procambarus* is more dynamic than in *Porcellio* and reflects the changing morphological boundaries observed through embryogenesis. At the early 15-20% stages *Antp* is detected throughout most of the trunk segments and all of the pereion. At the later 40-60% stages; however, the posterior boundary of expression retracts into the anterior part of the thorax and becomes restricted to the second maxillae and maxillipeds. This pattern may be the key to understanding the distinct morphologies of the posterior maxillipeds (Mxp2 and Mxp3) as they form a functionally and morphologically distinct food-handling assemblage. If so, the two maxilliped segments may represent a novel tagma with a genetic make-up distinct from both true mouthparts and pereionic legs. This observation also suggests that whatever the genetic/developmental function of *Antp* is in the isopod malacostracans, it is likely that several aspects are different in the decapods.

We can only speculate at this point on the developmental function of *Antp* in the species that we have studied. For example, in *Porcellio* the observation that *Antp* expression is limited to segments bearing uniramous walking legs suggests that *Antp* is important for their development (Abzhanov & Kaufman 1999a, 2000a). This is especially true at the at the 30% stage of embryogenesis when the T1/Mxp appendage is still leg-like and expresses *Antp* but not any other trunk *Hox* gene. Only at the onset of Scr protein accumulation does one see a transformation into a maxilliped. Additionally, the *Porcellio Antp* expression domain, although broadly overlapping with that of another trunk *Hox* gene, *Ubx,* is highly compartmentalized within the segments suggesting that this gene may be playing a unique role in the specification of the thorax/pereion in spite of an apparent complete overlap with the *Ubx* domain.

4.4   Ultrabithorax

The first evidence of the expression pattern of *Ubx* in malacostracans derived from studies performed using the monoclonal antibody FP6.87 that recognizes a conserved epitope in the protein products of both *Ubx* and *abd-A* (Kelsh et al. 1994; Averof & Patel 1997; Abzhanov & Kaufman 1999a). In branchiopod crustaceans, the Mab-detected signal correlates with the cephalon/trunk boundary (Averof & Akam 1995). Moreover, in *Artemia* this boundary was shown to belong to *Ubx* alone (Averof & Akam 1995). As already discussed, in many malacostracan crustaceans, including the Isopoda, the anterior trunk appendages evolved into maxillipeds (Schram 1986; Brusca & Brusca 1990; Richter & Scholtz 2001). In a landmark study Averof & Patel (1997) demonstrated that the anterior trunk segments and appendages that transformed into maxillipeds do not accumulate Ubx and Abd-A. These results suggest that Ubx might be an important trunk marker and was an important player in malacostracan morphological evolution. In good agreement with the results of Averof and Patel (ibid), the Ubx/Abd-A anterior boundary in *Porcellio*, as detected by the FP6.87 antibody, is in the second trunk segment which becomes the first adult pereionic segment (T2/P1) (Fig. 6; Abzhanov & Kaufman 1999a). Since early T1 appendages develop as uniramous legs, it is clear that neither *Ubx* nor *abd-A* is required for early walking leg identity or initiation (Abzhanov & Kaufman 1999a). Examination of the 50-60% stage embryos reveals that the Ubx/Abd-A domain extends anteriorly into the ventral T1 segment but that no Ubx/Abd-A accumulation is observed in the head, most of the T1/Mxp segment, the ventral portion of the pleon or uropods. In short, as in *Artemia* and other crustaceans, the Ubx and Abd-A proteins are detected in the developing trunk. We cannot determine however, if *Ubx* and *abd-A* are co-expressed within the trunk because the FP6.87 antibody cannot separate the individual expression domains of the two genes.

In order to address the latter issue we cloned Ubx fragments from *Porcellio*. Two apparent splicing variants with an identical homeobox but a different upstream region were recovered for *Ubx* (Abzhanov & Kaufman 2000a). Specific anti-sense *Ubx* RNA probes to both splice variants were used to analyze the expression of *Ubx*. The *Ubx* mRNA accumulation can be detected shortly before the morphological differentiation of the trunk (Abzhanov & Kaufman 2000a). Both *Ubx* transcripts are expressed exclusively in the locomotory pereion and overlap completely with each other. Consistent with the results obtained with the Mab, the anterior boundary of *Ubx* expression in *Porcellio* is in the posterior ventral T1/Mxp segment. The posterior boundary of expression is in the last pereionic leg-bearing segment in the adult (T8/P7 segment) and there is no expression detected in the pleon. These molecular boundaries are established very early in development and pereionic expression is clearly detectable at the 20% stage when only small pereiopods limb buds are present on the germ band. Thus, beginning in very early *Porcellio* development, the *Ubx* expression domain almost completely overlaps with that of *Antp* but its anterior boundary lies posteriorly by one segment (Fig. 6; Abzhanov & Kaufman 2000a). However, in early development only low levels of *Ubx* expression are detected in the ventral portion of the stage 50% embryos whereas high levels of *Antp* are found in this same part of the embryo. The differences in *Ubx* and *Antp* expression patterns may reflect distinct developmental functions for these genes in these portions of the germ band.

It was found that in *Procambarus* the anterior boundary of Mab signal is more dynamic than in either hexapods or isopods (Abzhanov & Kaufman 2000b). At the early 15-25% stage, some weak Mab signal is seen in the epidermis of the second trunk/second maxilliped (T2/mxp2) limb buds and a stronger signal in the T3/mxp3 with additional strong expression through the rest of the trunk (Fig. 6). This domain includes strong accumulation of Ubx/Abd-A in the developing pleon and its limb buds. At this stage of development, most limb buds in the posterior head and in the trunk are morphologically indistinguishable. No signal was detected in the antennal, gnathal, maxillary segments and T1/Mxp1 segment/limbs. Later at the 50-70% stages the anterior boundary of expression appears to retract towards the caudal end of the embryos as no signal is detected in the T2/Mxp2, while in contrast the signal becomes stronger in the T3/Mxp3 appendages. Ubx/Abd-A expression can still be detected at these later stages in the pereionic exopods and in the pleon. Importantly, it is at around the 50-70% stages that embryonic limbs start developing their unique morphological identities. By this time the T1/Mxp1 limbs are already developing as Mx2-like appendages. However, the T2/Mxp2 limbs begin to fall behind other trunk limbs in growth rate and take up a more mouthpart-like identity. In contrast, the T3/Mxp3 limbs continue to develop similarly to the more posterior walking legs. Shortly before the embryo hatches at the 90-95% stage, it becomes increasingly difficult to detect Ubx/Abd-A signal in the T3/Mxp3, and T4/P1 appendages. The loss of signal in the T3/Mxp3 limbs is in accord with the data available from other decapods (Averof & Patel 1998). Therefore, the T3/Mxp3 maxilliped morphology as the largest and most pediform of all the "supplementary" mouthparts correlates well with the expression of Ubx/Abd-A in this structure until late in development.

The *in situ* hybridization with a specific probe to *Procambarus Ubx* showed diffuse expression in early stage 15-35% embryos that appears to be weaker in the more anterior segments and stronger in the more posterior ones (Abzhanov & Kaufman 2000b). The anterior boundary of expression at these stages is in T2/Mxp2 and T3/Mxp3 (Fig. 6). The posterior boundary is not well defined and there is a gradient of expression that peaks around the T4-T7 appendages and then diminishes towards the posterior end of the trunk. Interestingly, at around the 50-70% stage of development, *Ubx* expression retracts both anteriorly and posteriorly into a smaller domain. At the 70-75% stage, the anterior-most expression boundary is in T3/Mxp3, in agreement with the Mab results. Strong *Ubx* expression continues more posteriorly into T4/P1 - T7/P4 and appears to be weaker in the T8/P5 segment. At the end of embryogenesis (90-95% stage), the last two pairs of pereiopods no longer show accumulation of *Ubx* transcript with distal podomeres loosing the expression first (Fig. 6). A decrease in *Ubx* expression is also detected in the distal-most part of the T4/P1 appendages. The T4/P1, T5/P2 and T6/P3 legs all develop as walking limbs armed with chelae at the distal ends. We suggest that these dynamic expression domains of *Ubx* are indicative of the intricate genetic/developmental programs taking place in the crayfish embryo, which shape the complex crayfish body plan.

4.5 abdominal-A

The trunk gene *abdominal-A* (*abd-A*) in hexapods is expressed in a pattern very similar to that of *Ubx*. It has also been demonstrated that the *abd-A* protein product in hexapods functions similarly to *Ubx* (Akam & Martinez-Arias, 1985; Karch et al. 1990; Macias et al. 1990; Casares et al. 1996). The *abd-A* expression domain in the abdomen is conserved from the zygentoman *Thermobia domestica* to the dipteran *Drosophila melanogaster* (Tear et al. 1990; Nagy et al. 1991; Peterson et al. 1999). The expression patterns of the three apparent splicing variants of *abd-A* have been reported for *P. scaber* (Fig. 6 Abzhanov & Kaufman

2000a). The *in situ* probes revealed very similar but not identical accumulation patterns for the three alternative splicing products and all are expressed primarily in the developing pleonic appendages. No expression is detected either in the cephalon, thorax, and uropods, or in the ventral part of pleon. The anterior boundary of two of the splicing variants (*abd-A1* and *abd-A2*) is in the first pleopod-bearing segment (T9/P1), while the posterior boundary is in the last T13/P5 segment (Fig. 6). The expression of these variants is strongest in the first two pairs of pleonic appendages (T9/P1 and T10/P2). Expression in the three more posterior pairs of pleopods is uniform but weaker than in P1 and P2. This pattern is established early in development and continues throughout embryogenesis. Both *abd-A1* and *A2* transcript accumulation match the receding anterior-posterior gradient in the pleon of the Mab stained embryos. The third splicing variant of *abd-A* (*abd-A3*) is detected mostly in the first two pairs of pleopods. However, unlike the *abd-A1* or *abd-A2* transcripts, the *abd-A3* domain after the 50% stage expands into the posterior part of pereion. The expansion begins as a spot of strong expression in the ventrolateral portion of the T8/P7 segment and somewhat weaker expression on the periphery of the coxa and in the mesoderm of the developing P6 and P7 pereiopods. This late expansion into a tagma that has already acquired a distinct morphology is highly reminiscent of the expansion of *Ubx* expression into the posterior thoracic segments in the apterygote hexapod *Thermobia domestica* (Peterson et al. 1999). Another novel expression domain specific to the *abd-A3* transcript lies in a single row of cells in the anterior part of the T6/P5, T7/P6 and T8/P7 segments in late embryonic stages. This expression is possibly associated with cells of the developing central nervous system. The absence of any *abd-A* expression in the ventral pleonic ectoderm might be due to the expression of *Antp* in this domain and a reciprocal negative regulation by that Hox gene. Accordingly, the Mab antibody detects no signal in the ventral part of pleon.

In summary, studies on the isopod *Porcellio scaber* have revealed a well-defined domain in the pleon. The anterior border of *abd-A* lies against the posterior border of the *Ubx* domain in the pereion and the combined *Ubx* and *abd-A* mRNA expression domains closely match the distribution of their protein products, as detected with the Mab antibody. Their individual domains although both restricted to the developing post-gnathal trunk segments are distinct and largely non-overlapping beginning in early development. This derived pattern of the trunk genes is novel for any arthropod studied thus far and correlates with the morphological border between thorax and pleon.

Similarly to *Porcellio,* the expression domain of the *abd-A* homolog of *Procambarus* is in the pleon (Fig. 6; Abzhanov & Kaufman 2000b). In 30% stage embryos, the expression seems to be strongest in the most anterior pleonic segment (T9/p1) and gradually diminishes toward the posterior end of the embryo. The pleonic expression domain does not seem to change during development. No *abd-A* expression is detected in the sixth (last) pleonic segment (T14/ur) or in the telson. At later stages of development, *abd-A* expression expands into the posterior part of pereion and into its appendages (T7/P4 and T8/P5). This expansion begins with a localized expression in the distal podomeres of the T8/P5 pereionic limbs, then in the tips of T7/P4 and continues to extend into the distal half of these legs. This later expression seems to be reciprocal with that of *Ubx* in the same appendages (Fig. 6). This dramatic shift of the anterior boundary of *abd-A* expression into posterior legs, as well as a similar late expression pattern modification in the isopod *Porcellio*, is analogous to the expansion of the *Ubx* expression into the posterior thorax and its derivatives, such as wings and legs, in higher hexapods. Although the morphology of the T7/P4 and T8/P5 legs does not change towards a more pleonic identity during late embryogenesis, they are modified relative to the other pereiopods. The fourth and fifth pairs of pereionic legs do not have chela on their distal tips and in adults they have primarily locomotory (walking) function. It is tempting to suggest that *abd-A* in isopods and decapods, analogously to *Ubx* function in the thorax and abdomen of higher hexapods, has two distinct developmental roles in the posterior pereion and in the pleon.

4.6 caudal

In *Drosophila*, *caudal* (*cad*) has been shown to act like a homeotic gene in that it is required for the identity of the analia, appendages derived from the most posterior, non-segmental part of the body (telson) (Moreno & Morata 1999). Mutations in vertebrate *caudal* homologues cause developmental homeotic transformations of the tail vertebrae and other abnormalities and thus an important role in posterior patterning seems to be conserved (Chawengsaksophak et al. 1997; Brooke et al. 1998; Schulz et al. 1998). We used *in situ* hybridization to reveal the *cad* transcript expression pattern in the crayfish embryos (Abzhanov & Kaufman 2000b). Interestingly, only the posterior half of the telson accumulates transcript suggesting that the anterior half has a different developmental origin. Leptostracans, the most basal malacostracans, are known to have seven segments in the pleon (Brusca & Brusca 1990; Richter & Scholtz 2001). The last, seventh pleonic segment is limbless and is thought to have been lost in the lineage leading to the eumalacostracans, possibly due to fusion with the sixth pleonic segment (Beklemishev 1964; Schram 1986; Richter & Scholtz 2001). Many basal crustacean groups, such as branchiopods, have multiple limbless segments posterior to the leg-bearing trunk (see Deutsch et al. 2003). Additionally in crayfish, multiple stripes of expression of the Engrailed protein product, a molecule required for segmentation in hexapods, are detected in part of the trunk immediately adjacent the telson (Patel et al. 1989; Scholtz 1995b, Abzhanov & Kaufman 2000c). Therefore, we propose that the anterior half of the decapod "telson" represents the remnants of at least one (the seventh) pleonic segment that was fused to the analia proper during decapod evolution. No *cad* expression is detected elsewhere in the developing embryos.

5 UNDERSTANDING MALACOSTRACAN EVOLUTION THROUGH DEVELOPMENT

5.1 *Higher crustaceans in evo-devo studies*

Higher crustaceans (class Malacostraca) represent one of the most species-rich and morpho-logically diverse groups of non-hexapod arthropods. We attempted in this work to review some data from our own and other laboratories on the deployment of various *Hox* genes in the Higher Crustacea (Malacostraca). Most observations have been reported from two key representatives of the Malacostraca - the common woodlouse *Porcellio scaber* (Peracarida, Isopoda, Oniscidea) and the red crayfish *Procambarus clarkii* (Decapoda, Pleocyemata, Astacida). Decapoda and Peracarida, two large groups that separated over 350 million years ago encompass most of malacostracan diversity and taken together encompass the majority of malacostracan species (Schram 1986). Decapods have evolved highly cephalized trunks as opposed to peracarids (isopods and allies). In decapods, the first three pairs of trunk appendages are maxillipeds, limbs modified for food handling/feeding. These limbs are morphologically distinct from pereionic walking legs and the swimming pleopods. Besides maxillipeds, decapods have a wide variety of other specialized appendages, which often have mixed or dual functions and morphologies. We hypothesize that the reported differences in isopod and decapod Hox deployment may be directly associated with the complex nature of the decapod body plan.

5.2   *Synopsis of the malacostracan data*

Most of the studies described here aimed to achieve a better understanding of the mechanisms involved in large-scale evolution of the diverse arthropod body plans through the comparative developmental analysis of a few chosen new "model" species of crustaceans. The primary focus was to understand how *Hox* genes are deployed in representatives of major arthropod groups with markedly different body plans, from primitive to complex. In each lineage studied, a good correlation was found between *Hox* expression patterns and prominent morphological features of higher crustaceans, such as tagmatic borders and certain types of appendages. Here we will attempt to link important changes in morphology that occurred during malacostracan evolution with the observed changes in *Hox* expression domains.

First, the results of PCR surveys in representatives of all arthropod classes and in recognized out-groups (Onychophora, Annelida, Chordata, Hemichordata, Nematoda) indicate that only a single cluster of ten genes likely existed in the arthropod ancestor (Lawrence & Morata, 1994; Akam 1995; Averof & Akam 1995; Grenier et al. 1997; Abzhanov & Kaufman 1999b). This ancestor, according to morphological studies of both extinct and extant protostomes, was probably a worm-like, segmented animal with a pair of antennae and a series of homonomous, leg-like appendages. Some major extinct arthropod groups, notably the class Trilobita, retained this primitive body plan essentially unchanged for most of their evolutionary history. Based on comparative studies of close arthropod relatives, we propose that such a trilobite-style body plan featured a single set of broad overlapping patterns of *Hox* gene expression with staggered and co-linear anterior boundaries. As no arthropod species with such a primitive body plan has survived to the present, we can only speculate about the exact segmental affinities of *Hox* expression domains in those animals. Interestingly, the dorsal side of the body in some trilobites appears to be more complex than the ventral: up to three groups of fused and modified segments are visible dorsally (Clarkson, 1998). However, in the same animals, little or no modification is seen in the appendages, suggesting separate patterning systems on the dorsal and ventral sides of the body. Thus for the purposes of this study, tagmata are defined by the morphologies and functional character of the appendages rather than the segments.

Less derived members of both the Chelicerata and Mandibulata have at least two tagmata, which could be indicative of the existence of some degree of tagmatization in their last common ancestor. Since appendage morphologies in the corresponding tagmata are so different, we suspect that primitively both had leg-like limbs. The chelicerate/mandibulate anterior and posterior tagmata appear to comprise a largely separate deployment of two different sets of Hox genes. The anterior tagma (the prosoma in spiders and the head in mandibulates which are not homologous) is the domain of five genes: *lab, pb, Hox3/zen, Dfd* and *Scr* (Damen et al. 1998; Telford & Thomas 1998; Peterson et al. 1999; Abzhanov et al. 1999; Abzhanov & Kaufman 1999). The posterior tagma, the chelicerate opistho-soma and mandibulate trunk, contain the expression domains of *Antp, Ubx, abd-A* and *Abd-B* (Damen et al. 1998; Telford & Thomas 1998; Peterson et al. 1999; Abzhanov et al. 1999; Abzhanov & Kaufman 2000). It is likely that the *Hox* genes of the arthropod ancestor that gave rise to both the Chelicerata and Mandibulata had redundant and/or combinatorial patterning functions allowing later utilization of individual *Hox* genes for the specification of increasingly more specialized appendages. For example, the appendages in the opistho-soma of arachnids are suppressed or highly modified, suggesting that the posterior set of *Hox* genes has evolved different developmental functions from the anterior set and was important in setting up a unique arachnid body plan.

The next significant step was the evolution of the jawed head in the Mandibulata including crustaceans. All of the anterior genes are known to be required in head development in the fruit fly *Drosophila melanogaster,* a highly derived hexapod (Kaufman et al. 1990). Unlike chelicerates, however, hexapods and malacostracans deploy their head genes in a series of distinct, mostly non-overlapping domains, which divide the head into developmentally and genetically distinctive territories (Kaufman et al. 1990; Peterson et al. 1999; Abzhanov & Kaufman 1999). The well-defined and small expression domains of these *Hox* genes mirror the morphology of the mandibulate head where each segment has a unique identity. Comparative analysis suggests that the mandibulate ancestor may not have had such a modular head Hox code, instead relying on a chelicerate-style pattern. Thus a more specifically defined deployment of the anterior *Hox* genes may have facilitated evolution of the specialized head seen in the hexapods and higher crustaceans (Abzhanov & Kaufman 1999b)

5.3 *Future of* Hox *genes studies*

The expression domains of the trunk genes in the decapod *P. clarkii* are similar in principle but differ in particular details from those in the isopod *P. scaber.* The expression patterns of both malacostracans have common features that suggest conserved functions. For example, *Scr* expression in the second maxillae and the first pair of maxillipeds in *Porcellio* and *Procambarus* may be indicative of a conserved function for this gene in development of these mouthparts. Assuming that the first pair of maxillipeds evolved independently in isopods and decapods, it is likely that *Scr* was recruited into the T1 limbs independently (Brusca & Brusca 1990). Alternatively, our findings may support a view that the first pair of maxillipeds is homologous in all eumalacostracans (Richter & Scholtz 2001).

Similarly, it can be predicted that the overall expression patterns of the trunk *Hox* genes in myriapods and non-malacostracan crustaceans will be similar to those of the branchiopod *Artemia franciscana,* i.e., overlapping and restricted to the trunk (see Deutsch et al. 2003). However, in the Hexapoda and Malacostraca, the mandibulate trunk underwent further tagmatization by subdividing into an anterior locomotory section (the hexapod thorax and malacostracan pereion) and a posterior portion (the hexapod abdomen and malacostracan pleon) (Siewing 1963; Lauterbach 1975; Walossek & Müller 1997; Scholtz 2001). In both lineages, a striking correlation is observed: the anterior and posterior boundaries of the trunk *Hox* gene expression domains coincide with the tagmatic borders (Peterson et al. 1999; Abzhanov & Kaufman 2000). Specifically, *Antp* and *Ubx* are largely co-expressed in the thorax of *Porcellio* and *Procambarus* with their anterior boundaries of expression being offset by one segment. In contrast, *abd-A* is the only trunk gene expressed in the pleon.

Thus the malacostracan trunk *Hox* genes are more restricted in their expression domains than the broadly overlapping pattern seen in *Artemia*. Moreover, despite a superficial resemblance to the discrete domains of their hexapod homologues, their anterior and posterior expression domain boundaries are quite different from those in hexapods (Figs. 5, 6; Abzhanov & Kaufman 2000). These findings support the hypothesis that similar to the case of the mandibulate head genes, the trunk genes were co-expressed and performed highly redundant or combinatorial roles in the homonomous trunk in the last common ancestor of hexapods and crustaceans (Averof & Akam 1995). We suggest that the ancestral trunk has independently differentiated in hexapods [thorax + abdomen] and in malacostracans [thorax/pereion + pleon] due to specialized deployment of individual *Hox* genes in each clade (Lauterbach 1975; Walossek & Müller 1997; Abzhanov & Kaufman 2000). Indeed, this may in part explain how the homologous *Hox* genes of the trunk have evolved to acquire different developmental functions in the closely related Hexapoda and Malacostraca (Ballard et al. 1992; Boore et al. 1995, 1998; Friedrich & Tautz 1995; Whitington & Bacon 1997; Nilsson & Osorio 1997; Dohle 1997; Zrzavy et al. 1997; reviewed in Gilbert & Raunio 1997). Thus it would appear, as is the case with the malacostracan and other mandibulate heads, that a more restricted deployment of the *Hox* genes is associated with the evolution of novel body plans and it is likely that this narrowing of regulatory gene expression domains played an important role in the evolution of the new specialized regions of the arthropod trunk. It is also clear that this process continued within the Malacostraca as the significant morphological changes between isopod and decapod trunk designs are reflected in the expression patterns of their homologous *Hox* genes. A more thorough sampling of different malacostracan species with more intermediate phylogenetic positions will be very useful for a better understanding of the roles that *Hox* genes play in their evolution.

In summary, it is becoming increasingly clear that although the complete *Hox* cluster predates the divergence of the arthropods, the evolution of individual patterns and specific homeotic functions took place after their divergence. As more complex developmental environments emerged in the otherwise genetically identical segments, new avenues were opened for the evolution of body plans. Thus, tagmatization of higher crustaceans is best explained as a gradual process governed by natural selection. We support the view that no *Hox*-driven saltations in morphology or "hopeful monsters" were necessary for generating the wealth of extant and extinct crustacean body plans (Goldschmidt 1940; Akam et al. 1994; Akam 1998).

REFERENCES

Abzhanov, A. & Kaufman, T.C. 1999a. Novel regulation of the homeotic gene *Scr* in crustacean leg-to-maxilliped appendage transformation. *Development* 126: 1121-1128.

Abzhanov, A. & Kaufman, T.C. 1999b. Homeotic genes and the arthropod head: expression patterns of the *labial, proboscipedia,* and *Deformed* genes in crustaceans and insects. *Proc. Natl. Acad. Sci. USA* 96: 10224-10229

Abzhanov, A. & Kaufman, T.C. 2000a. Crustacean (Malacostracan) Hox genes and the evolution of the arthropod trunk. *Development* 127: 2239-2249.

Abzhanov, A. & Kaufman, T.C. 2000b. Embryonic Expression patterns of the Hox genes in the crayfish *Procambarus clarkii* (Decapoda, Crustacea). *Evol. Dev.* 2: 271-283.

Abzhanov, A & Kaufman, T.C. 2000c. Evolution of distinct expression patterns for *engrailed* paralogues in higher crustaceans (Malacostraca). *Dev. Genes Evol.* 210: 493-506.

Abzhanov, A & Kaufman, T.C.. 2000d. Homologs of *Drosophila* appendage genes in the patterning of arthropod limbs. *Dev. Biol.* 227: 673-689.

Abzhanov, A., Holtzman, S. & Kaufman, T.C. 2001. The *Drosophila* proboscis is specified by two Hox genes, *proboscipedia* and *Sex combs reduced*, via repression of leg and antennal appendage genes. *Development* 128: 2815-2822.

Abzhanov, A., Popadic, A. & Kaufman, T.C. 1999. Chelicerate Hox genes and the homology of arthropod segments. *Evol. Dev.* 2: 77-89.

Akam, M. & Martinez-Arias, A. 1985. The distribution of *Ultrabithorax* transcripts in *Drosophila* embryos. *EMBO* 4: 1689-1700.

Akam, M. 1995. Hox genes and the evolution of diverse body plans. *Philos. Trans. R. Soc. Lond. B* 349: 313-319.

Akam, M. 1998. Hox genes, homeosis and the evolution of segment identity: no need for hopeless monsters. *Int. J. Dev. Biol.* 42: 445-451.

Akam, M., Averof, M., Castelli-Gair, J., Dawes, R., Falciani, F. & Ferrier, D. 1994. The evolving role of Hox genes in arthropods. *Development* 1994 Supplement: 209-215.

Averof, M. & Akam, M. 1995. Hox genes and the diversification of hexapod and crustacean body plans. *Nature* 376: 420-423.

Averof, M., & Patel, N.H. 1997. Crustacean appendage evolution associated with changes in Hox gene expression. *Nature* 388: 607-698.

Ax, P. 1999. *Das System der Metazoa II.* Stuttgart: Gustav Fischer Verlag.

Ballard, J.W., Olsen, G.J., Faith, D.P., Odgers, W.A., Rowell, D.M. & Atkinson, P.W. 1992. Evidence from 12S ribosomal RNA sequences that onychophorans are modified arthropods. *Science* 258: 1345-1348.

Beeman, R.W., Brown, S.J., Stuart, J.J. & Denell, R.E. 1990. Homeotic genes of the red flour beetle, *Tribolium castaneum.* In: Hagedorn, H.H. (ed), *Molecular Hexapod Science:* 21-29. New York: Plenum Press.

Beeman, R.W., Stuart, J.J., Brown, S.J., & Denell, R.E. 1993. Structure and function of the homeotic gene complex (HOM-C) in the beetle, *Tribolium castaneum. BioEssays* 15: 439 444.

Beklemishev, W.N. 1964. *Principles of the comparative anatomy of invertebrates.* Chicago: The University of Chicago Press.

Boore, J.L., Collins, T.M., Stanton, D., Daehler, L.L., & Brown, W.M. 1995. Deducing the pattern of arthropod phylogeny from mitochondrial DNA rearrangements. *Nature* 376: 163-165.

Boore, J.L., Lavrov, D.V. & Brown, W.M. 1998. Gene translocation links insects and crustaceans. *Nature* 392: 667-668.

Briggs, D.E.G., Fortey, R.A. & Wills, M.A. 1992. Morphological disparity in the Cambrian. *Science* 256: 1670-1673.

Brooke, N.M., Garcia-Fernandez, J. & Holland, P.W.H. 1998. The paraHox gene cluster is an evolutionary sister of the Hox gene cluster. *Nature* 392: 920-922.

Brown, S.J., J. P. Mahaffey, J.P., Lorenzen, M.D., Denell, R.E., & Mahaffey, J.W. 1999. Using RNAi to investigate orthologous homeotic gene function during development of distantly related insects. Evol. Devel. 1: 11-15.

Brusca, R.C. & Brusca, G.J. 1990. *Invertebrates.* Sunderland, Massachusetts: Sinauer Associates Inc.

Carroll, S.B. 1995. Homeotic genes and the evolution of arthropods and chordates. *Nature* 376: 479-485.

Carroll, S.B., Dinardo, S., O'Farrell, P., White, R., & Scott, M. 1988. Temporal and spatial relationships between segment segmentation and homeotic gene expression in *Drosophila* embryos: distribution of the fushi tarazu, engrailed, Sex combs reduced, Antennapedia, and Ultrabithorax proteins. *Gen. Dev.* 2: 350-360.

Casares, F., Calleja, M. & Sanchez-Herrero, E. 1996. Functional similarity in appendage specification by the *Ultrabithorax* and *abdominal-A Drosophila* HOX genes. *EMBO* 15: 3934-3942.

Chawengsaksophak, K., James, R., Hammond, V.E., Kontgen, F. & Beck, F. 1997. Homeosis and intestinal tumours in *Cdx2* mutant mice. *Nature* 386: 84-87.

Clarkson, E.N.K. 1998. *Invertebrate palaeontology and evolution.* Blackwell Science.

Cook, C.E., Smith, M.L., Telford, M.J., Bastianello, A. & Akam, A. 2001. *Hox* genes and the phylogeny of the arthropods. *Curr. Biol.* 11: 759-763.

Cribbs, D.L., Pultz, M.A., Johnson, D., Mazzulla, M. & Kaufman, T.C. 1992. Structural complexity and evolutionary conservation of the *Drosophila* homeotic gene *proboscipedia*. *EMBO* 11: 1437-1449.

Damen, W.G.M., Hausdorf, M., Seyfarth, E-A. & Tautz, D. 1998. A conserved mode of head segmentation in arthropods revealed by the expression pattern of Hox genes in a spider. *Proc. Natl. Acad. Sci. USA* 95: 10665-10670.

Denell, R.E., Brown, S.J. & Beeman, R.W. 1996. Evolution of the organization and function of hexapod homeotic complexes. *Sem. Cell Dev. Biol.* 7: 527-538

Deutsch, J.S., Mouchel-Vielh, E., Quéinnec, É. & Gibert, J.-M. 2003. Genes, segments and tagmata in cirripedes. In: Scholtz, G. (ed), *Crustacean Issues 15, Evolutionary Developmental Biology of Crustacea*: 19-42. Lisse: Balkema.

Diederich, R.J., Merrill, V.K.L., Pultz, M.A. & Kaufman, T.C. 1989. Isolation, structure, and expression of *labial*, a homeotic gene of the Antennapedia complex involved in *Drosophila* head development. *Genes Dev.* 3: 399-414.

Diederich, R.J., Pattatucci, A.M. & Kaufman, T.C. 1991. Developmental and evolutionary implications of *labial, Deformed,* and *engrailed* expression in the *Drosophila* head. *Development* 113: 273-281.

Dohle, W. 1997. Myriapod-hexapod relationships as opposed to an hexapod-crustacean sister group relationship. In: Fortey, R.A. & Thomas, R.H. (eds), *Arthropod Relationships:* 305-316. London: Chapman & Hall.

Dohle, W. 2001. Are the insects terrestrial crustaceans? *Ann. Soc. Entomol. Fr. (N.S.)* 37: 85-103.

Duncan, I. 1987. The Bithorax complex. *Ann. Rev. Genet.* 21: 285-319.

Falciani, F., Hausdorf B., Schröder R., Akam M., Tautz D., Denell R., Brown S. 1996. Class 3 Hox genes in insects and the origin of zen. *Proc. Natl. Acad. Sci. USA.* 93: 8479-8484.

Friedrich, M. & Tautz, D. 1995. Ribosomal DNA phylogeny of the major extant arthropod classes and the evolution of myriapods. *Nature* 376:165-167.

Gilbert, S.F. & Raunio, A.M. 1997. *Embryology: Constructing the Organism.* Sunderland, Massachusetts: Sinauer Associates Inc.

Goldschmidt, R. 1940. *The Material Basis for Evolution.* New Haven: Yale University Press.

Grenier, J.K., Garber, T.L., Warren, R., Whitington, P.M. & Carroll, S.B. 1997. Evolution of the entire arthropod Hox gene set predated the origin and radiation of the onychophoran/arthropod clade. *Curr. Biol.* 7: 547-553.

Hayward, D.C., Patel, N.H., Rehm, E.J., Goodman, C.S. & Ball, E.E. 1995. Sequence and expression of grasshopper Antennapedia: Comparison to *Drosophila. Dev. Biol.* 172: 452-465.

Hegner, R. W. & Engemann, J. G. 1968. *Invertebrate Zoology,* 2nd Edition. New York: Macmillan Publishing Co.

Kaestner, A. (1970) *Invertebrate Zoology. Volume III.* Interscience Publishers.

Karch, F., Bender, W. & Weiffenbach, B. 1990. *abd-A* expression in *Drosophila* embryos. *Genes Dev.* 4: 1573-1587.

Kaufman, T.C., Seeger, M.A. & Olsen, G. 1990. Molecular and genetic organization of the Antennapedia gene complex of *Drosophila melanogaster. Adv. Genet.* 27: 309-362.

Kelsh, R., Weinzierl, R.O.J., White, R.A.H. & Akam, M. 1994. Homeotic gene expression in the locust *Schistocerca*: an antibody that detects conserved epitopes in Ultrabithorax and Abdominal-A proteins. *Dev. Genet.* 15: 19-31.

Kourakis, M.J., Master, V.A., Lokhorst, D.K., Nardelli-Haefliger, D., Wedeen, C.J., Martindale, M.Q. & Shankland, M. 1997. Conserved anterior boundaries of Hox gene expression in the central nervous system of the leech *Helobdella. Dev. Biol.* 190: 284-300.

Kukalova-Peck, J. 1991. Fossil history and the evolution of hexapod structures. In: Naumann, I.D., Carne, P.B., Lawrence, J.F., Nielsen, E.S., Spradbery, J.P., Taylor, R.W., Whitten, M.J. & Littlejohn, M.J. (eds), *Insects of Australia:* 141-179. Ithaca: Cornell University Press.

Kukalova-Peck, J. 1997. Arthropod phylogeny and "basal" morphological structures. In: Fortey, R.A. & Thomas, R.H. (eds), *Arthropod Relationships:* 97-108. London: Chapman & Hall.

Lauterbach,. K-E. 1975. Über die Herkunft der Malacostraca (Crustacea). Zool. Anz. 194: 165-179.

Lawrence, P. & Morata, G. 1994. Homeobox genes: their function in *Drosophila* segmentation and pattern formation. *Cell* 78: 181-189.

Lewis, E.B. 1978. A gene complex controlling segmentation in *Drosophila*. *Nature* 276: 565-570.

Lewis, E.B. 1981. *Developmental Genetics of the Bithorax Complex in Drosophila.* ICN-UCLA Symposia on Molecular and Cellular Biology. New York: Academic Press.

Macias, A., Casanova, J. & Morata, G. 1990. Expression and regulation of the *abd-A* gene of *Drosophila. Development* 110: 1197-1207.

Merrill, V.K.L., Turner, F.R. & Kaufman, T.C. 1987. A genetic and developmental analysis of mutations in the *Deformed* locus in *Drosophila melanogaster. Dev. Biol.* 122: 379-395.

Merrill, V.K.L., Turner, R.F., Diederich, R.J. & Kaufman, T.C. 1989. A genetic and developmental analysis of mutations in *labial,* a gene necessary for proper head formation in *Drosophila melanogaster. Dev. Biol.* 135: 376-391

Morata, G. & Kerridge, S. 1981. Segmental functions of the bithorax complex in *Drosophila. Nature* 290: 778-781.

Morata, G. & Sanchez-Herrero, E. 1999. Patterning mechanisms in the body trunk and the appendages of *Drosophila. Development* 126: 2823-2828.

Moreno, E. & Morata, G. 1999. *Caudal* is the Hox gene that specifies the most posterior *Drosophila* segment. *Nature* 400: 873-877.

Nagy, L.M., Booker, R. & Riddiford, L.M. 1991. Isolation and embryonic expression of an *abdominal-A-like* gene from the lepidopteran, *Manduca sexta. Development* 112: 119-129.

Nielsen, C. 2001. *Animal Evolution: Interrelationships of the Living Phyla* 2nd Ed. Oxford: Oxford University Press.

Patel, N.H., Kornberg, T.B. & Goodman, C.S. 1989a. Expression of *engrailed* during segmentation in grasshopper and crayfish. *Development* 107: 201-212.

Patel, N.H., Martin-Blanco, E., Coleman, K.G., Poole, S.J., Ellis, M.C., Kornberg, T.B. & Goodman, C.S. 1989b. Expression of *engrailed* proteins in arthropods, annelids, and chordates. *Cell* 58: 955-968.

Pattatucci, A.M., Otteson, D.C. & Kaufman, T.C. 1991. A functional and structural analysis of the *Sex combs reduced* locus of *Drosophila melanogaster. Genetics* 129: 423-441.

Peterson, M.D., Rogers, B.T., Popadic, A., & Kaufman, T.C. 1999. The embryonic expression pattern of labial, posterior homeotic complex genes and the *teashirt* homologue in an apterygote hexapod. *Dev. Genes Evol.* 209: 77-90.

Randazzo, F.M., Cribbs, D.L. & Kaufman, T.C. (1991). Rescue and regulation of *proboscipedia*: a homeotic gene of the Antennapedia complex. *Development* 113: 257-271.

Richter, S. & Scholtz, G. 2001. Phylogenetic analysis of the Malacostraca (Crustacea). *J. Zool. Syst. Evol. Research* 39: 113-136.

Roch, F. & Akam, M. 2000. *Ultrabithorax* and the control of cell morphology in *Drosophila* halteres. *Development* 127: 97-107.

Rogers, B.T. & Kaufman, T.C. 1997. Structure of the hexapod head in ontogeny and phylogeny: a view from *Drosophila. Int. Rev. Cytol.* 174: 1-84.

Rogers, B.T., Peterson, M.D. & Kaufman, T.C. 1997. Evolution of the hexapod body plan as revealed by the *Sex combs reduced* expression pattern. *Development* 124: 149-157.

Scholtz, G. 1995a. Head segmentation in Crustacea - an immunocytochemical study. *Zoology* 98: 104-114.

Scholtz, G. 1995b. Expression of the *engrailed* gene reveals nine putative segment-anlagen in the embryonic pleon of the freshwater crayfish *Cherax destructor* (Crustacea; Malacostraca, Decapoda). *Biol. Bull.* 188: 157-165.

Scholtz, G. 1997. Cleavage, germ band formation and head segmentation: the ground pattern of the Euarthropoda. In: Fortey, R.A. & Thomas, R.H. (eds), *Arthropod Relationships:* 317-332. London: Chapman & Hall.

Scholtz, G. 2001. Evolution of developmental patterns in arthropods - the contribution of gene expression to morphology and phylogenetics. *Zoology* 103: 99-111.

Schram, F.R. 1986. *Crustacea.* New York: Oxford University Press.

Schram, F.R. & Koenemann, S. 2003. Developmental genetics and arthropod evolution: on body regions of Crustacea. In: Scholtz, G. (ed), *Crustacean Issues 15, Evolutionary Developmental Biology of Crustacea*: 75-92. Lisse: Balkema.

Schulz, C., Schröder, R., Hausdorf, B., Wolff, C. & Tautz, D. 1998. A *caudal* homologue in the short germ band beetle *Tribolium* shows similarities to both, the *Drosophila* and the vertebrate *caudal* expression patterns. *Dev. Genes Evol.* 208: 283-289.

Shippy, T.D., Brown, S.J., & Denell, R.E. 2000. *Maxillopedia* is the *Tribolium* ortholog of *proboscipedia*. *Evol. Dev.* 2: 145-51.

Siewing, R. 1963. Studies in Malacostracan morphology, results and problems. In: Whitington, H.B. & Rolfe, W.D.I. (eds), *Phylogeny and Evolution of Crustacea:* 85-104. Cambridge, Massachusetts: Museum of Comparative Zoology.

Spears, T. & Abele, L.G. 1997. Crustacean phylogeny inferred from 18S rDNA. In: Fortey, R.A. & Thomas, R.H. (eds), *Arthropod Relationships:* 169 -188. London: Chapman & Hall.

Tear, G., Akam, M. & Martinez-Arias, M. 1990. Isolation of an *abdominal-A* gene from the locust *Schistocerca gregaria* and its expression during early embryogenesis. *Development* 110: 915-925.

Telford, M.J. & Thomas, R.H. 1995. Demise of the Atelocerata? *Nature* 376: 123-124.

Telford, M.J., & Thomas, R.H. 1998. Expression of homeobox genes shows chelicerate arthropods retain their deutocerebral segment. *Proc. Natl. Acad. Sci. USA* 95: 10671-10675.

Vachon, G., Cohen, B., Peifle, C., McGuffin, M.E., Botas, J. & Cohen, S.M. 1992. Homeotic genes of the bithorax complex repress limb development in the abdomen of the *Drosophila* embryo through target gene *Distal-less*. *Cell* 71: 437-450

Walossek, D. & Müller, K.J. 1997. Cambrian "Orsten"-type arthropods and the phylogeny of Crustacea. In: Fortey, R.A. & Thomas, R.H. (eds), *Arthropod Relationships:* 139-154. London: Chapman & Hall.

Warren, R. & Carroll, S.B. 1995. Homeotic genes and diversification of the hexapod body plan. *Curr. Op. Genet. Dev.* 5: 459-465.

Whitington, P.M., Leach, D. & Sandeman, R. 1993. Evolutionary change in neural development within arthropods: axogenesis in the embryos of two crustaceans. *Development* 118: 449-461.

Whitington, P.M. & Bacon, J.P.. 1997. The organization and development of the arthropod ventral nerve cord: insights into arthropod relationships. In: Fortey, R.A. & Thomas, R.H. (eds), *Arthropod Relationships:* 349-370. London: Chapman & Hall.

Wills, M.A. 1997. A phylogeny of recent and fossil Crustacea derived from morphological characters. In: Fortey, R.A. & Thomas, R.H. (eds), *Arthropod Relationships:* 189-209. London: Chapman & Hall.

Zheng, Z., Khoo, A., Fambrough, D., Garza, L. & Booker, R. 1999. Homeotic gene expression in the wild-type and a homeotic mutant of the moth *Manduca sexta*. *Dev. Genes Evol.* 209: 460-472.

Zrzavy, J., Hypsa, V., Vlaskova, M. 1997. Arthropod phylogeny: taxonomic congruence, total evidence and conditional combination approaches to morphological and molecular data sets. In: Fortey, R.A. & Thomas, R.H. (eds), *Arthropod Relationships:* 97-108. London: Chapman & Hall.

Developmental genetics and arthropod evolution: on body regions of Crustacea

FREDERICK R. SCHRAM & STEFAN KOENEMANN
Institute for Biodiversity and Ecosystem Dynamics, University of Amsterdam, Netherlands

INTRODUCTION

The evolution of arthropod body plans remains an irksome issue for the field of arthropod systematics. For example, Gould believed it was illusory to use patterns of tagmosis as a basis for higher-level classification of arthropods and that "... all modern creatures called Crustacea..." displayed the "... stereotyped tagmosis of this class ..." regardless of their "... variety in size and ecological specialization" (Gould 1991: 414). Yet, the distinctness of the various arthropod body plans has been an important basis for the higher classification of arthropod subphyla and classes, e.g., Snodgrass (1952). For most arthropod groups above the class level we can clearly assign taxonomic ranks based on anatomical characteristics of the head, location of gonopores, numbers and arrangements of trunk limbs, and so forth. The Crustacea stand as a sole exception to this in that they exhibit no single body plan but actually possess several *Baupläne* (structural or archetypal body plans indicative of higher taxa), some more or less distinguished from the others (Schram 1986).

There have been several noteworthy attempts to analyze and order the complexity of arthropod body plans. Minelli & Bortoletto (1988: 331) concluded that there was an underlying pattern that made it possible to subsume the variation in centipede body plans within a single model of segment formation. In their model, the number of segments could be described by "a set of even values, all multiples of eight!" Minelli and colleagues further elaborated this idea in a series of papers (Minelli 1996, 2000, 2001; Minelli et al. 2000; Minelli & Fusco 1995; Minelli & Peruffo 1991) with implications for all myriapods. Minelli (2001) proposed a three-phase model of repeatedly duplicating post-naupliar "eosegments" (primary segmental units) that form the final number of segments in the adult (merosegments). While Minelli's Naupliar/Post-Naupliar/Merosegmental model (NPM) may easily apply to epimorphic myriapods, whether it can be extended to other arthropods remains to be seen.

Schram & Emerson (1991) and Minelli & Schram (1994) approached the issue of patterns of segmentation in arthropods from a different angle. They postulated target areas, respectively denoted as node segments or hot spots, at regularly spaced positions along the arthropod body. Minelli and Schram believed that these hot spots were under control of key genes. The authors argued that expression patterns of these genes could facilitate recognition of structural and positional homologies among *Baupläne* of higher taxa. Emerson & Schram (1997) actually tested this hypothesis cladistically. They concluded that information about positional patterns of different arthropod body plans was a fundamental component for reconstructing arthropod phylogeny.

All these discussions, however, have taken place largely within the framework of classical comparative anatomy. It is only relatively recently that interesting insights regarding the genesis of arthropod *Baupläne* has begun to emerge from the field of developmental genetics. For example, the *Hox* genes play a pivotal role during the early ontogenetic formation of the arthropod structural plans (see Cook et al. 2001). In *Drosophila*, these homeotic genes are organized into two families, the *Antennapedia* gene complex (ANT-C), and the *Bithorax* series of genes (BX-C). The genes within both complexes are expressed co-linearly, i.e., the expression along the anterior-posterior axis of the developing body corresponds with the linear order of these genes on the chromosomes (Fig. 1). Based on a comparison of *Hox* gene expression patterns and structural constraints during ontogenesis, e.g., gonopore locations, we propose to redefine crustacean body regions. We also discuss an evolutionary/developmental hypothesis of arthropod segmentation including predictions about the role of *Abd-B* in specifying a key node segment in arthropods.

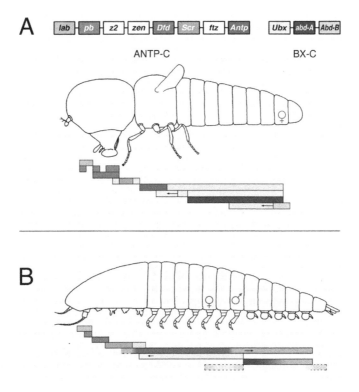

Figure 1. [See Plate 7 in color insert.] (A) Co-linear chromosomal arrangement of *Drosophila Hox* genes of the Antennapedia and Bithorax complexes (ANTP-C and BX-C). Abbreviations: *lab* = labial; *pb* = proboscipedia; *z2* = gene similar to *zerknüllt*; *zen* = *zerknüllt*; *Dfd* = *Deformed*; *Scr* = *Sex combs reduced*; *ftz* = *fushi tarazu*; *Antp* = *Antennapedia*; *Ubx* = *Ultrabithorax*; *abd-A* = *abdominal-A*; *Abd-B* = *Abdominal-B*. (B) *Hox* gene expression patterns of *Porcellio scaber* relative to adult body plan. Colors coordinated with other figures.

2 ON BODY PLANS

2.1 *Spider heads and the homology of cerebral segments*

One of the most exciting discoveries to emerge from developmental genetics is the simultaneous and independent discovery of a true deutocerebral segment in chelicerates based on coincident mapping of *Hox* genes between insects, and spiders and mites (Damen et al. 1998; Telford & Thomas 1998). Classic phylogenetic discussions, extending well back into the nineteenth century, had always assumed that the "deutocerebral segment" was absent in the Cheliceriformes (chelicerates and pycnogonids). Consequently, the chelicerae, or chelifores, were supposedly borne on a somite homologous with both the intercalary segment of atelocerates (myriapods and insects), and the second antennal segment of crustaceans (see Schram 1978; Schram & Hedgpeth 1978).

Nevertheless, Damen et al. (1998) and Telford & Thomas (1998) independently and simultaneously mapped *Hox* gene expressions in a spider (*Cupiennius*) and a mite (*Archegozetes*), respectively, and compared them to those of *Drosophila* (Fig. 2). Their astounding conclusion (Averof 1998) was that the supposedly missing deutocerebral segment of cheliceriforms was not missing at all. It clearly emerged from the *Hox* gene expression patterns that the cheliceral segment is not the homologue of the intercalary segment of the Atelocerata, but rather the homolog of the antennal segment, i.e., the segment bearing the antennae in atelocerates and the so-called first antennae in crustaceans.

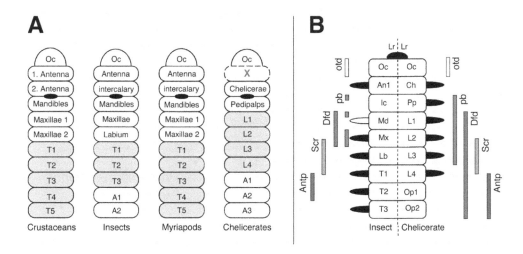

Figure 2. Alternative views of deutocerebral segment in arthropods. (A) Traditional view with deutocerebral segment supposedly missing in chelicerates (marked with "x"); shaded areas indicate leg-bearing segments (modified from Damen et al. 1998). (B) Deutocerebral segment present in chelicerates; this view based on developmental genetics implies a homology of antennae and chelicerae (from Telford and Thomas 1998). Abbreviations: Oc = ocular (pre-antennal) segment, Ch = chelicerae, An = antenna, Ic = intercalary segment, Pb = pedipalp, Md = mandible, L1-4 = legs, Mx = maxilla, Lb = labium, T1-3 thoracic segments, Op1-2 = opisthosomal segments; *otd* = *orthodenticle*; other abbreviations as in Fig. 1.

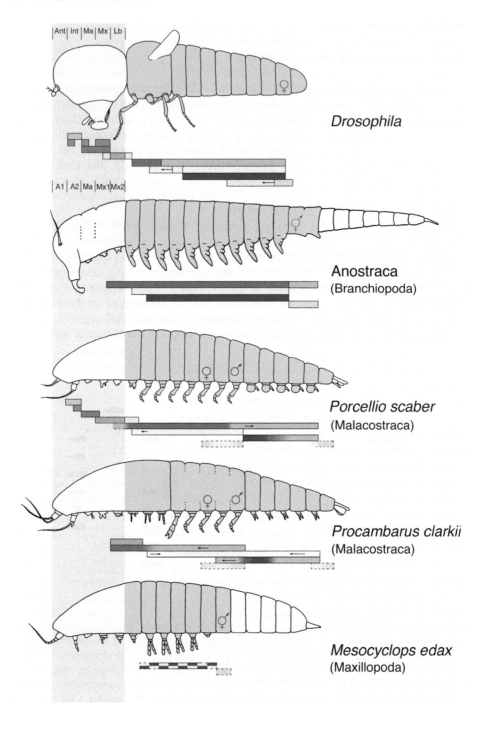

Drosophila

Anostraca
(Branchiopoda)

Porcellio scaber
(Malacostraca)

Procambarus clarkii
(Malacostraca)

Mesocyclops edax
(Maxillopoda)

This conclusion was possible because of careful mapping and comparison of *Hox* gene expression in several arthropod taxa. It completely overturned a traditional, but previously untested, supposition in comparative anatomy of the Arthropoda. These benchmark studies demonstrate that *Hox* genes can prove useful in identifying homologous regions across different arthropod body plans (Averof 1998; Scholtz 2001).

2.2 Abd-B *and the relocation of gonopores*

In *Drosophila*, there are only three genes in the BX-complex: *Ultrabithorax* (*Ubx*), *abdominal-A* (*abd-A*), and *Abdominal-B* (*Abd-B*). It has been shown that deletion of the BX-C results in a mutant larva that develops all segments posterior to the 2nd thoracic segment (T2) as identical repeats of T2 (Arthur 1997). Thus, these three genes, *Ubx*, *abd-A*, *Abd-B*, control the segment identity of the insect thorax and abdomen. However, the expression of the posterior-most gene of ANTP-C, *Antp*, can overlap with that of the posterior *Hox* genes. The degree of overlap shows characteristic variation that can be correlated with arthropod body plans (Fig. 3).

Averof & Akam (1995) compared the expressions of *Hox* genes between *Drosophila* and *Artemia franciscana*. In this branchiopod crustacean, the three "trunk genes" *Antp*, *Ubx*, and *abd-A* are expressed in overlapping domains, whereas in *Drosophila* they tend to be expressed in distinct segments. Averof and Akam explicitly proposed that the branchiopod thorax is homologous to the entire pre-genital trunk (thorax and abdomen) of the insect (Fig. 4).

Subsequently, several workers demonstrated for a number of malacostracans and two cyclopoid copepods that variations in the expression patterns of *Ubx* and *abd-A* accord with the varying specialization of thoracic limbs, e.g., maxillipeds on the anterior thoracic segments (Averof & Patel 1997; Browne & Patel 2000; Abzhanov & Kaufman 2000a). Patel and associates (pers. comm.) also apparently detected some slight *Abd-B* expression in the genital region of the peracarid *Mysidium columbiae* and in the posterior segments of the malacostracan pleon; however, these patterns remain to be confirmed. Nevertheless, it seems clear that variations in body plans are linked to variations in *Hox* gene regulation.

Figure 3. [See Plate 9 in color insert.] *Hox* gene expression patterns of *Drosophila* and 4 crustaceans relative to adult body plans. The area shaded in gray delineates the boundary of a 5-segmented head. Segments with color shading indicate comparable regions of *Hox* expression; position of gonopores marked by red male/female symbols; arrows show directional shifts during development; solid *Hox* gene colors indicate areas of strong expression, or main regions of expression; faded colors show faint expression patterns; dashed *Hox* boxes are areas of predicted expression patterns. The checkered pattern in *Mesocyclops* displays the undifferentiated expression of *Ubx* and *abd-A*. Abbreviations: Ant = antenna, Int = intercalary segment, Ma = mandible, Mx = maxilla, Lb = labium. See Fig. 1 for color coding of *Hox* genes. *Hox* data from various sources (see text).

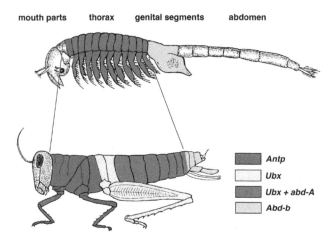

Figure 4. [See Plate 8 in color insert.] Proposed homology of *Artemia* (above) thorax and insect pre-genital trunk (grasshopper, below) based on *Hox* gene expression patterns.   See Fig. 1 for abbreviations (from Akam 1995).

We believe that *Abd-B* remains a particularly interesting gene, with great relevance to issues of positional homologies and the genesis of *Baupläne*. Kelsh, et al. (1993) looked at the expression of *Abd-B* in the grasshopper *Schistocerca gregaria*. By the time segmentation in the embryos extends the end of the trunk, *Abd-B* is expressed only in segment A11. Subsequently, expression expands anteriorly to eventually include segment A8. The anal cerci do not stain with *Abd-B* antibody. This expression pattern corresponds with the regulatory transcripts *(r*-function) of *Abd-B* in *Drosophila* that suppress embryonic ventral epidermal structures in the 8th and 9th segments of the abdomen. Thus, *Abd-B*, the last gene in the co-linear *Hox* sequence, has come to be viewed as a sort of marker for the end of the arthropod body. These segments are, coincidentally, concerned with gonopore expression as well.

However, this conceptualization of *Abd-B* as an end-of-body marker does not apply universally. *Abd-B* in the crustacean *Artemia* is the most posteriorly expressed *Hox* gene but detected on the genital segments located in the middle of the body at the posterior end of the trunk. It is not expressed at the body terminus (see Fig. 3).

Correspondingly, Damen & Tautz (1999) note that there is a residuum of *Abd-B* expression in embryos of the spider *Cupiennius salei* that coincides with the location of the adult genital segment near the anterior end of the opisthosoma. This anterior expression is separated from the main expression of *Abd-B* at the posterior end of the body. As a result, Damen & Tautz concluded that the expression of *Abd-B* on the genital segment in the spider is independent from its function in the posterior domain of the *BX* cluster. Deutsch (1997: 334) came to a similar conclusion based on preliminary results of *Abd-B* expression in cirripede crustaceans. Furthermore, Damen & Tautz note that the gene *egl-5*, an *Abd-B* ortholog from the nematode *Caenorhabditis elegans,* is also expressed more anteriorly in the gonads and sex muscles of males.

Damen & Tautz (1999) suggested that the specification of genital openings is the original function of *Abd-B*, and that the gene was only secondarily recruited into the *Hox* cluster. This hypothesis implies that the ancestral position of the gonopores was at the posterior end of the body. According to Damen & Tautz the later recruitment of *Abd-B* into the BX cluster then coincided with an anteriad movement of the gonopores (Fig. 5A). These authors noted that the hypothesized positional mechanism involving such a relocation of genital openings is not yet understood. However, Damen & Tautz did not consider an alternative: namely, that the separation of *Abd-B* expressions along the crustacean bodies might be the result of a rather different developmental pathway. It is possible that genitalia did not relocate, rather that the body termination moved further posteriad (Fig. 5B) leaving the gonopores "behind."

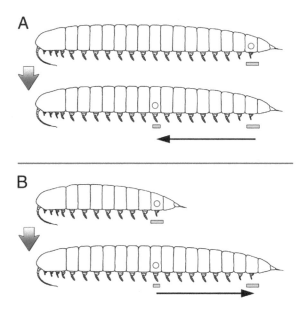

Figure 5. Alternative hypotheses of shifting *Abd-B* expression. (A) Relocation of gonopores (open circle) according to Damen and Tautz (1999). The original function of *Abd-B* (shaded box) was the specification of gonopores at the posterior end of an hypothesized, ancestral arthropod. During the course of evolution (filled arrow), developmental changes resulted in an anterior movement of the gonopores (thin black arrow). (B) Alternative model herein proposing an unchanged position of the gonopores, while the body termination is extended posteriorly.

2.3 *What are abdomens?*

The orthodox view of most arthropods possessing abdomens is another example of how well-established hypotheses about segmental homologies constrain thinking. However, Lauterbach (1975) showed that traditional definitions of body regions in crustaceans could be interpreted differently. He proposed a homology of the entire limb-bearing body region in malacostracans to the limb-bearing segments of non-malacostracans.

Walossek & Müller (1997) in their Tagmosis Model further elaborated this idea. They graphically compared various *Baupläne* and proposed that the crustacean "thorax"

("Thorax I" in their terms) was the limb-bearing region of the trunk, whereas the "abdomen" was that part of the trunk for which limbs were lacking (Fig. 6). The interpretations of these workers stood in strong contrast to long held views that the thorax in crustaceans was simply the area 'in front' of the gonopores, while the abdomen (with or without limbs) was defined as region posterior of the gonopores. For example, the entire limb-bearing trunk of malacostracans was traditionally divided into an anterior thorax equipped with walking legs and a posterior abdomen that bore pleopods (swimmerets). Both Lauterbach's proposed homologies, and Walossek & Müller's definitions imply that the entire limb-bearing malacostracan trunk is homologous to the thorax of a branchiopod like *Artemia* (Scholtz 2001).

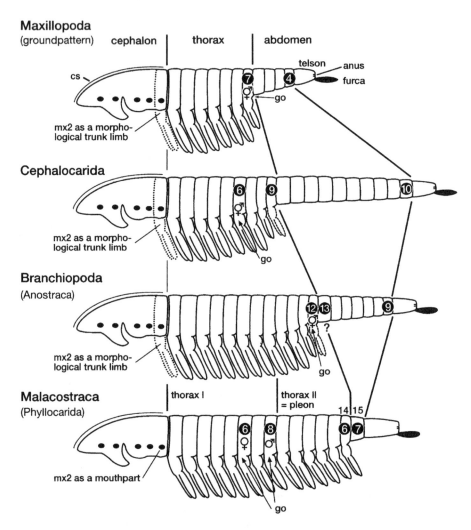

Figure 6. Model of tagmosis in crustaceans after Walossek and Müller (1997). Numbers in filled circles mark individual segments within a tagma, for example, in Maxillopoda, 7 = 7th thoracic segment, 4 = 4th abdominal segment. Abbreviations: cs = cephalic shield, mx2 = second maxilla, go = gonopore.

Interestingly, Akam (1995) suggested that the thorax of *Artemia* was the homolog of the entire trunk (thorax + abdomen) of insects based on expression patterns of *Antp, Ubx, abd-A,* and *Abd-B* (Fig. 4). However, if we accept both the *Artemia*/malacostracan homology proposed by Walossek & Müller, and Akam's *Artemia*/insect homology, then we might logically conclude by extension that the limb-bearing trunk of malacostracans must be homologous to the insect trunk. Currently available *Hox* data found in several studies do not sustain a definite conclusion in this case (Averof & Patel 1997; Browne & Patel 2000; Abzhanov & Kaufman 2000a, 2000b, this volume; Deutsch et al. 2003). A comparison of *Hox* expression patterns of *Drosophila* with both *Artemia* and malacostracans does not unequivocally support either one of the proposed homologies, nor does it reject them (Fig. 3). Abzhanov & Kaufman (2000b) advocate the view that trunk domains defined by *Antp, Ubx* and *abd-A* evolved independently in insects and the malacostracan isopod *Porcellio scaber*. At present, all we can conclude convincingly from a comparison of available *Hox* gene expressions in these taxa is that (1) a true abdomen does not show any *Hox* expression, and therefore (2) the insect "abdomen" should be redefined.

2.3.1 *Control of limb-bearing trunk segments*

However, can we speak of strict homology between various regions in different animals, regardless of absolute segment number? Point-by-point comparisons between body plans are neither easy, nor straightforward to make (see Schram & Emerson 1991; Emerson & Schram 1997). Nevertheless, based on the increasing availability of developmental genetic data we can certainly conclude that variations in the interaction of arthropod *Hox* genes are at least in part responsible for the morphologic differentiation of trunk limbs and segments.

For example, we know from the work of Averof & Akam (1995) that *Antp, Ubx,* and *abd-A* are expressed throughout the appendage-bearing thorax of *Artemia*, while *Abd-B* is expressed only in the modified thoracic segments 12 and 13 associated with the gonopores (see Fig. 3). There is no *Hox* expression in the *Artemia* abdomen, and we as yet know very little about *Hox* gene expression in the branchiopod head.

In malacostracan crustaceans, the expressions of *Ubx* and *abd-A* seem to be separated into well-defined domains. More posteriorly, this separation appears to accord with the boundary between different sets of appendages on the thorax and pleon (Abzhanov & Kaufman 1999, 2000a). On the other hand, the withdrawal of *Ubx/abd-A* expression anteriorly appears to correlate with maxilliped development in combination with posteriorly changing patterns of expression for *Scr* (Abzhanov & Kaufman 1999, 2000a; Averof & Patel 1997; Browne & Patel 2000).

Although overlap occurs in immature *Procambarus* in the expression of *Antp, Ubx,* and *abd-A*, contiguous but distinct fields of expression become established as the adult stage is approached with *Antp* restricted to the anterior thorax, *Ubx* occurring through most of the thorax, and *abd-A* in the pleon. Similar patterns occur in *Porcellio* in regard to *Ubx* and *abd-A*, while *Antp* occurs throughout the thorax and has a weak or transient expression in the pleon (Fig. 3).

Although to date only a very limited sampling of all the available body plans from within crustaceans has occurred, from the examples above we could begin to draw some general observations. For example, the results of these several studies suggest that the expression of *Hox* genes, especially *Antp, Ubx* and *abd-A*, is needed in crustaceomorph arthropods to modulate the development of limbs on a trunk segment. Certainly in insects, thoracic leg development in adults is connected with the expression of *Antp* and *Ubx*. In chelicerates, the more posterior expression of these two genes corresponds with the development of spinnerets on the opisthosoma, which appear to be modified appendages.

Similarly, it appears that the development of appendages on the arthropod body seems to occur anterior to the most terminal expression of *Abd-B*. This is true for insects, chelicerates, and crustaceans, for example, *Artemia*. However, little has been done to explore patterns of *Abd-B* expression in crustaceans beyond *Artemia*. To do so will be critical (see below).

Although the currently available Hox data do not support convincing conclusions about homologous crustacean body regions, we believe that the thorax of *Artemia* is comparable to the thorax and pleon of malacostracans as proposed by Lauterbach (1975) and Walossek & Müller (1997). This is not the point-for-point positional homology of Minelli & Schram (1994), but a manifestation of serial homology (Owen 1843) that coincides with domains under the control of *Hox* genes.

2.3.2   *Crustacean body regions redefined*

Based on the comparison of *Hox* gene expression patterns, we propose to redefine the terms relevant to body regions, at least regarding crustaceans, that have been used all too loosely for 200 years. A "trunk" in an arthropod was traditionally regarded as the region posterior to the head in which the segments were all rather alike (homonomous) and generally (but not always) bore limbs. Based on this definition, the term "trunk" fails to distinguish between body regions of a myriapod and a remipede crustacean. A "thorax" has been commonly defined as the anterior, leg-bearing part of the trunk, the "abdomen" as the region posterior to the thorax, which may or may not have limbs. In malacostracans, the term "pleon" ("Thorax II" in the terminology of Walossek & Müller 1997) serves as an alternative designation for the abdomen. However, we now have the means to specify more clearly the various regions of a crustacean. We suggest the following definitions (see also Fig. 7):

trunk
  - the region posterior to the head, in which generally all segments bear limbs linked with the coincident expression of *Antp*, *Ubx*, and *abd-A*; and
  - which is anterior to or coincident with the posterior-most expression of *Abd-B*.

thorax
  - the region just posterior to the head, when differentiated from posterior regions by a structurally distinct set of limbs; and
  - which is anterior to or coincident with the anterior-most expression of *Abd-B*.

pleon
  - the region posterior to the thorax when differentiated by a structurally distinct set of limbs (which may be reduced or absent); and
  - which is typically posterior to the gonopores (and the postulated anterior-most expression of *Abd-B*; see section 3.2).

abdomen
  - the region posterior to the trunk that lacks limbs;
  - which is posterior to the expression of *Abd-B*; and
  - which also exhibits no expression of *Hox* genes.

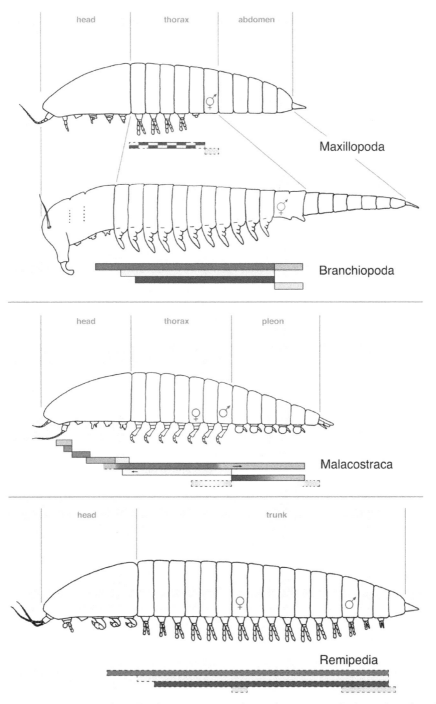

Figure 7. [See Plate 10 in color insert.] Comparison of crustacean body regions based on morphological characteristics and *Hox* gene expression patterns (*Hox* expression patterns for Remipedia are predicted; see Fig. 3 for legend of *Hox* genes and text for detailed definitions).

These new definitions should facilitate the identification of homologous body regions. It is now possible to determine more easily whether a particular body region of a crustacean is modified or lacking, especially in regard to the posterior aspects of the body. Furthermore, the application of these definitions to pancrustacean hexapods requires that we no longer refer to an insect "abdomen". Rather the adult insect "abdomen" is comparable with what we see in some crustaceans, for example Bathynellacea, that lack appendages on posterior segments of the thorax and pleon. While this definition does in principal accord with the proposed homology of the entire insect trunk to the thorax of *Artemia* by Akam (1995), it disgarees with Walossek & Müller (1997, p. 150), who believe it is "evidently critical to use the term abdomen for the posterior part of insects".

3   CONCLUSIONS

3.1   *The significance of gonopore locations*

The mid-body location of crustacean gonopores has always figured prominently in delineating the group as a whole, and variations thereon served in defining the major higher taxa within the sub-phylum. The paradox in this is that strict adherence to the location of gonopores in other arthropods helps to determine subphylum membership, whereas in crustaceans wide variations in the exact location of the gonopore has only been considered significant at class or ordinal level.

However, if we applied the same standard to crustaceans that we do to other arthropods we might be inclined to recognize at least three major higher taxa represented by distinct body plans. These are as follows (Fig. 8):

- The Mystacocarida and Branchiura with gonopores on thoracic segment 4 (T4). [The position of the Ostracoda within this scheme remains problematic. Olesen (2001) concludes that ostracode gonopores occur on T3. However, Schulz (1976) advanced evidence that ostracode gonopores were originally on T7. Most recently, Tsukagoshi & Parker (2000) surmised that the gonopores in female podocopid ostracodes originated from T5, while the position of male gonopores on the second most terminal segment (T10) was probably a derived feature. Given all these conflicting observations, no safe conclusion can be drawn for the Ostracoda at this time.]

- The Eucrustacea with gonopores on post-maxillary segments 6-8. These include: Malacostraca (female and male gonopores on T 6 and T8, respectively), Cephalocarida (gonopore on T6), most maxillopodans (both sexes with gonopores on T7, except in female cirripedes with the gonopore on T1), and perhaps also Remipedia (female pore on post-maxillary segment 8; but the male gonopore occurring seven segments further posteriad on segment 15).

- The Branchiopoda with gonopores on T11 or T12. [Deviations from this body plan are found in *Polyartemia* (gonopores on T20/T21), *Polyartemiella* (gonopores on T18/T19), and the extremely reduced Cladocera (gonopores near/at body terminus).]

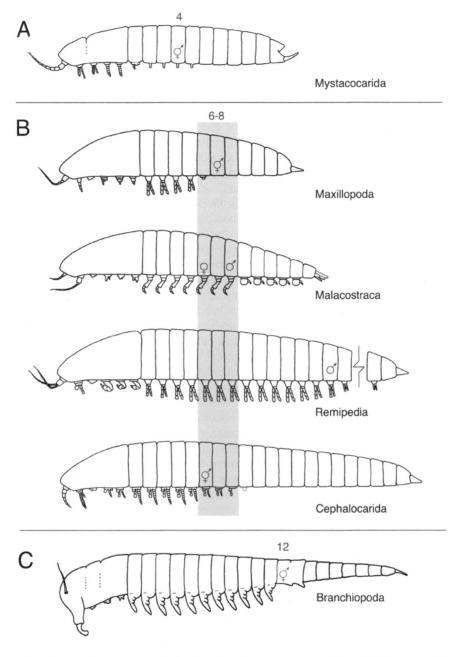

Figure 8. Three major crustacean body plans based on gonopore position. (A) Mystacocarida and Branchiura (Branchiura not shown), with gonopores on T4. (B) Eucrustacea with gonopores on post-maxillary segments 6-8 (marked by area shaded in gray). (C) Branchiopoda (*Artemia* as a representative), with gonopores on T11/T12. The jagged line in Remipedia indicates variability in number of trunk segments.

In any other arthropods, these differences would be the basis for placing possessors of such distinctly different gonopore locations in separate higher taxa or subphyla. At this point it is not relevant which of these locations might be the most primitive. It is sufficient to note that each forms a critical component of a distinctive body plan.

A comparison of these three crustacean body plans based on gonopore locations appears to reveal a basic underlying pattern. The positions of gonopores on the crustacean thorax are characterized by distinct intervals that are approximate multiplies–of four, e.g., gonopores on T4 in Mystacocarida and Branchiura, on T8 in Eumalacostraca, and on T12 in Branchiopoda. As stated above, there are, of course, exceptions within each of the three major body plans. However, all positional deviations apparently occur within one or two segments of the "main" (most frequently occurring) position. Therefore, we think that a *quantum addition* of four segments may constitute one important mechanism that gave rise to distinct, basic crustaceomorph body plans. This hypothesis is compatible with our alternative model of a posterior extension of arthropod trunks (Fig. 5; see section 2.2).

Furthermore, a pattern of quantum segment addition can also be observed in the expression of *engrailed* (*en*). The segment-polarity gene *en* delineates the anterior parts of the parasegments throughout the body. Scholtz (1995) noted eight stripes of *en* throughout the pleon in the crayfish *Cherax* (a 9th, most terminal stripe fades relatively early during embryogenesis). This might indicate that a primeval condition in Malacostraca is at least an 8-segment pleon (thorax II), i.e., a doubling of the eight thoracic stripes. We do not see at this time how these patterns can be accommodated with the NPM model of Minelli (2001). While Minelli's analysis of merosegmentation patterns applies readily to the epimorphic Myriapoda, a quantum segmentation model might better characterize crustaceomorph arthropods with a predominantly anamorphic development. Furthermore, it is possible to apply the process of quantum segmentation to the hexapods, which could be included with the branchiopod body plan (gonopores on T12). Indeed, a positional homology of genital segments in insects and branchiopods has been previously proposed by Averof & Akam (1995). Together with the insect/*Artemia* linkage based on *Hox* gene expression patterns (see section 2.3), the location of gonopores seems to provide further support for a relationship of hexapods and branchiopod crustaceans (see Spears & Abele 1997 for a linking of collembolans and branchiopods based on molecular data).

This quantum hypothesis leaves unexplained, however, how abdomens came to be. Abdomens apparently lack *Hox* gene expression as well as limbs. The length of abdomens varies. Maxillopoda have four abdominal segments in the classic 5-7-4 ground pattern delineated for the group (Newman 1983). Cephalocarida possess 12 abdominal segments, Anostraca eight, Mystacocarida six, while Branchiura have only an unsegmented stub. Whether or not Notostraca and Leptostraca have both a pleon as well as an abdomen must await the results of detailed *Hox* gene studies in members of these groups.

We suggest the possibility that some process other than quantum segmentation is at work concerning the origin and function of abdomens. Some kind of asynchrony between *Hox* and *ParaHox* genes could be invoked, which may be linked to the need of a hydrodynamic stabilizer in swimming animals, e.g., comparable with the tail on a kite. *Hox* genes control ecto- and mesodermal tissues that, in our model, develop into body regions devoted to locomotion and feeding, with concomitant appendage and central nervous system specializations. *ParaHox* genes are expressed in the endoderm, perhaps helping to generate a "trailing" region that lacks limbs and in which nerve tissues are generally characterized by poorly differentiated neurons and ganglia.

3.2 *What to do and what to expect*

Two objectives are crucial for a more complete understanding of crustacean body plans. First, a more comprehensive selection of taxa has to be analyzed. It will be critical to obtain complete *Hox* data for cephalocarids, mystacocarids, ostracodes, branchiurans (along with pentastomids), non-anostracan branchiopods, and eventually even remipedes. Representatives of all these groups, except the last, are readily available and merely require the time and effort to study them. Second, we need expression patterns for *all* of the *Hox* genes in crustaceans, since to date more attention has been paid to the posterior *Hox* genes, *Ubx*, *abd-A*, *Abd-B*, or their epitopes than to the more anterior genes. For example, while *Artemia* was the first crustacean for which *Hox* expressions were determined, these data were restricted to the three posterior genes and to date little has been published on *Antp* and more anterior genes in that readily available model species.

Nevertheless, from what is known, we might predict that we would find the following patterns of gene expression.

- We would expect that in the limb-less abdomens of maxillopodans, cephalocarids, and mystacocarids there would be no *Hox* gene expression. Naturally, the *ParaHox* gene *caudal* will mark the body terminus on the body of these groups, but *Abd-B* will probably not occur terminally (Rabet et al. 2001).

- We should see *Ubx* and *abd-A* occurring throughout the entire trunk in remipedes (Fig. 7), if we ever succeed in probing for *Hox* expressions in these crustaceans.

- We predict that some form of *Abd-B* expression will be seen in the gonopore segments of T4 in mystacocarids and branchiurans, and in T6-8 in cephalocarids, malacostracans, and maxillopodans. In addition, we suspect that Abd-B will be found in at least T8 and T15 in short-bodied remipedes, respectively the sites of the female and male gonopores, and, possibly, in other selected posterior trunk segments of longer-bodied forms (see Yager & Carpenter 1999).

- We would also look for terminal *Abd-B* expression in pentastomids, branchiurans, remipedes, cladocerans, and all malacostracans, but not expect to find terminal expression in any of the other crustacean groups.

- We expect that *Abd-B* will also occur in segments 11-12 of notostracans and conchostracans, as has been already noted in *Artemia*.

4   ARE THE CRUSTACEA MONOPHYLETIC?

Based on our analysis we must accept that crustaceomorph arthropods do not conform to a single body plan as is the case in all other arthropod monophyla, and this implies that the concept of a "monophylum Crustacea" is questionable. It would appear that the postulated concepts of a "Pan-Crustacea" (insects and crustaceans), or "Crustaceomorpha" (Crustacea *sensu stricto* and some Cambrian stem-fossils), may have some considerable usefulness. Accepting this will be especially critical towards identifying not only the relationships amongst the various living and Cambrian fossil forms, but also the placement of the Hexapoda. In any case, it seems likely that determining the relationships of the various crustaceomorph body plans to the other body plans within the Arthropoda will require comprehensive approaches towards the ultimate solution (Schram & Jenner 2001; Schram & Koenemann in prep.).

ACKNOWLEDGEMENTS

We would like to thank Jan van Arkel for the final preparation of all figures. We also wish to acknowledge several people for discussions, personal communications, and exchanges of information that have helped immeasurably to shape our ideas on this and related subjects in the last few years. These include in this instance Ronald Jenner, Michalis Averof, and especially Dieter Waloszek. We also wish to acknowledge the great stimulus provided by the other members of the Dutch national program in systematics and developmental biology, including Jo van den Biggelaar, Peter Damen, Wim Dictus, Lex Nederbragt, and Andre van Loon, supported by the Earth and Life Sciences Foundation (ALW) of the Netherlands Organization for Scientific Research (NWO), grant number 805-33-430, of which this paper is the thirteenth contribution.

REFERENCES

Abzhanov, A. & Kaufman, T.C. 1999. Novel regulation of the homeotic gene *Scr* associated with a crustacean leg-to-maxilliped appendage transformation. *Development* 126: 1121-1128.

Abzhanov, A. & Kaufman, T.C. 2000a. Crustacean (malacostracan) *Hox* genes and the evolution of the arthropod trunk. *Development* 127: 2239-2249.

Abzhanov, A. & Kaufman, T.C. 2000b. Embryonic expression patterns of the *Hox* genes of the crayfish *Procambarus clarkii* (Crustacea: Decapoda). *Evol. Dev.* 2: 271-283.

Abzhanov, A. & Kaufman, T.C. 2003. HOX genes and tagmatization of the higher Crustacea (Malacostraca). In: Scholtz, G. (ed), *Crustacean Issues 15, Evolutionary Developmental Biology of Crustacea*: 43-74. Lisse: Balkema.

Akam, M. 1995. Hox genes and the evolution of diverse body plans. *Phil. Trans. R. Soc. Lond. B* 349: 313-319.

Arthur, W. 1997. *The Origin of Animal Body Plan: A Study in Evolutionary Developmental Biology*. Cambridge University Press, Cambridge.

Averof, M. & Akam, M. 1995. *Hox* genes and the diversification of insect and crustacean body plans. *Nature* 376: 420-423.

Averof, M. & Patel, N.H. 1997. Crustacean appendage evolution associated with changes in *Hox* gene expression. *Nature* 388: 682-686.

Averof, M. 1998. Origin of the spider's head. *Nature* 395: 436-437.

Browne, W.F. & Patel, N.H. 2000. Molecular genetics of crustacean feeding appendage development and diversification. *Sem. Cell Dev. Biol.* 11: 427-435.

Cook, C.E., Smith, M.L., Telford, M.J., Bastianello, A. & Akam, A. 2001. *Hox* genes and the phylogeny of arthropods. *Curr. Biol.* 11: 759-763.

Damen, W.D. & Tautz, D. 1999. *Abdominal-B* expression in a spider suggests a general role for *Abdominal-B* in specifying genital structures. *J. Exp. Zool.* 285: 85-91.

Damen, W.D., Hausdorf, M., Seyfarth, E.-A. & Tautz, D. 1998. A conserved mode of head segmentation in arthropods revealed by the expression pattern of *Hox* genes in spider. *Proc. Nat. Acad. Sci. USA* 95: 10665-10670.

Deutsch, J. 1997. The origin of Hexapoda: a developmental genetic scenario. *Mem. Mus. Nat. Hist. Nat.* 173: 329-340.

Deutsch, J. 2001. Are Hexapoda members of the Crustacea? Evidence from comparative developmental genetics. *Ann. Soc. Entomol. Fr.* 37: 41-49.

Deutsch, J.S., Mouchel-Vielh, E., Quéinnec, É. & Gibert, J.-M. 2003. Genes, segments and tagmata in cirripedes. In: Scholtz, G. (ed), *Crustacean Issues 15, Evolutionary Developmental Biology of Crustacea*: 19-42. Lisse: Balkema.

Emerson, M.J. & Schram, F.R. 1997. Theories, patterns, and reality: game plan for arthropod phylogeny. In: Fortey, R.A. & Thomas, R.H. (eds) *Arthropod Relationships*: 67-86. London: Chapman & Hall.

Gould, S.J. 1991. The disparity of the Burgess Shale arthropod fauna and the limits of cladistic analysis: why we must strive to quantify morphospace. *Paleobiology* 17: 411-423.

Kelsh, R., Dawson, I. & Akam, M. 1993. An analysis of *Abdominal-B* expression in the locust *Schistocerca gregaria*. *Development* 117: 293-305.

Lauterbach, K.-E. 1975. Über die Herkunft der Malacostraca (Crustacea). *Zool. Anz.* 194: 165-179.

Minelli, A. 1996. Segments, body regions, and the control of development through time. *Mem. Calif. Acad. Sci.* 20: 55-61.

Minelli, A. 2000. Holomeric vs. meromeric segmentation: a tale of centipedes, leeches, and rhombomeres. *Evol. Devel.* 2: 35-48.

Minelli, A. 2001. A three-phase model of arthropod segmentation. *Dev. Genes Evol.* 211: 509-521.

Minelli, A. & Bortoletto, S. 1988. Myriapod metamerism and arthropod segmentation. *Biol. J. Linnean Soc. London* 33: 323-343.

Minelli, A. & Fusco, G. 1995. Body segmentation and segment differentiation: the scope for heterochronic change. In: McNamara, K.J. (ed) *Evolutionary change and heterochrony*: 49-63. London: John Wiley.

Minelli, A. & Peruffo, B. 1991. Developmental pathways, homology and homonomy in metameric animals. *J. Evol. Biol.* 3: 429-445.

Minelli, A. & Schram, F.R. 1994. Owen revisited: a reappraisal of morphology in evolutionary biology. *Bijdr. Dierk.* 64: 65-74.

Minelli, A., Foddai, D. Pereira, L.A. & Lewis, J.G.E. 2000. The evolution of segmentation of centipede trunk and appendages. *J. Zool. Syst. Evol. Res.* 38: 103-117.

Newman, W.A. 1983. Origin of Maxillopoda: urmalacostracan ontogeny and progenesis. *Crustacean Issues* 1: 105-119.

Olesen, J. 2001. External morphology and larval development of *Derocheilocaris remanei* Delamare-Deboutteville & Chappuis 1951 (Crustacea, Mystacocarida), with a comparison of crustacean segmentation and tagmosis patterns. *Biologiske Skrifter, The Royal Danish Academy of Sciences and Letters,* 53: 1-59.

Owen, R. 1843. *Lectures on the Comparative Anatomy and Physiology of the Invertebrate Animal.* Delivered at the Royal College of Surgeons in 1843. London: Longman.

Rabet, N., Gibert, J.-M., Quéinnec, É., Deutsch, J.S. & Mouchel-Vielh, E. 2001. The caudal gene of the barnacle *Sacculina carcini* is not expressed in its vestigial abdomen. *Dev. Genes Evol.* 211: 172-178.

Scholtz, G. 1995. Expression of the *engrailed* gene reveals nine putative segment-anlagen in the embryonic pleon of the freshwater crayfish *Cherax destructor* (Crustacea, Malacostraca, Decapoda). *Biol. Bull.* 188: 157-165.

Scholtz, G. 2001. Evolution of developmental patterns in arthropods - the analysis of gene expression and its bearing on morphology and phylogenetics. *Zoology* 103: 99-111.

Schram, F.R. 1978. Arthropods, a convergent phenomenon. *Fieldiana, Geology* 39: 61-108.

Schram, F.R. 1986. *Crustacea.* New York: Oxford University Press.

Schram, F.R. & Emerson, M.J. 1991. Arthropod Pattern Theory: a new approach to arthropod phylogeny. *Mem. Queensl. Mus.* 31: 1-18.

Schram, F.R. & Hedgpeth, J.W. 1978. Locomotory mechanisms in some Antarctic pycnogonids. In: Fry, W.D. (ed) *Sea Spiders (Pycnogonida). Zool. J. Linnean Soc. Lond.* 63: 145-169.

Schram, F.R. & Jenner, R.A. 2001. The origin of Hexapoda: the crustacean perspective. In: Deuve, T. (ed) *Origin of the Hexapoda*; *Ann. Soc. Entomol. France* 37: 243-264.

Schulz, K. 1976. *Das Chitinskelett der Podocopida und die Frage der Metamerie dieser Gruppe.* Unpublished doctoral dissertation, University of Hamburg: 167 pp.

Snodgrass, R.E. 1952. *A Textbook of Arthropod Anatomy.* Ithaca, Comstock Publishing Associates.

Spears, T. & Abele, L.G. 1997. Crustacean phylogeny inferred from 18S rDNA. In: Fortey, R.A. & Thomas, R.H. (eds) *Arthropod Relationships*: 169-187. London: Chapman & Hall.

Telford, M.J. & Thomas, R.H. 1998. Expression of homeobox genes shows chelicerate arthropods retain their deutocerebral segment. *Proc. Nat. Acad. Sci. USA* 95: 10671-10675.

Tsukagoshi, A. & Parker, A.R. 2000. Trunk segmentation of some podocopine lineages in Ostracoda. *Hydrobiologia* 419: 15-30.

Walossek, D. & Müller, K.J. 1997. Cambrian 'Orsten'-type arthropods and the phylogeny of Crustacea. In: Fortey, R.A. & Thomas, R.H. (eds) *Arthropod Relationships*: 139-153. London: Chapman & Hall.

Yager, J. & Carpenter, J.H. 1999. *Speleonectes epilimnius* new species (Remipedia, Speleonectidae) from surface waters of an anchialine cave on San Salvador Island, Bahamas. *Crustaceana* 72: 965-977.

II CELLS AND SEGMENTS

Cell lineage, segment differentiation, and gene expression in crustaceans

WOLFGANG DOHLE
Freie Universität Berlin, Institut für Biologie, Zoologie, Berlin, Germany

MATTHIAS GERBERDING [1], ANDREAS HEJNOL [2] AND GERHARD SCHOLTZ
Humboldt-Universität zu Berlin, Institut für Biologie, Vergleichende Zoologie, Berlin, Germany.

[1] *present address: University of California Berkeley, USA*

[2] *present address: Technische Universität Braunschweig, Institut für Genetik, Braunschweig, Germany*

1 INTRODUCTION

The Crustacea is a highly diverse group. This diversity is seen in the various adult morphologies consisting of different overall shapes, tagmatization, limb types, and segment numbers (Schram 1986; Gruner 1993; Schram & Koenemann 2003). The patterns of ontogeny are also very diverse (Siewing 1969; Anderson 1973; Fioroni 1987; Scholtz 1997). On the one hand, we find yolk-poor eggs with holoblastic cleavage, a hollow blastula, and an invagination gastrula (Kühn 1912; Zilch 1978, 1979). On the other hand, there are yolky eggs, which undergo intralecithal cleavage with subsequent formation of a blastoderm, a germ disc and a germ band (Dohle 1976). There are direct developers and others passing through a variety of larval stages such as nauplius, cypris or zoea (Gruner 1993; Scholtz 2000). Furthermore, we find different embryonic and larval structures such as lateral and dorsal organs (Fioroni 1987; Meschenmoser 1996) and a great variety of modes of limb development (Williams & Müller 1996; Williams & Nagy 1996; Olesen et al. 2001; Schram & Koenemann 2001; Williams 2003). The variety of adult shapes and developmental pathways is also reflected in differences of gene expression patterns (Scholtz et al. 1993, 1998; Panganiban et al. 1995; Averof & Patel 1997; Abzhanov & Kaufman 1999, 2000b, c). However, one has to keep in mind that there is no direct correlation or causal connection between the patterns of development and of specific adult morphologies. Similar adult structures in closely related groups can arise via different ontogenetic pathways, whereas highly divergent organisms or characters can start at quite similar ontogenetic stages (Remane 1952; Sander 1983; Dohle 1989; Wagner & Misof 1992; Scholtz & Dohle 1996).

A prominent characteristic of the ontogeny of some crustacean taxa is the occurrence of stereotyped cell lineages and cell division patterns. Stereotyped cell lineages are found in the germ band of malacostracans (Dohle & Scholtz 1988; Scholtz & Dohle 1996) and during early total cleavage of several malacostracans and non-malacostracans (Kühn 1912; Anderson 1969; Hertzler & Clark 1992; Gerberding et al. 2002; Hertzler 2002; Scholtz & Wolff 2002). These stereotyped cell lineages exhibiting an invariant sequence of mitoses and division products is a widespread phenomenon within the animal kingdom, probably the most famous examples for cell lineages are spiral cleavage (van den Biggelaar et al. 1997) and the development of nematodes (Schierenberg 1997; Schnabel 1997). Ever since the cell lineage mode of development has been detected more than a century ago, hypotheses have been put forward to explain the putative functional role of "determinate" development (Conklin 1896; Stent 1985, 1998). The most widespread view is that cell lineage plays a causal role in cell fate determination. Cell fate is supposed to be influenced by factors inherited from precursor cells and distributed via characteristic cell divisions. According to this view, cells act predominantly autonomously without communicating with their neighbors. There is increasing evidence that this view is an oversimplification.

Even in organisms that serve as standard examples for determinate and mosaic development such as nematodes, ctenophorans, and spiralians, there is no strictly autonomous differentiation of cells (Martindale & Henry 1997; Schnabel 1997; van den Biggelaar et al. 1997). This view is supported by comparative and experimental approaches such as cell ablation, vital staining, 4D microscopy, genetic studies etc. There is much more regulation and cell-cell interaction than previously suspected. This raises the question: What is the function of cell lineage? This question cannot be satisfactorily answered at present. All we can say is that even in animals with highly fixed cell lineages these are no indispensable prerequisites for normal development (Dohle 1989). Against this background, it is astonishing how stable some of these patterns remain during evolution. This stability provides us with another aspect of the study of cell lineage. Cell lineages can provide characters as important as highly complex structures for the reconstruction of phylogenies. The specific characteristics of mitoses, spindle directions, temporal sequences, equality and spatial arrangement of the division products can be included in the evaluation of homology and synapomorphy. Again, spiral cleavage is a well-known example (van den Biggelaar et al. 1997; Dohle 1989, 1999).

In recent times, gene expression patterns have been used to draw phylogenetic and evolutionary conclusions (Patel 1994; Scholtz 2001; Hughes & Kaufman 2002b). This is a new and exciting field as these patterns provide us with additional characters. In conjunction with cell lineage and morphogenesis, these expression patterns can be characters of high complexity. Nevertheless, they are no panacea for the solution of controversial relationships. Their value for phylogenetic reconstructions has to be discussed as carefully as for any other set of characters.

The general body organization of crustaceans is characterized by the occurrence of metameric body units, the segments. This is true for all arthropods but the stereotyped cell lineages found in several crustacean (malacostracan) groups offer a much more integrative approach to the study of segment formation in crustaceans than in chelicerates, myriapods, and insects where no comparable cell lineage patterns occur. In the following, we review the present knowledge about formation of segments in crustaceans at the levels of cell lineage patterns, gene expression, and morphogenesis, in particular, the establishment of segmental boundaries, early neurogenesis and limb bud formation in the embryonic germ band. The conceptual framework of the comparative approach is phylogenetic systematics (cladistics). Ontogenetic characters of all levels can either be used to infer phylogeny or they can be interpreted in a phylogenetic framework to draw conclusions about their evolutionary transformations. The phylogenetic background for the following treatment of crustacean development is discussed by Scholtz (2002b) for the Articulata and Arthropoda;

Dohle (2001), Richter (2002) for the Arthropoda, Mandibulata, and Crustacea; and Richter & Scholtz (2001) for Malacostraca (Fig. 1).

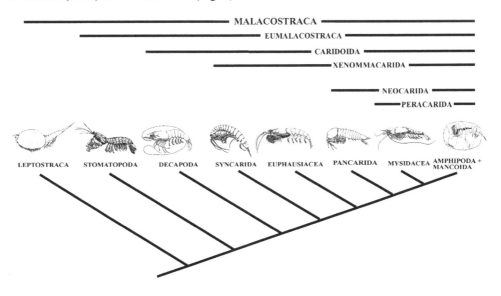

Figure 1. The phylogenetic tree of the Malacostraca as favored by the authors (after Richter & Scholtz 2001)

2 FORMATION OF THE GERM BAND

In insects, a distinction is made between short-germ and long-germ types of development. On the one hand, there are germ bands that at their first appearance on the egg surface do not show more than part of the head. Posterior to it there is an undifferentiated growth zone that successively contributes material for further segments. On the other hand, there are long germ bands that - shortly after cellularization of the blastoderm - develop all segments almost at the same time, as in *Drosophila*. Between these two extremes there is a number of intermediate types (Krause 1939; Sander 1983; Tautz et al. 1994; Davis & Patel 2002).

In crustaceans, nothing equivalent to extreme long-germ types occurs. In nearly all cases, a few anterior segments are laid down at the beginning (Scholtz 1992). The first appearance of differentiated structures is restricted to the protocerebral head lobes, the first and second antennae and the mandibular segment. Behind the mandibular segment, there are only comparatively few cells responsible for the generation of all trunk segments (Fig. 2A). The cellular material for the segments is provided either by randomly distributed mitoses or by the orderly proliferation of cells by teloblasts.

In many crustacean groups, the embryo hatches at an incomplete stage. The resulting free-swimming larva is called a nauplius. However, even in groups without a free nauplius, a stage comparable to a nauplius is formed in the egg that is called an egg-nauplius (Fig. 2B). This stage occurs in most malacostracans. In the few malacostracans with a free-swimming nauplius, this precocious larva is interpreted as the secondarily derived state (Scholtz 2000).

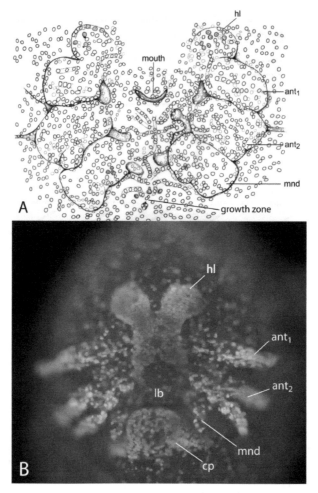

Figure 2. Egg-nauplius in *Anaspides tasmaniae*. (A) Early stage with buds of first and second antennae and mandibles and undifferentiated growth zone (Original). (B) Later stage with naupliar appendages in course of differentiation and formation of a caudal papilla (Original). ant_1 = first antenna; ant_2 = second antenna; cp = caudal papilla; hl = head lobe; mnd = mandible; lb = labrum.

In non-malacostracans, a free-swimming nauplius is typical, and epimorphic development - hatching with the full complement of segments - is rare. In the best-studied genus, *Artemia*, the larva hatches with differentiated naupliar limbs exposing bristles and spines and with a growth cone with undifferentiated scattered cells. Differentiation is initially not connected to cell multiplication. The gastrula stage has nearly as many cells as the nauplius so that differentiation is more a process of organization and specialization of already existing cells (Olson & Clegg 1978).

Whereas most non-malacostracans are thought to have such an undifferentiated growth zone, the production of cellular material for segmentation in malacostracans is connected to the action of teloblasts.

3 TELOBLASTS

3.1 *Formation of ectoteloblasts in malacostracans*

In the typical case, a great part of the post-naupliar germ band in malacostracans is generated by orderly divisions of teloblasts. Teloblasts can be defined as large stem cells that give rise to files of smaller derivatives by repeated asymmetric divisions in one direction only (Siewing 1969; Stent 1985; Dohle & Scholtz 1997). In malacostracans, ectoderm as well as mesoderm cells are proliferated by teloblasts. The ectoteloblasts differentiate in the germ disc at an early stage. At this stage, the material for the whole trunk region of the embryo is still missing. The germ corresponds to a nauplius within the egg shell (egg-nauplius).

The formation of the ectoteloblasts (ETs) follows different pathways: divisions of precursors, migration of cells, or differentiation *in situ*. Some earlier descriptions were not precise enough so that it was not clear whether differences had been based simply on inaccurate observations or not. More meticulous investigations in recent times have shown that real variation exists.

Oishi (1959, 1960) was the first to show that in decapods, the ectoteloblasts are formed by a sequence of elaborate cleavages from precursor cells. The result is a ring of 19 ectoteloblasts (ETs) consisting of 9 pairs and an unpaired median ET (Fig. 3A). Oishi confirmed such a cleavage pattern for three species, a caridean, a pagurid, and a brachyuran (Fig. 3B). As these species are not closely related within the decapods (Scholtz & Richter 1995), it can be concluded that the cleavage pattern represents the original mode at least in the ancestor of the Pleocyemata.

A ring of 19 ETs has also been reported in a large selection of different malacostracans such as leptostracans, stomatopods, syncarids, euphausiaceans, and pancarids (reviews and discussions in Dohle 1972; Zilch 1974; Scholtz 1984, 2000; Hertzler 2002). As leptostracans are regarded as the sister group of all other malacostracans (= Eumalacostraca) (Richter & Scholtz 2001) (Fig. 1), a ring of 19 ETs must inevitably have been present in the common ancestor of all malacostracans. As studies into the cell lineage of the groups named are still missing, it remains unknown whether the complicated pattern of cleavages of precursor cells as described for decapods was also a feature of the ur-malacostracan.

The presence of a ring of ETs is a prerequisite for the formation of a caudal papilla in the strict sense (Fig. 3B). The ETs encircle the site of the developing proctodaeum and the future anus. Through repeated generation of smaller cells, they become erected above the surface of the egg and form a finger-like projection, the caudal papilla. This caudal papilla folds towards the anterior end of the germ and thus obscures large parts of the ventral side. To analyze the cleavages, the papilla must be folded back.

Several deviations from the original pattern have been found with regard to the number, arrangement and formation of ETs. Even within the decapods, the freshwater crayfish show a much higher number of ectodermal stem cells arranged in a ring around the end of the caudal papilla. There are up to over 40 ETs, the number varies slightly (from 39 to 46 in *Cherax destructor*) (Scholtz 1992). They are formed by *in situ* differentiation and not by a complex division pattern as in other decapods. It has been suggested that the high number of ETs supports the argument for a monophyletic origin of the freshwater crayfish (Astacoidea and Parastacoidea). Before this character was introduced in the debate (Scholtz 1993, 2002a), several authors had advocated a polyphyletic origin and a convergent immigration of crayfish into freshwater in the Northern and Southern Hemispheres.

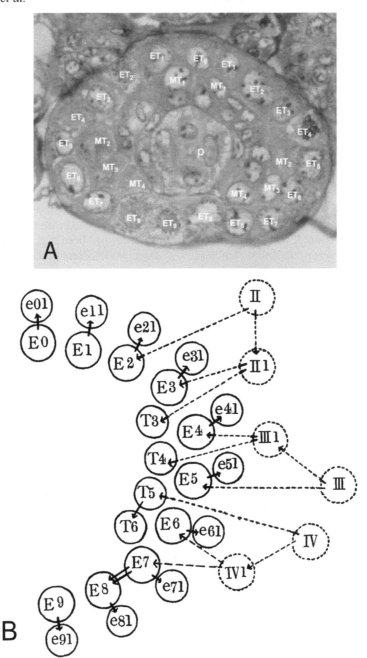

Figure 3. Formation of ectoteloblasts. (A) Transverse section through the caudal papilla of *Homarus americanus* in the teloblast region. ET_0 = median ectoteloblast; $ET_1 - ET_9$ = paired lateral ectoteloblasts; $MT_1 - MT_4$ = mesoteloblasts; p = proctodaeum (same preparation as in Scholtz 1993, fig. 2). (B) Diagrammatic representation of the differentiation of ectoteloblasts in several decapod embryos (after Oishi 1959).

In the Peracarida, no ring of ETs and no "true" caudal papilla are formed. If one takes for granted that the formation of a ring of 19 ETs is the original mode in malacostracans, this ring must have been secondarily disrupted during evolution forming a transverse band of ETs that only propagate cells for the ventral germ band and do not contribute material for the dorsal side. Accordingly, the transverse arrangement of ETs is an apomorphy of the Peracarida (see Richter & Scholtz 2001) (Fig. 1). In peracaridans, ETs are, as far as is known, never formed by stereotyped divisions of precursors, but are differentiated from scattered blastoderm cells. In the mysid *Neomysis integer* as well as in tanaidaceans and isopods (Scholtz 1984; Dohle 1972; Hejnol 2002), the ETs are differentiated *in situ,* which means that at that point in time when they can first be discerned as larger cells with large and faintly staining nuclei they are already arranged in a transverse and slightly crescent-like row in front of the blastopore area. Augmentation in number of ETs comes about by the lateral addition of adjacent cells. In the cumacean *Diastylis rathkei*, the ETs first appear in a crescent behind the blastopore and migrate towards the front thus recapitulating a ring-like arrangement (Dohle 1970). The "dorsal" side of the crescent disrupts and the ring flattens to form a transverse row. A similar migration has also been reported for the mysids *Hemimysis lamornae* (Manton 1928) and *Mesopodopsis orientalis* (Nair 1939).

It is noteworthy that mysids and lophogastrids still form a projection which resembles a caudal papilla and which covers the anterior ventral side of the embryo. However, ecto-teloblasts as well as mesoteloblasts are only arranged at the ventral side of this papilla. Therefore, the lumen of the papilla is crammed with yolk and yolk cells, which is in contrast to the "true" caudal papilla that is nearly devoid of yolk. A ring-like arrangement of the ETs in mysids is only achieved when the last pleonic segments are formed and the ETs are pushed back to the narrow pointed tip of the papilla. In amphipods, the germ band becomes folded ventrally in the region of the sixth thoracic segment. Whether this ventral furrow can be regarded as a remnant of a papilla formation or not has been discussed by Scholtz (1984). In cumaceans, isopods and tanaidaceans all traces of a caudal papilla are obliterated. There is no ventral furrow either. The germ band expands around the egg periphery so that in relatively small eggs, head lobes and telson tend to touch each other. In these cases, the dorsal side of the embryo must be formed anew by infolding processes. This formation of the germ band can be taken as a synapomorphic character of the three groups that are combined as Mancoidea (Ax 1999; Richter & Scholtz 2001) (Fig. 1).

The most derived case is found in amphipods. In two genera of Gammaridea (*Gammarus*, *Orchestia*), which have been carefully investigated, ETs are lacking completely (Scholtz 1990; Scholtz et al. 1994). All cells of the post-naupliar germ band are provided through unorderly adequal divisions of scattered cells. These cells are arranged in a pattern of transverse rows. As early as 1894, Bergh recognized that amphipods lack ETs. Weygoldt (1958), in his study of embryogenesis of *Gammarus pulex*, also could not find individualized ETs. He assumed that there is a whole field of cells with teloblastic activity ("Ektoteloblastenfeld"). This was a misinterpretation of the cleavages of the ectodermal rows with ordered longitudinal spindles. As no ectoteloblasts could be found in a representative of the Hyperiidea, *Hyperia galba* (Scholtz 1990), it can be deduced that all amphipods lack ectoteloblasts and that this deficiency is an apomorphy of the amphipods (Scholtz & Wolff 2002).

3.2 Are there ectoteloblasts in non-malacostracans?

It has been claimed that ectoteloblasts occur in several non-malacostracan crustaceans. In his seminal book "Embryology and phylogeny in annelids and arthropods", Anderson (1973) writes: "Ectoteloblasts which bud off the segmental, post-naupliar ectoderm have been described in branchiopods (Anderson, 1967) ..." (Anderson 1973: 343). However, in the original paper on two branchiopod species, the conchostracan *Limnadia stanleyana* and the anostracan *Artemia salina*, no mention of ectoteloblasts is made. On the contrary, it is stressed: "The components of this growth zone are paired vertical rows of six to seven mesoteloblasts laterally placed just in front of the telson, and an overlying ring of generalized ectoderm cells lacking distinct teloblasts" (Anderson 1967: 88). Why Anderson changed his mind remains obscure. He knew and cited the voluminous paper by Benesch (1969) who explicitly denies the existence of ETs in *Artemia*. We have seen and appreciated the preparations of Benesch and have investigated *Artemia* ourselves and can confirm that there are no ETs in the brine shrimp (Fig. 4). Cells at the end of the developing nauplius and metanauplius are neither larger nor more ordered than cells in front and do not show higher mitotic activity. The investigations of Freeman (1986, 1989), Manzanares et al. (1993, 1996) and Harzsch (2001) come to the same result.

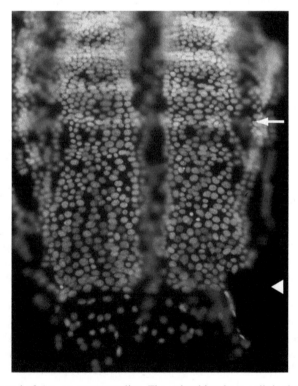

Figure 4. Posterior end of *Artemia* metanauplius. The animal has been split in the dorsal midline, the gut removed and the body wall flattened. No regular transverse rows and no larger cells with teloblastic activity (small arrow) can be detected at the border between trunk and telson (arrowhead). Cells are concentrated at the anterior trunk segments (I – VI). The large arrow points to the rudiment of the 7th thoracic segment. Stained with Bisbenzimid (Original).

In the raptorial water flea *Leptodora kindtii*, the ectodermal germ band grows by scattered divisions and without the action of ETs (Gerberding 1997). This study was explicitly designed to detect ETs. However, there are no specialized stem cells; no preferred orientation of spindles could be found; only randomly distributed cell divisions increase cell density. We suspect that there are no ETs in the branchiopods.

Another group where teloblasts have been described are the cirripedes (Anderson 1965, 1969, 1973). The formation of 7 ETs has been reported. This observation urgently needs a reinvestigation.

The emerging picture of ET evolution is: whereas the ancestor of crustaceans had a generalized growth zone, the ancestor of the Malacostraca has developed a ring of exactly 19 ETs (1 median and 9 pairs). Within the Decapoda, the ring was conserved but the number of ETs has more than doubled in freshwater crayfish. In Peracarida, the ring was transformed into a transverse row only contributing to the ventral side of the embryo. In the Amphipoda, ETs were eliminated altogether as a source of ectodermal cell proliferation.

3.3 The pattern of mesoteloblast formation in malacostracans

The existence and number of mesoteloblasts (MTs) are much more conservative than ETs in Malacostraca. In all groups, the number of 4 pairs of MTs has been found. In embryos with a „true" caudal papilla, they are arranged in a ring (Fig. 3A), and in the Peracarida they form a transverse row beneath the ETs. The number of 4 pairs is also retained in freshwater crayfish where the number of ETs is augmented and in amphipods where the ETs are lacking.

In decapods, Oishi (1959) recognized a pattern of formation of MTs from 2 pairs of precursor cells (Fig. 5B). One of them is transformed directly into the most laterodorsal MT4. Through a complicated cleavage pattern, the medioventral precursor gives rise to the three more median MTs ($MT_1 – MT_3$). This pattern has been verified and supplemented in the cumacean *Diastylis* (Fig. 5A) where the cleavages have been reconstructed on the grounds of a large number of mitotic figures (Dohle 1970). The only question mark concerns the fate of the two cells that are first given off by the precursors and migrate in an anterior direction and most probably represent the 2 pairs of primary mesoderm cells underlying the 2nd maxillae. A similar cleavage pattern was observed in tanaidaceans (Dohle 1972) and amphipods (Scholtz 1990). It must be conceded that in the latter named examples it is more difficult to trace the precursors within the immigrating cell mass. Nevertheless, the cleavages are so similar in decapods, cumaceans and amphipods that it can be said with confidence that the formation of 2 pairs of precursors is a trait of the caridoid ancestor (see Richter & Scholtz 2001) (Fig. 1).

There are several cases in the literature where another cleavage pattern is given or the 4 pairs of MTs are deduced from one pair of precursors (Strömberg 1965, 1967, 1971; Hertzler 2002). In our opinion, these cases are not well documented so that at present the most convincing hypothesis is that the formation of the MTs in the ancestor of the Malacostraca followed the pattern described for decapods and peracarids. In the holoblastic eggs of dendrobranchiate shrimps, the MTs are said to go back to a single large median precursor (Zilch 1978, 1979; Hertzler 2002).

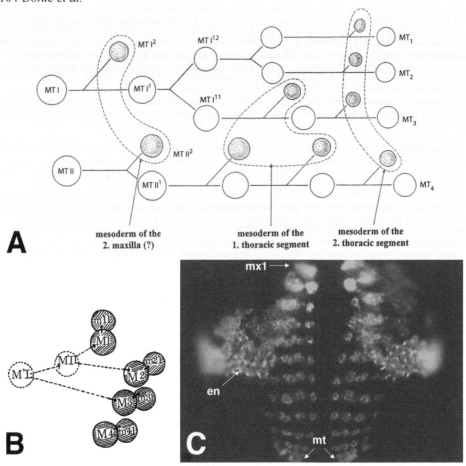

Figure 5. [See Plate 11 in color insert.] Mesoteloblast formation in malacostracans. (A) Peracarida (*Diastylis rathkei*): There are two pairs of mesoteloblast precursors. The posterior and more lateral mother cell gives rise to the lateral mesoteloblast. The anterior and more median mother cell gives rise to the 3 median mesoteloblasts. Only one side is represented (after Dohle & Scholtz 1997). (B) The differentiation of mesoteloblasts in several decapod embryos. Only one side is represented (after Oishi 1959). (C) The post-naupliar mesoderm (and entoderm (en)) in the germ band of *Orchestia cavimana* stained with DiI (after Wolff & Scholtz 2002). The 4 pairs of large mesoteloblasts (mt) mark the posterior end. Their smaller derivatives lie anterior to the mesoteloblasts with those of the 1st maxilla (mx1) at the anteriormost position. All stained cells are the outcome of an injection of DiI into two micromeres at the 16-cell stage (see Wolff & Scholtz 2002).

3.4 *Mesoteloblasts in non-malacostracans?*

Comparable to the situation in ectoteloblasts, the existence of mesoteloblasts (MTs) in non-malacostracans is doubtful. Anderson remains the only investigator who describes MTs in branchiopods (Anderson 1967) and cirripedes (Anderson 1965, 1969). In *Limnadia* and *Artemia*, a variable number of six or seven MTs on each side are described which increase to nine or ten as development proceeds. Benesch (1969) explicitly denies the existence of MTs in *Artemia*. We also could not find an orderly array of mesodermal stem cells in this species. Gerberding (1997) has also not found MTs in *Leptodora*.

In cirripedes, Anderson (1965, 1969) describes a number of exactly 4 pairs of MTs giving rise to a crescent of 8 mesodermal primordial cells for each trunk segment. Anderson points to the fact that the Malacostraca have the same specialization of 8 MTs, and he assumes that this number "... has evolved convergently ..." (Anderson 1973: 336). Walley (1969), who made a thorough investigation of histogenesis of barnacle larvae before and after metamorphosis, cites Anderson's descriptions but does not actually confirm them. Therefore, it must be repeated what has been said concerning the ETs: cirripedes urgently need a reinvestigation in order to clarify the situation.

On the whole, we have great reservations regarding the claim that ETs or MTs form part of the trunk in non-malacostracans.

4 FORMATION OF ECTODERMAL CELL ROWS

The ETs, irrespective of their arrangement in a ring or in a transverse row, begin to proliferate small derivatives in the anterior direction. Though there are reports that these derivatives are given off simultaneously, we conclude from our own experience that there is always a successive generation of ET derivatives following a mediolateral mitotic wave.

In the classic case that has been worked out in *Diastylis*, twelve rows of smaller cells are given off by the ETs in an anterior direction (Dohle 1970). These cells are not only arranged in neat transverse rows but also in exact longitudinal files with an ET at their ends, respectively. In principle, this grid-like pattern could be brought forth by the longitudinal direction of spindles. However, the precision of the network supports the idea that another influence must be responsible for maintaining the order. This idea is corroborated by two observations. In the tanaidacean *Leptochelia*, the first three rows given off by ETs do not initially fit exactly into the grid. Only later are they arranged in a precise order.

The most convincing argument is that scattered cells in front of the first ET derivatives are also forced into a grid. This observation, first made in *Diastylis* (Dohle 1970), has been verified in tanaidaceans, mysids, pancarids and decapods. There are two to four rows that are arranged in this way. The most striking case is that of amphipods. No ectoteloblasts occur. All cells behind the mandibles and in front of the telson are forced into an orderly grid. The future destiny of these cells corresponds exactly to the rows of ectoteloblastic or non-ectoteloblastic origin in other malacostracans (Figs. 6-11).

In order to differentiate between the cell rows of ectoteloblastic origin and cell rows, which have been arranged by formerly scattered blastoderm cells, a different designation has been introduced. Rows of ectoteloblastic origin (e-rows) are designated with Roman numerals in the order of their generation: eI, eII, eIII or simply I, II, III and so on. Rows formed by scattered blastoderm cells (E-rows) are designated with Arabic numerals in brackets: E(1), E(2), E(3) or simply (1), (2), and (3). The cells of the midline are numbered $E(3)_0$, eI_0 and so on. The lateral cells are counted from the cell nearest to the midline: $E(3)_1$, eI_1 to the lateral side. These cells are arranged in strictly symmetrical order so that it is sufficient to describe and figure the events on one side. The two sides are, however, independent. If one side is slightly ahead of the other, this advantage is seen in all hemi-segments of one side.

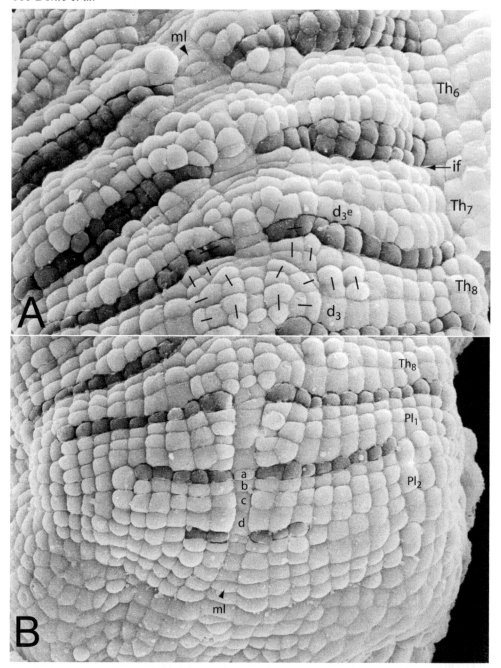

Figure 6. [See Plate 12 in color insert.] Posterior part of a germ band of the amphipod *Orchestia cavimana* (after Hejnol 2002). SEM picture. The cells of row a have been tinged with blue for orientation. Two views (A, B) of the same preparation. In th8, cells after the first differential cleavage are connected by lines. Abbreviations: a, b, c, and d = cells and rows belonging to one genealogical unit; if = intersegmental furrow; ml = midline; Pl = pleonic segments; Th = thoracic segments.

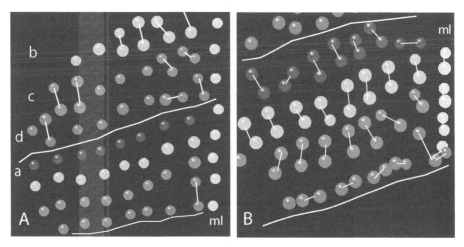

Figure 7. [See Plate 13 in color insert.] The first differential cleavages in the germ band of the isopod *Porcellio scaber* as revealed by 4D microscopy (after Hejnol 2002). Row a: blue; row b: green; row c: pink; row d: red; midline (ml): yellow; genealogical border: white line. (A) The cleavages in two genealogical units. The differential cleavage begins in cell d_1 (arrow head) of the posterior genealogical unit. In the preceding rows the cells d_1, d_2, d_7, c_1, c_2, c_6, c_7 and $b_1 - b_4$ are already divided. Row a of this anterior unit is not shown. (B) Nearly all cells have performed the first differential cleavage. Cell d_1h has already divided for the second time (arrow head).

Can it be excluded that cells of the anterior rows which are thought to be of non-ectoteloblastic origin are sister cells of ETs? It is possible that some cells of the E-rows have descended from precursors of ETs. However, this must have taken place before their visible differentiation and arrangement into rows. It can be said that ETs, after they are recognizable by size, arrangement in a row, and the beginning of asymmetrical divisions, do not contribute to the E-rows.

The number of twelve rows of small ET derivatives generated in an anterior direction has also been found in tanaidaceans, isopods and mysids. The twelfth cleavage is followed by another cleavage. This cleavage is also unequal. The row of small cells is, however, given off in a posterior direction. This row of small cells directly abuts on the tightly packed cells of the future telson. The row of larger cells is counted as row XIII and the posterior row of small cells as row XIV. With this division, the action of the ETs comes to an end in the named species. In the decapod *Cherax*, a fourteenth cleavage of the ETs follows so that one more row of cells is formed (Scholtz 1992).

5 FIRST DIVISIONS OF ECTODERMAL CELL ROWS

The cells of the rows generated by ectoteloblasts divide in a typical manner. They divide twice with spindles oriented longitudinally and parallel to the midline so that the result of these two divisions is 4 rows of cells. These cells are arranged in a precise grid-like pattern. Once more, the accuracy of this cell lattice cannot only be brought forth by the same spindle directions. An influence forcing the cells into the lattice must still be in action.

The question is whether the first two divisions are completely equal resulting in equipotential cells. A careful examination reveals that the divisions are slightly unequal. In *Diastylis*, the posterior cell row cd is slightly larger than the anterior one, named ab. In the next division, the cd cells have a slight temporal advantage over the anterior ones (Dohle 1976). Once again, the cells of row d are slightly larger than those of row c, and the cells of row a are larger than those of row b. The nuclei of rows b and c are placed more to the surface, and the cells of these rows are more elevated than their sister cells. These relations are not the same in all species investigated. In *Cherax*, row ab is always in advance of row cd and has almost completed its mitotic wave when cd begins to divide (Scholtz 1992). Also in different rows in the same species, the mitotic advantage can be reversed. In *Neomysis*, E(2)ab always divides before E(2)cd, whereas row e1cd divides before e1ab (Scholtz 1984).

The observation that the first division is already unequal is corroborated by the fact that in most species, the anterior row ab is distinguished by *engrailed* expression. This observation cannot be generalized as the amphipods show *en* expression only after the second division (Scholtz et al. 1993, 1994) (see below).

As described in the preceding section, two rows anterior to the ectoteloblastic rows which had been arranged by migration and ordering of scattered blastoderm cells, divide twice in the same manner as described, the result being 4 rows respectively. After their formation, these two sets of 4 rows can only be distinguished from the following sets of rows of ectoteloblastic origin by their slightly larger nuclei. The influences triggering the divisions and forcing the cells into the grid-like pattern must be the same and cannot be coupled with the ectoteloblast activity. This observation led to the consideration that the grid-like formation of rows could in principle be possible without the action of ectoteloblasts. This consideration was made before an actual case in nature had been verified. Fortunately, such a case exists in the form of the amphipods (Scholtz 1990). So to speak, nature has performed this crucial experiment in the course of evolution. Amphipods do not develop ectoteloblasts. Nevertheless, all cells of the germ band are forced into rows and divide twice, in a manner absolutely comparable to the ectoteloblastic and non-ectoteloblastic cell rows in other species.

An equivalent surgical experiment would be to eliminate ectoteloblasts and see whether more rows are added in front of the destroyed ETs by the arrangement of formerly scattered blastoderm cells. We cannot expect, however, that the cells in front of the disturbance can regenerate the eliminated part. Preliminary experiments, which have been performed in isopods, show that the eliminated cells cannot be replaced (Hejnol 2002).

In the cumacean *Diastylis* and the mysid *Neomysis*, two rows of cells are arranged in front of the ectoteloblastic derivatives and divide twice to form 4 rows each, as has been described in the preceding section (Dohle 1970; Scholtz 1984). These 8 rows of non-teloblastic origin are named E(2)a, b, c, and d and E(3)a, b, c, and d. In the tanaidacean *Leptochelia* and the isopod *Porcellio*, these 8 rows go back to one single row that divides three times (Dohle 1972; Hejnol 2002). This might be another synapomorphy of Tanaidacea and Isopoda supporting their sister group relationship (see Richter & Scholtz 2001). Despite this difference, the 8 rows in all the species are homologous with regard to their future fate. In front of row E(2), there are additional rows that form the main part of the 1st maxillary segment. Depending on the species, these rows can be traced back to one or two primary rows – E(1) and E(0). The result is always a number of 4 rows. These rows consist of only a few cells on each side. Their division patterns are completely different from the subsequent rows and differ widely in different species so that there is nearly no means to homologize the patterns and the cells (Fig. 8). Thus, there is a transition zone between mandibles and 2nd maxillae that exhibit stereotyped divisions as in the subsequent rows but does not obey the same division patterns. The cells forming the posterior border of the mandibular segment still follow a certain order in arrangement and cleavage direction, the more anterior cells, however, divide "wildly" without any detectable sequence or direction in equivalent stages or on the two sides of the germ band. In the two antennal segments as well as in the part in front belonging to a procephalon, lineage and arrangement of cells lack all signs of stereotypy (Fig. 11).

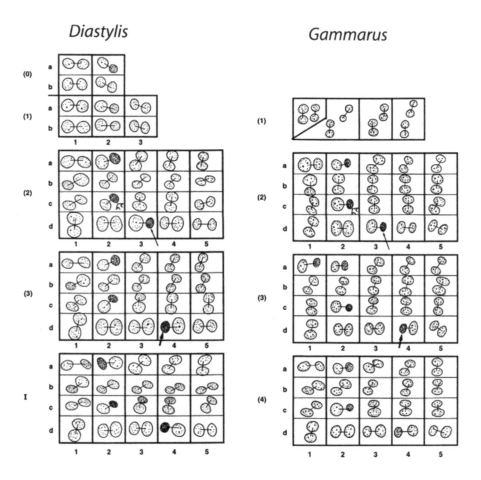

Figure 8. Schematic representation of the anterior postnaupliar part of the germ band of two peracaridan species, *Diastylis rathkei* (Cumacea) and *Gammarus pulex* (Amphipoda) (after Dohle & Scholtz 1997). Explanation in the text.

6 FIRST DIVISIONS OF THE MESODERMAL CELL ROWS

The naupliar segments are underlain by mesoderm cells (m-cells) that immigrate at the site of the blastopore and migrate to the corresponding segments (Gerberding et al. 2002, Wolff & Scholtz 2002). The post-naupliar segments of malacostracans are equipped with a few cells originating from the MTs (Fig. 5C). The origin of the MTs has already been described. They are differentiated successively. Thus, the most anterior post-naupliar segments do not have the full complement of MT derivatives. In *Diastylis*, we find in the 1st maxillary segment (beneath row E(1)) a pair of m-cells the origin of which is doubtful. The 2nd maxillary segment has 2 pairs of m-cells that are most probably derived from the 2 pairs of MT precursors. The 1st thoracic segment - the ectodermal part of which is mainly formed by row E(3) - is underlain by 3 pairs of m-cells which originate from MT_3 and MT_4 (Dohle 1970). The subsequent segment, which is mainly formed by the ectoteloblastic row eI, is equipped with the full complement of 4 pairs of m-cells (Fig. 5C). These cells divide in a stereotyped manner. The patterns in this and the following segments are clearly homonomous. One interesting point should be mentioned. The 4 pairs of m-cells lie beneath the ectodermal rows c and d and thus do not straddle later segment borders. The 1st thoracic segment makes an exception. This segment first contains only 3 pairs of m-cells but later on exhibits 4 pairs of cell clusters. The innermost cluster has been contributed by the preceding 2nd maxillary segment from, where a pair of cells has emigrated and filled the median gap in the 1st thoracic segment. This migration has been observed in *Neomysis* and *Gammarus* (Dohle & Scholtz 1988; Scholtz 1990).

7 GRADIENTS

In insects, many species show an anteroposterior gradient of differentiation in the germ band (Bock 1939). If one disregards the most anterior part of the head in front of the gnathal appendages, a segment can be defined where differentiation is ahead of subsequent and even of more anterior segments. This advantage implies demarcation by intersegmental grooves, limb bud formation and detachment and differentiation of ganglia. The peak of such a gradient can be located in varying segments in different species: mandibles, first or second maxillae, first or even second thoracic segment. This region was regarded by Seidel et al. (1940) as an essential physiological organizer and was called the differentiation centre (Differenzierungszentrum). It seems to operate even in extreme long-germ insects such as *Drosophila* in which all segments appear nearly simultaneously.

In crustaceans with stereotyped division patterns, this differentiation center can be determined very precisely. Even if a row is only minimally ahead of another, say 1/5 of a division cycle, it can easily be ascertained. It is not always the same row that divides first.

Though the cells of the non-ectoteloblastic rows E(2) and E(3) only assembled when the first teloblastic rows were already formed, these non-ectoteloblastic rows are normally the first to divide. In *Diastylis*, row E(2) is ahead of row E(3), in *Gammarus*, the rows E(2), (3) and (4) cleave nearly at the same time. An exception is the tanaidacean *Leptochelia* where row eI is ahead of subsequent as well as preceding rows. In this species, rows E(2) and E(3) are generated by the division of the cells of one precursor row. Therefore it seems reasonable to deduce that they lag behind because of this additional division. However, in isopods also the rows E(2) and E(3) are generated by one precursor row (Vehling 1994; Hejnol 2002). Nevertheless, row E(2) is ahead as in most other known examples. This shows that the differentiation center is autonomous and can switch from one row to the other in evolution without disturbing the general pattern.

The gradients of differentiation in anteroposterior direction are also different in their steepness in different species. In *Cherax*, the gradient is very steep. A given row is nearly one cell cycle in advance of the subsequent row. In amphipods, the gradient is rather flat. One cell cycle stretches over 4 rows (Scholtz 1992).

Another gradient that is obvious in the early germ band is a mediolateral gradient. A very careful examination in insects has been made by Bock (1939). In crustaceans, this gradient has been described as early as 1893 by Bergh. In the cumacean *Diastylis*, Dohle (1970) tried to formalize this gradient. It has a sink in the midline and a peak between the first and second cells adjacent to the midline. It gently slopes to the lateral cells. An equivalent gradient could be found in all other species investigated. In principle, one has to postulate a stimulating effect which triggers an adjacent cell to divide, and an antagonistic retarding effect preventing the next but one cell from dividing. This operates when the field unit of the gradient matches more or less one cell. This seems to be the case in *Diastylis*. In other species, there is a field covering 2 or 3 cells so that cells within this field behave stochastically: an outer cell can divide before an inner one. In isopods, the gradient is more like a double peak so that lateral cells divide first. Even in cases where the gradient is textbook-like at first, it breaks down after the first or second differential cleavage. Then the division centers lie in the ganglion and/or in the tip of a limb bud.

By using 4D microscopy, it is possible to define the mediolateral gradient exactly. In *Porcellio*, which has been investigated in this way (Hejnol 2002), the gradient is rather flat which means that the time lapse between the mitosis of an inner cell and the adjacent outer cell is not very accentuated. It may even be that an outer cell is in advance of an inner one in individual cases. What is general in this species is a second peak in lateral cells. In contrast, the gradient of *Diastylis* is steep. There is no outer sub-peak.

Also the gradients within the descendants of one row vary in different species. In *Diastylis*, it is always row d that cleaves first and is ahead of the other rows. In *Porcellio*, it is row b that is ahead of the other rows.

8 DIFFERENTIAL CLEAVAGES

After completion of the first two cleavages of the cells of one row which result in the formation of 4 rows (a, b, c, and d), the similarities between cell and division patterns of different rows and species are not at an end. The cells proceed into rounds of divisions that are peculiar for each cell. The divisions can be characterized by the orientation of the spindles relative to the midline, by the equality or inequality of the daughter cells and their nuclei and by the timing of the divisions. As the cleavages are heterochronous and mostly unequal, the pattern of the cleavages can be reconstructed from fixed and mounted preparations. Recently, direct observations and verifications of the reconstructed patterns have been made possible by use of 4D microscopy - a technique developed for the assessment of the cell lineage of single embryos in nematodes in space and time by Schnabel et al. (1997) and extended to crustaceans by Hejnol (2002) (Fig. 7). These peculiar cleavages have been called "differential cleavages" in line with the terminology of Kühn (1965).

In the cell row E(2) as well as in subsequent cell rows, the first differential cleavage takes place in the cell d_1. The spindle is nearly parallel to the midline, slightly oblique so that an anterior cell d_1v and a posterior cell d_1h that is nearer to the midline are generated

(Figs. 6-8). The posterior cell is slightly larger than the anterior one. In the sequence of the following cleavages, there is slight variation. In most cases, d_2 is the next cell to cleave. The spindle is at right angles to the midline, the two resulting cells are equal in size. The next cell to cleave is c_1, the division is adequal, the orientation of the spindle being 45^0 relative to the midline. d_3 cleaves with a spindle once more forming right angles to the midline producing an inner cell d_3i and an outer cell d_3e (Figs. 6, 8). In this differential cleavage, there is an obvious distinction in equality between different rows. The cells generated after the first differential cleavage can be arranged in a scheme regardless of their time sequence (Fig. 8). The same scheme can of course be established for the second differential cleavage. A complete scheme for the third differential cleavage has not yet been feasible. Though the cells nearest to the midline could be followed to the fifth or even the seventh differential cleavages, this is not possible for the more lateral cells because of their delay which becomes more pronounced in later stages. It is to be hoped that the further fate of different cells can be traced by the application of markers such as DiI as has been performed in the midline cells (Gerberding & Scholtz 1999, 2001).

When comparing the schematized patterns of cleavages in different rows of the same species and of corresponding rows in different species, detailed similarities and slight differences can be worked out. This method is equivalent to comparisons on the molecular level. The corresponding segments in different species are like orthologs, whereas the homonomous segments are like paralogs after a duplication event. We are convinced that the patterns share an overall homology, which means that all these segmental patterns go back to a common and finally identical origin. The question is in how far different segments and different species have deviated from this postulated putative common origin. Thus, the similarities and differences lend themselves to questions about functional inter-pretation, to considerations about homology and about phylogenetic divergence.

As an example, we compare the schematized first differential cleavages of the des-cendants of anterior rows in the cumacean *Diastylis* and the amphipod *Gammarus* (Fig. 8). The overall similarity between homonomous rows of one species and between the rows of different species is more than obvious. In homonomous rows of one species, there are slight differences. In *Diastylis*, the daughter cells of $(2)a_2$ and $(3)a_2$ are similar whereas in Ia_2 the spindle direction is different and the equality is reversed. In contrast, the daughter cells of $(3)d_3$ and Id_3 are similar whereas the cleavage of $(2)d_3$ is different: the lateral daughter cell $(2)d_3e$ (thin arrow) is far smaller than the median one. Both differences are obviously related to the formation of limb buds. The first and second maxillae are formed more medially than the thoracic limbs. The tip of the appendage, which is marked by the small cell $(2)d_3e$ (thin arrow) in the second maxilla, is marked by the cell d_4i (fat arrow) in the 1st and subsequent thoracic limbs. The same relation can be found in *Gammarus*. It must be borne in mind that row I in *Diastylis* is the 1st row generated by the ectoteloblasts whereas in *Gammarus* the equivalent row (4) has been formed by the arrangement of scattered blastoderm cells.

There are also differences in equality between the daughter cells of a_1 in different seg-ments. These cannot be explained in functional terms. As far as is known, the two daughter cells of a_1i become neuroblasts in all segments. A remarkable difference pertains to *Diastylis* and *Gammarus* regarding the cell c_2. In *Diastylis*, this cell in all segments divides unequally. The spindle is oriented at a 45^0 angle, the more anterior and lateral daughter cell (white hollow arrow) is much smaller than the more median and posterior one. In *Gammarus*, the division of c_2 is also highly unequal. However, the spindle is oriented at a 90^0 angle so that the smaller daughter cell (white hollow arrow) lies exactly lateral to the larger sister. This position is invariant in all segments. It could also be found in other amphipods (*Orchestia, Calliopius*), and an early germ band of an amphipod can, apart from the lack of ectoteloblasts, be easily recognized by this arrangement. The 45^0 spindle orientation of *Diastylis* is not unique but can be found in *Leptochelia* (Tanaidacea), *Neomysis* (Mysidacea), *Porcellio* (Isopoda) and even Decapoda. It is thus a plesiomorphic feature. The orientation in *Gammarus* is the apomorphic state and is probably an autapomorphy of Amphipoda (Scholtz & Wolff 2002).

In this sense, a detailed comparison between germ bands of different malacostracans is at present possible but has still to be completed for the hitherto known species. Such a comparison will further underpin two general statements:

- Highly similar and homologous division patterns can come about in rows of different origins.
- Slight variations and deviations are possible without disturbance to most parts of the pattern and are thus revealed to be genetically independent.

The differences in the rows in front of row E(2) are much more pronounced. They also await a careful comparative analysis.

9 NEUROGENESIS

9.1 *The lateral neuroblasts in malacostracans*

In insects, neurogenesis is accomplished by the action of neuroblasts (Wheeler 1893; Bate 1976; Tamarelle et al. 1985; Truman & Ball 1998). Both in the head and in the trunk segments, a well-defined number of neuroblasts in each segment are differentiated in several steps. Each neuroblast (NB) gives rise to a chain of ganglion mother cells (GMCs) which divide once to generate two cells; these differentiate into neurons or glia cells. In our opinion, the term neuroblast should be restricted to cells cleaving asymmetrically following the stem cell mode (Stent 1985). We are aware that many neurobiologists also call cells directly transforming into neurons or equally dividing precursor cells "neuroblasts". We do not approve this usage as it obscures fundamental differences (see also Whitington 2003).

Concentrating on the post-naupliar segments, a stereotyped pattern of NB formation could be revealed in malacostracan embryos (Dohle 1976; Dohle & Scholtz 1988; Scholtz 1990, 1992). When identifying a cell as a neuroblast several characteristics have been considered. The NB is a relatively large cell. During division, the spindle is oriented perpendicular to the germ band surface, a small daughter cell is given off to the interior, the peripheral cell soon gaining the size of the mother cell. Further cleavages lead to a column of smaller cells like in insects. As only one further division has been found in these smaller cells, it has been concluded that the chain of cells are GMCs and that the two daughter cells are differentiated into neurons and/or glia cells. This last assumption has not been verified conclusively in the lateral NBs. This will only be possible by individual injection or application of markers and observation of the further development.

Most of the NBs of the post-naupliar segments are arranged in strict symmetrical order lateral of the midline. From the segment of the 2nd maxilla onwards, the patterns of generation of NBs are very much alike. The first cell to become a neuroblast is the posterior division product of d_1, namely d_1h. This cell divides into a large peripheral NB, d_1hn, and a small interior GMC, d_1hg. 12 further NBs could be traced in the genealogical scheme. This scheme is based on the reconstruction of the sequence of heterochronous mitoses and has a very high degree of confidence. Such schemes could until now only be established for the cumacean *Diastylis* and the amphipod *Gammarus* (Fig. 9) and in a simplified form for the crayfish *Cherax destructor* (Scholtz 1992; Scholtz & Gerberding 2002). Further cleavages and additions of NBs could not be traced with certainty by this method. Therefore, it is not clear whether this is already the final number or whether further NBs are added. Nevertheless, some interesting results can be gained from this scheme:

- Neuroblasts do not differentiate simultaneously in steps or in "waves" (Doe 1992) as in insects, but one after the other. d_1h gives off two GMCs before d_2ie, d_2ii, c_2hi, c_2he and d_1vi divide successively to give off the next GMCs.
- Both products of an adequal division of a NB precursor at the surface do not necessarily become NBs. d_1ve cleaves adequally; the more lateral daughter cell d_1vee becomes a NB whereas the more median cell d_1vei divides once more adequally.
- With the generation of a small GMC, the larger cell is not definitely differentiated into a NB but can divide once more adequally at the surface with tangentially oriented spindles. The NB d_1hnn after having produced two GMCs divides adequally on the surface. The median cell once more becomes a NB whereas the lateral cell d_1hnne divides adequally and is probably an epidermoblast.

It is common belief that in insects NB precursors, or functional NBs, no longer divide equally. In older papers, claims about equally dividing and multiplying neuroblasts are not rare. In recent times, equal divisions of NBs have been suspected only in a few cases (Tamarelle et al. 1985). Probably the differentiation of NBs in insects is much more flexible than commonly described.

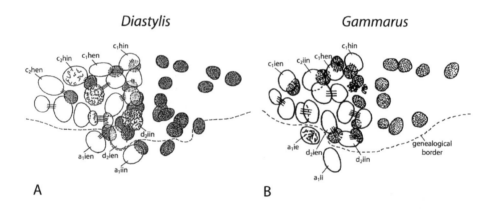

Figure 9. The early pattern of neuroblasts (NBs) in malacostracans. On the left (morphological right side) NBs, GMCs (= ganglion mother cells) and normal epidermal cells are shown; on the right (morphological left side) only GMCs are shown. (A) The rudiment of a thoracic ganglion in *Diastylis* (Cumacea) where 13 GMCs have been produced by NBs (Full details of cell lineage in Dohle 1976, and Dohle & Scholtz 1988). Only two early NBs are contributed by cells of the En-positive row a, posterior to the genealogical border (dotted line). (B) The rudiment of a thoracic ganglion in *Gammarus* (Amphipoda) in a slightly earlier stadium, with 9 GMCs formed (adapted from Scholtz 1990, and Scholtz & Gerberding 2001).

Scholtz (1992) has assumed a rather high number of NBs in the crayfish *Cherax destructor*, namely 25 – 30 NBs per hemisegment. This corresponds closely to the number that is found in insects. Harzsch (2001) describes 5 rows comprising 3 to 4 NBs within each hemineuromere in embryos of *Palaemonetes argentinus*. This number of 13 to 15 NBs agrees rather well with the number of NBs present in the stadium depicted for *Diastylis* and *Gammarus* (Fig. 9). Unfortunately, the putative NBs of *Palaemonetes* have not been brought into relation to the original 4 rows and to the stereotyped division pattern. The question of the final number of NBs has not been settled. In crustaceans, the persistence of NBs into the adult phase has been demonstrated (Harzsch & Dawirs 1994; Schmidt 2001).

9.2 *The median neuroblast*

In addition to the lateral NBs, which are arranged in strict symmetrical order, there is one median NB per genealogical unit. It differentiates in the midline of the germ band (Gerberding & Scholtz 1999, 2001; Scholtz & Gerberding 2002). This midline has not been analyzed in detail in most previous investigations as the cleavages could not be verified by their counterparts on the contralateral side. Gerberding & Scholtz (1999, 2001) have traced the lineage in the amphipod *Orchestia* by applying the fluorescent marker DiI *in vivo*. The median NB is the most posterior cell after the first two cleavages of the primary midline cell. At hatching the median NB-clone comprises about 10 neurons. The three more anterior midline cells divide only once with the progeny transforming into glia cells. This median NB can be compared to NBs in the midline in insects (Bate & Grunewald 1981; Bossing & Technau 1994). It is not clear whether non-malacostracans have an equivalent neuronal midline differentiation. However, some figures in the papers of Manzanares et al. (1993) on *Artemia* and of Gerberding (1997) on *Leptodora* suggest at least the existence of a comparable midline.

9.3 *Neuroblasts in naupliar segments*

The scheme of NB differentiation as described in the preceding sections and depicted in Fig. 9 applies more or less to the segments from the 2nd maxilla onwards. The more anterior segments have a completely different scheme. The mitoses on the first maxillary segment are stereotyped, the first NBs could also be traced (Dohle 1976). However, they do not fit into the scheme for the subsequent segments. The naupliar segments do not have an unvaried cell division pattern. Nevertheless, NBs can be identified by their size and by the groups of small interior cells obviously originating from them (Scholtz 1992). There are clusters of NBs median of the mandibles, of the two pairs of antennae and of the head lobes and NB-like cells can even be found in the midbrain of advanced embryos (Benton & Beltz 2002). In insects, the NBs of the brain have been mapped (Zacharias et al. 1993). An equivalent effort has not yet been made in crustaceans.

9.4 *Neuroblasts in non-malacostracans?*

It is a matter of much debate whether non-malacostracan crustaceans possess NBs homologous to malacostracan NBs. Many old and recent investigations are not conclusive. Gerberding (1997) was the first to demonstrate unequal mitoses in the germ band of the raptorial water flea *Leptodora kindtii* with spindles perpendicular to the surface producing a chain of small cells. He depicts 23 putative neuroblasts per hemisegment. Duman-Scheel & Patel (1999) identified seven rows of larger-sized cells on the ventral side of *Artemia franciscana*. Without further specification they "speculate that these large cells are NBs". Neither asymmetrical divisions nor the arrangement of GMCs are demonstrated.

In a recent paper, Harzsch (2001) presents investigations on *Triops cancriformis* (Notostraca) and *Artemia salina* (Anostraca) and suggests "... that in the ventral neurogenic zone of Branchiopoda, neuronal stem cells with cellular characteristics of malacostracan neuroblasts are present" (p. 155). This suggestion is based on incubation of larvae in BrdU that is incorporated in cleaving cells. The method does not distinguish between different types of cells. The only indication of the existence of NBs is that large labeled nuclei were associated by smaller labeled nuclei that are assumed to be GMCs. A segmentally iterated cleavage pattern could not be recognized. To summarize, the existence of NBs in the form of stem cells, their number and distribution in non-malacostracans needs closer investigation and confirmation.

10   LIMB BUD FORMATION

One of the astounding results of the early cell lineage studies in peracarids was the observation that the limb buds are formed as complex structures consisting of derivatives of cells from different genealogical units. For instance, the thoracic legs in *Diastylis* are composed of cells originating from cells $c_3 - c_5$ and $d_3 - d_5$ of one row and $a_2 - a_5$ of the subsequent row (Dohle 1976, Dohle & Scholtz 1988) (Fig. 10). In contrast, the second maxillae, which are placed more medially, go back to derivatives of cells $E(2)c_3 - c_4$, $d_3 - d_4$ and $E(3)a_2 - a_3$ with a certain contribution of $E(2)c_2$, d_2, and $E(3)a_1$. The limb buds of the first maxillary segment, which are very much like the second maxillae in position and extension, have a different origin. The formation of the mandibles cannot be attributed to a stereotyped cell division pattern, nor can the vestiges of the second and first antennae (Fig. 11).

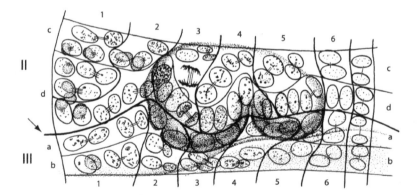

Figure 10. Formation of the third thoracic limb bud in *Diastylis* (after Dohle & Scholtz 1988). Only the left morphological side is shown. Medially, 9 GMCs are formed; compare with Fig. 9. Arrow points to genealogical border between the rows II and III of ectoteloblastic origin.

Despite completely different structures in the legs of a variety of species, the first appearance of limb buds and the underlying cell pattern is much the same (Dohle 1972; Scholtz 1984, 1990, 1992; Hejnol 2002). The cell pattern is even the same in abdominal limbs and in segments that are not fitted with limb buds as in the 8th thoracic segment in representatives of the 'mancoids' (Cumacea, Tanaidacea and Isopoda, see Hessler 1983; Richter & Scholtz 2001). Some more details on limb bud formation will be treated in the section on *Distal-less* expression.

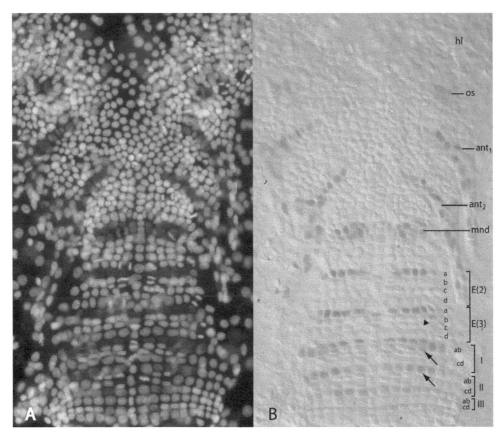

Figure 11. Anterior part of a germ band of *Neomysis integer* (same preparation as in Dohle & Scholtz 1997, fig. 3). The ectodermal row IX has been formed by the ETs. The posterior part of the germ band which forms a papilla-like projection has been obliterated. Anterior is up. (A) Staining with Bisbenzimid. The cells which are En-positive are darker. (B) The same preparation in bright field. En-positive cells are brown. Abbreviations: hl = head lobe; os = rudiment of ocular-protocerebral stripe; ant₁ = first antenna; ant2 = second antenna; mnd = mandible. Lower arrow: cell IIab₄ in row ab is stained; upper arrow: cell Ib₄ shortly after division, both daughter cells still retain En; arrowhead: cell E(3)b₄ has lost En-staining.

11 INTERSEGMENTAL FURROWS

Closely connected to the complex nature of ganglia and limb buds is the fact that the intersegmental furrows do not match the boundaries of the genealogical units but separate cells of the same origin. The intersegmental furrow runs transversely and slightly obliquely through the descendants of a primary cell row (Figs. 6, 10). Medially, the incision is placed behind descendants of row b whereas more laterally it passes between descendants of row a and b and at the margin of the germ band, it is obvious between descendants of row a and nearly merges with the genealogical boundary (Dohle & Scholtz 1988). It must be said that this clear-cut border is the intersegmental furrow of the germ band. Nobody can say whether there is a posterior shift when ventral sclerites and intersegmental membranes are formed and which cells contribute to dorsal structures such as tergites and their intersegmental membranes. If the parasegmental subdivision of the germ is a primary segmentation, the embryonic metamerization can be taken as a secondary segmentation which is superimposed by sclerite formation and folding of intersegmental membranes as a sort of tertiary segmentation.

12 GENEALOGICAL UNITS AND PARASEGMENTS

The embryonic segments are essentially homonomous and can be characterized by several traits: they have a pair of developing ganglia, a pair of mesodermal colonic sacs, a pair of limb buds, and are demarcated by intersegmental furrows (Heymons 1901; Tiegs 1941; Dohle 1974; Scholtz 2002b). In *Drosophila*, units have been revealed that predate the visible embryonic segments. These units have been termed parasegments by Lawrence & Martinez Arias (1985). They are shifted in phase in comparison to the embryonic segments. The anterior part of a parasegment constitutes the posterior part of a segment whereas the posterior part of the same parasegment gives rise to the anterior part of the subsequent segment. The idea of the parasegment rests on several observations: (1) the first transversal grooves in the early *Drosophila* embryo are parasegmental; (2) clones of cells are mixed within the parasegment but they respect the parasegmental borders; (3) anterior boundaries of expression of many *Hox* genes are parasegmental (Lawrence 1988; Hughes & Kaufman 2002a, b). Especially the second criterion can also be related to malacostracans. The descendants of one primary row can be regarded as the constituents of a parasegment. They contribute to posterior structures of an anterior segment including a small part of the developing ganglion and the posterior part of the limb bud, and they form the anterior and middle part of the subsequent segment. The intersegmental furrow separates cells belonging to one genealogical unit. The anterior genealogical boundary is a sharp boundary for *engrailed* expression. We never saw cells of the preceding genealogical unit (in cell row d) with *en* expression. On the other hand, cells of the b-row gain or regain *en* expression (Scholtz et al. 1993).

The parasegmental organization of the early germ band before formation of the intersegmental furrows seems to be a conserved trait in insects and crustaceans and is probably even a heritage from the euarthropod stem species as indications of parasegments have also been found in spider embryos (Damen 2002).

The importance of the phase shift for the construction of composed structures still seems to be obscure. However, it is obvious that it must have a functional value (Hughes & Krause 2001). We find the phenomenon in leeches and oligochaetes where the segments are composed of cells which straddle the intersegmental boundary (Shankland 1999; Shimizu

& Nakamoto 2001). And it is a well-known fact that vertebrae of the vertebrates are composed structures consisting of the posterior part of an embryonic somite and of the anterior part of the subsequent somite.

13 GENE EXPRESSION

The progress in the characterization of genes and gene products in *Drosophila* has made it possible to reveal and isolate equivalent genes in other arthropods. Crustaceans have not been investigated as amply as some insect groups. However, a growing wealth of know-ledge is accumulating. Genes that have been investigated in different crustacean species are the segment-polarity genes *engrailed* (see below), *wingless* (Nulsen & Nagy 1999; Duman-Scheel et al. 2002), the limb patterning genes *Distal-less* (see below), *extradenticle, dachshund, apterous,* and *pdm* (González-Crespo & Morata 1996; Averof & Cohen 1997; Abzhanov & Kaufman 2000d; Williams et al. 2002), *fushi tarazu* (Mouchel-Vielh et al. 2002), *trachealess* (Mitchell & Crews 2002), or the genes of the Hox series (Averof & Akam 1995; Mouchel-Vielh et al. 1998; Abzhanov & Kaufman 1999, 2000 b, c; Rabet et al. 2001; Shiga et al. 2002) (see also Abzhanov & Kaufman 2003; Deutsch et al. 2003). Unfortunately, not many investigators take advantage of the fact that in the developing germ band of malacostracans, expression of genes can be described with the resolution of single cells of which the origin is known. In the following we describe the expression patterns of the genes *engrailed* and *Distal-less* because these have been related to the cell division pattern in malacostracan germ bands.

13.1 *Engrailed*

One of the most intensively studied genes is the gene *engrailed* (*en*). It has a crucial function in the segmentation process (Hidalgo 1998; Gibert 2002). It was first taken as a gene of the pair rule class as lethal *engrailed* alleles in *Drosophila* lead to a substantial deletion of the posterior region of even-numbered segments (Nüsslein-Volhard & Wieschaus 1980). This phenomenon could not be found in other insects. *en* is now addres-sed as a segment polarity gene. It encodes a protein (En), which acts as a transcriptional repressor and is accumulated in cells defining the posterior border of a segment. *en* is highly conserved in arthropods (Damen 2002; Hughes & Kaufman 2002a).

Abzhanov & Kaufman (2000a) have revealed that there are two paralogs of *en*-related genes both in the woodlouse *Porcellio scaber* (Ps-*en*1 and Ps-*en*2) and in the crayfish *Procambarus clarkii* (Pc-*en*1 and Pc-*en* 2). This is in contrast to the brine shrimp *Artemia* where only one gene belonging to the *en* class has been found. Only one gene has also been found in some insects, e.g., *Tribolium, Achaeta, Schistocerca.* On the other hand, the thy-sanuran *Thermobia* has two *en*-related genes that show close similarities to the two malaco-stracan genes. The most plausible interpretation is that in the common ancestor of crusta-ceans and insects only one *en*-class gene was present which duplicated in the last common ancestor of malacostracans and hexapods. In Pterygota, one of these genes was lost so that in several insects only one *en*-class gene persists. Convergently in several lineages this remaining gene duplicated once more so that in *Drosophila, Bombyx* and *Periplaneta* there are once more two *en*-class genes. In *Oncopeltus* (Peterson et al. 1998) and, in a phylo-genetically very distant case, in cirripedes (Gibert et al. 1997, 2000; Quéinnec et al. 1999), the *en* gene has been duplicated more than once. In cirripedes, the situation is rather complicated (Gibert et al. 1997, 2000). At least two independent duplication events of the original *engrailed* gene must have occurred. The first duplication (into *en.a* and *en.b*) must have predated the radiation of the cirripedes. One of the *en* genes is assumed to have undergone a second duplication (into *en.a1* and *en.a2*) in a single lineage. However, since some species only had a single *en* gene, several losses of the *en.b* paralog must have taken place. These events are reviewed by Gibert (2002).

The expression patterns of *en* are widely studied in arthropods including crustaceans since monoclonal antibodies, Mab 4D9 and 4F11, have been raised against the En protein which are cross-reactive with a variety of species (Patel et al. 1989a, b). Abzhanov & Kaufman (2000a) have studied the expression patterns of the two *en*-related genes in *Porcellio* and *Procambarus* by *in situ* hybridization and found that they are fairly dissimilar both in pattern and in dynamics. They compared their data with En accumulations as detected by the antibody 4D9. The 4D9 antibody obviously binds to both gene products and therefore represents the sum of both protein accumulations. Another antibody, 4F11, does not match the expression pattern neither of Ps-*en*1 nor of Ps-*en*2. The reason is unclear. Probably, there is a third *en* locus that has not yet been found. Unfortunately, these comparisons have not been extended to the cellular level. Accordingly, all single cell analyses of the *en* expression pattern are based on studies using the 4D9 antibody.

13.1.1 *Post-naupliar segments in malacostracans*
In the post-naupliar part, from row E(2) onwards, the expression patterns are rather similar. In decapods, mysids and isopods, the first concentration of En can be detected after the first cleavage of the primary row. It is the anterior row ab that shows expression (Fig. 11B, small arrow). Expression follows the mediolateral wave of mitoses. Cells that are 3 to 4 cells more median than the cleaving primary cell become En-positive. Shortly after the following cleavage of the row ab, both nuclei are still stained (Fig. 11B, long arrow). However, the cells belonging to row b soon lose expression (Fig. 11B, arrowhead) so that only the cells of row a retain the En protein (Scholtz et al. 1993; Patel 1994). The amphipods are a remarkable exception. Here, no En can be detected after the first cleavage in row ab. Only after the following cleavage *en* is expressed in the most anterior row a (Scholtz et al. 1994; Scholtz & Wolff 2002). In spite of the difference, we ascertain the same situation in all species examined before the differential cleavages start: the cells of the row a are En-positive, the cells of the rows b, c, and d are En-negative (Scholtz & Dohle 1996). During the first differential cleavages (as far as has been analyzed) all descendants of row a express *en*. In addition, anterior derivatives of row b that had formerly lost en expression regain expression (Fig. 12). This proves that *en* expression is not exclusively clonally transmitted but is more dependent on cell-cell interactions. The widening of stripes is the result of transmission and recruitment.

As described above, concomitant with the differential cleavages, morphogenetic differentiation of segments begins and is triggered by the incision of the intersegmental grooves and the outgrowth of the limb bud *Anlagen*. Furthermore, neuroblasts indicate early segmental neurogenesis. The *en* expression pattern during segment formation corresponds well with the results gained from the cell lineage studies. This is especially true concerning the clonal composition of segmental structures, namely that the segmental boundaries do not match the clonal borders. The anterior border of the genealogical units is respected. We have in no case found the slightest accumulation of En in cells descendant from the d-row. Also the posterior border of the developing segment is not straddled. There is no En accumulation posterior to the intersegmental groove. The interactions in the gene expression cascades are of course unknown. They can only be inferred by analogy from insights in *Drosophila*. There is no indication whatsoever that could corroborate the suggestion of Emerson & Schram (1990) that the segments of crustaceans are fused double segments.

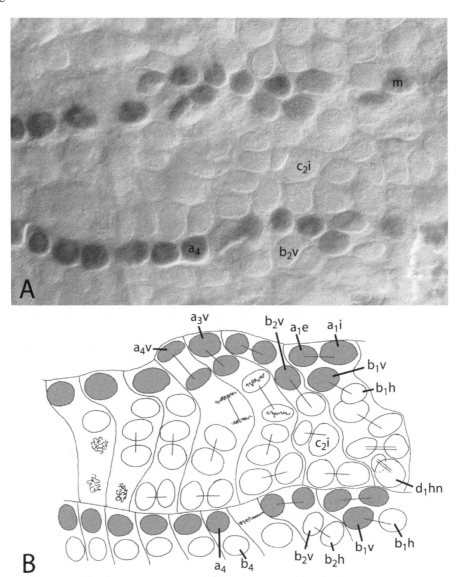

Figure 12. En stripes in the thoracic region of *Orchestia cavimana* (Amphipoda) (after Scholtz et al. 1994). Anterior is up, midline to the right. The posterior En-positive row consists of derivatives of row a and one derivative of row b (b_1v). The anterior En-positive row includes two derivatives of row b (b_1v and b_2v). (A) micrograph using differential interference contrast. m midline cell. (B) Camera lucida drawing of same preparation. One line connects sister cells after the first differential cleavage; two lines connect cells after the second differential cleavage.

13.1.2 Engrailed *and neurogenesis*

Duman-Scheel & Patel (1999) have performed a comparative analysis of En-positive cells on the ventral side of insect and crustacean embryos. They claim that two rows of En-positive cells in *Procambarus* (approx. 13 cells in each hemisegment) are to be regarded as NBs and are homologous to the two rows of En-positive neuroblasts in pterygote insects. From our own experience, we cannot subscribe to this claim. The figure given for *Procambarus* (Fig. 3B in Duman-Scheel & Patel 1999) is equivalent to our own figures for *Cherax* (e. g., fig. 3D in Scholtz & Dohle 1996). The En-positive cells mostly represent normal epidermal cells as well as cells involved in limb bud formation. In this stadium, we can only spot one En-positive putative NB precursor, i.e., cell a_1i, which will eventually divide into the two NBs a_1ii and a_1ie. It is also highly improbable that the crustacean median NB (= MNB) is En-positive. The MNB in *Orchestia* and in other malacostracans is the cell d_0 that is En-negative. The En-positive cell of the midline is a_0 that is a glia precursor and not a NB (Gerberding & Scholtz 1999). Therefore, we suggest that attempts to homologize the neural stem cells in insects and crustaceans should not exclusively rely on gene expression while disregarding origin and fate. Former efforts to trace the genealogy of cells should not be completely ignored.

The gene *engrailed* also plays a role in the differentiation of GMCs and neurons. Only a small subset of neurons is En-positive in *Drosophila* (Patel et al. 1989c) and the same is true for other insects and arthropods. It is interesting how similar the expression patterns of single neurons are in *Schistocerca*, *Procambarus*, *Porcellio*, *Triops* and *Artemia* (Dumann-Scheel & Patel 1999). It is more than tempting to homologize these expression patterns. However, in this case as well as in the former one, homologization should not only be based on number and position of the stained cells within the scaffold of ventral nerve fibers. The knowledge of origin and fate of the cells as well as additional characteristics such as direction of axon outgrowth or relation to other unstained cells will be useful. Otherwise, the conclusions are liable to be precocious. We would like to draw attention to the difficulties and the errors in identifying the origin of the aCC and pCC neurons in *Drosophila* (Broadus et al. 1995).

13.1.3 *Naupliar segments*

In front of the En-positive row E(2)a, the cells descendant of the rows E(1) and E(0) remain En-negative. This marks a discontinuity insofar as 4 rows remain En-negative instead of 3 rows in all subsequent segments. The posterior border of the mandibles is defined by bands of cells that are En-positive and show certain regularity (Fig. 11). In amphipods, this mandibular stripe is the first to appear on the germ band. It is completely contiguous. In mysids, the first En-positive cells are those of the first and second antennal segment. It is not possible to detect a cellular order in these stripes. Both of them have a wide median gap. They are approximately two cells wide from the beginning. The first indication of an additional anterior stripe that has been called the ocular-protocerebral stripe is shown by En concentration in but one cell on each side. Lateral cells are added successively to these (Fig. 11B, os). When the head lobes are distinctly formed, En accumulation can be seen surrounding the lobes in a circular manner (Scholtz 1995). This circular En accumulation has not been found in insects. In still later stages, some cells in front of the ocular-protocerebral stripe show En concentration. These cells can be compared to the secondary head spot (shs) cells in insects (Schmidt-Ott & Technau 1992; Boyan & Williams 2001). Whereas we agree with Boyan & Williams (2001) that the shs cells cannot be taken as indications of a segment because of their very late appearance in germ band formation, the assessment of the ocular-protocerebral stripe is different. It recalls the long-standing debate on the existence of a putative preantennal segment between acron and the first antennal segment. However, quite a different view seems possible. If the descendants of one row constitute a parasegmental unit, the En-positive stripe marks the anterior border of a parasegment. In analogy, the ocular-protocerebral stripe can be interpreted as part of the anterior compartment of a first antennal parasegment and not as the posterior border of an additional preantennal segment.

In this connection, another fact must not be forgotten. As many other genes, the *en* gene has many different functions, and the involvement in the subdivision process of the embryo and the task to define the anterior compartment of a parasegment and the posterior border of a segment is only one of these functions. It has additional functions in neurogenesis that seem to be phylogenetically much older than those in the segmentation process (Gibert 2002). In this context, it becomes comprehensible that genes homologous to *engrailed* are already present in unsegmented animals such as nematodes and that other *en*-class genes play a role in unsegmented and segmented deuterostomes. Although on the one hand the *engrailed* gene is indispensable for the subdivision and patterning of the embryo into segments, on the other hand not any En accumulation can be taken as an indication of an additional segment.

13.1.4 *Non-malacostracans*

Two non-malacostracan groups have been investigated with the purpose of showing the existence and the expression patterns of *en* namely the anostracans and the rhizocephalans. In the brine shrimp *Artemia franciscana* (Manzanares et al. 1993, 1996), only one single *en*-class gene could be found. The expression patterns in the first larval stages have been revealed by *in situ* hybridization as well as by a specially raised antibody. Both patterns agree. As could be expected, in the nauplius the posterior border of the two antennal and the mandibular segments are marked. No expression anterior to the first antennal stripe could be detected. The posterior borders of the trunk segments are stained successively in further larval stages. The two maxillary segments show En accumulation with a considerable delay. In *Artemia*, there is no stereotyped cell division pattern. The cells arrange themselves within the segment in ill-defined rows (Fig. 4).

The expression patterns of two *en*-class genes (*en.a* and *en.b*) have been studied in the rhizocephalan cirripede *Sacculina carcini* (Quéinnec et al. 1999; Gibert et al. 2000). When the neuroectoderm differentiates into epidermis and neuroderm, *en.a* expression fades in the segmental epidermis whereas it is subsequently detected in reiterated putative neural cells. On the contrary, *en.b* expression increases in the epidermis, which makes it an excellent segmentation marker. Behind the sixth thoracic segment, five tiny stripes are observed that are interpreted as molecular indications of vestigial abdominal segments. Neither teloblasts, nor an orderly array of cells have been reported by the authors.

13.2 Distal-less

Distal-less (*Dll*) is a gene which plays a key role in the differentiation of appendages. One can expect that its gene product (Dll) is always concentrated in the telopodite of arthropod legs. However, the establishment of a P/D axis in legs is not its exclusive role. Other formations can be influenced as well. We find Dll concentration in some structures the homology of which with appendages is possible but doubtful: labrum, furca. Other structures are stained which are clearly not equivalent to telopodites of appendages: endites, exites, border of carapace, optic lobes, sensillae (Mittmann & Scholtz 2001; Williams et al.

2001). Since an antibody against Dll of *Drosophila* has been raised which is cross-reactive with many other arthropods and also many other metazoans (Panganiban et al. 1995), much comparative work on expression patterns has been done.

13.2.1 Distal-less *and cell lineage in malacostracans*

In malacostracans, *Dll* expression has been investigated in germ bands of leptostracans, decapods, mysids, isopods, and amphipods and in the nauplius larva of a penaeid decapod (Panganiban et al. 1995; Scholtz & Gerberding 1997; Scholtz et al. 1998; Popadic et al. 1998; Williams 1998; Abzhanov & Kaufman 2000b, c, d; Hejnol 2002). However, detailed data on *Dll* expression at the level of cell lineage are only available for amphipods and isopods (Scholtz & Gerberding 1997; Hejnol 2002).

In isopods and amphipods, *Dll* expression is first visible in the naupliar segments. It is concentrated in stripes consisting of a few cells which cannot, however, be assigned to a fixed cell lineage. The first maxilla has a traceable cell lineage which corresponds neither to that of the mandibles nor of the subsequent segments. From the second maxilla onwards, cell lineage and expression are comparable. The first cell in which Dll can be detected is the cell d_3. This corresponds in isopods and amphipods. Further expression, however, is different. At first, *Dll* expression and limb formation in thoracic segments shall be treated.

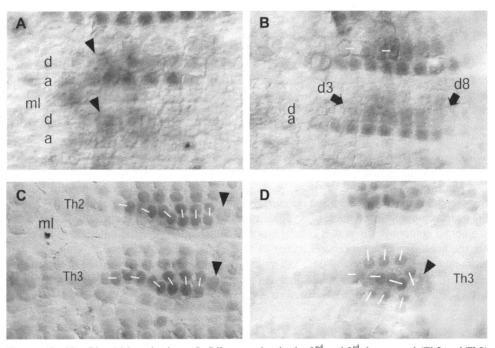

Figure 13. [See Plate 14 in color insert.] *Dll* expression in the 2nd and 3rd thoracopod (Th2 and Th3) in the amphipod *Orchestia cavimana* (from Hejnol 2002). (A) Double staining for En (blue) and Dll (brown). *Dll* is expressed first in cell d_3 (arrow head), then in the adjacent cells d_4, d_5 and a_4. (B) Same staining as A, germ band more advanced. In the anterior genealogical unit, the cells $d_3 - d_5$ are divided. d_3 to d_8 (arrows) are Dll positive. (C) Staining for Dll only (brown). Dll staining extends to d_9 (arrow head). The inner cell d_3i loses Dll expression. (D) Same staining as C. Dll staining is reduced; d_6h and d_7 (arrow head) are no longer stained.

In amphipods, many lateral cells are included so that *Dll* expression is arranged in a stripe from d_3 to d_9 (Fig. 13). When the d-cells divide, both daughter cells of $d_4 - d_8$ retain *Dll* expression. The outer cell d_3e expresses *Dll* whereas the inner cell d_3i loses expression. In d_9, Dll also fades away. At the same time, the posterior daughter cells of $c_3 - c_8$, namely c_3h – c_8h, gain *Dll* expression (Hejnol 2002). The anterior cells c_3v – c_8v are not stained. Some cells in row a are stained, too. The cells with Dll concentration form a stretched oval field in the embryonic segment. This corresponds to a long elevated limb rudiment. The main growth and elevation of the limb takes place in the median part.

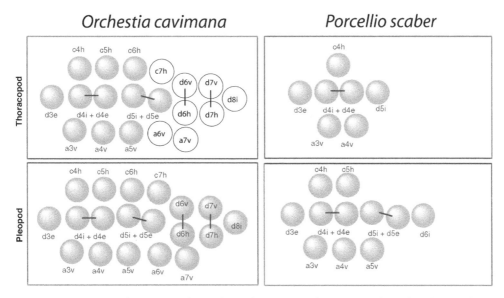

Figure 14. *Dll* expressing cells in the vestiges of a thoracopod (upper panels) and a pleopod (lower panels) in *Orchestia cavimana* (Amphipoda) and *Porcellio scaber* (Isopoda) (from Hejnol 2002). In *Orchestia*, many cells in pleonic segments express *Dll*. At first, the same subset of cells in the thoracic segments is Dll-positive. Many of the lateral cells lose *Dll* expression later (white circles). In *Porcellio*, few cells express *Dll* in the thoracic segments. *Dll* expression is a bit more extended in pleonic segments.

In the isopod *Porcellio scaber*, *Dll* expression does not have a comparable lateral extension. Expression arises a little later than in amphipods. The cells d_3 and d_4 gain expression before outer cells and shortly before their differential divisions (Hejnol 2002). Both daughter cells of d_4 retain *Dll*, as well as the outer cell of d_3 whereas the inner cell d_3i loses it. The result are three Dll-positive cells d_3e, d_4i and d_4e. Further cells that are added to the *Dll*-positive cell cluster in thoracic segments are c_4h, d_5i, a_3v and a_2v. Thus, *Dll* expression does not affect cells more lateral than d_5, and even d_5i later loses expression. Though the orientation and equality of differential cleavages of the cells are nearly identical in isopods and amphipods, *Dll* expression is much more restricted in isopods (Fig. 14). The *Anlagen* of the walking limbs in isopods grow in a finger-like manner while in amphipods they are elongated elevations.

The expression patterns in the pleopods are different from the thoracic limbs (though the cleavage patterns are essentially the same). In amphipods, the most lateral parts retain their initial *Dll* expression. In isopods, additional lateral cells, namely d_5e and d_6i, gain *Dll* expression (Fig. 14). All these observations tend to show that the stereotyped differential cleavages of the post naupliar germ band which are nearly identical in thoracic and abdominal segments (and even in the limbless 8th thoracic segment of the manca stage of isopods) are not a prerequisite for limb differentiation but that the "order" for leg formation is superimposed on a field of cells in equivalent positions.

This view is corroborated by limb formation in the first maxilla, the mandibles and the two pairs of antennae. In these appendages, a cell division pattern homologous with the rows in the remaining post-naupliar region cannot be detected. Nevertheless, the differentiation of the gnathal appendages has much more in common than those with the thoracic or with the pleonic limbs.

13.2.2 Distal-less *in non-malacostracans*

Non-malacostracans have been favorite objects in the investigation of *Dll* expression. A whole series of branchiopods have been studied including representatives of the anostracans, the notostracans, the conchostracans, and the cladocerans (Panganiban et al. 1995; Williams 1998; Williams et al. 1999; Popadic et al. 1998; Olesen et al. 2001; Richter 2002). Most of these examples are equipped with phyllopodous limbs and do have, in addition to an endopod and an exopod, a whole series of endites and a few exites as well. All these structures show *Dll* expression with the exception of an exite-like projection acting as a gill. These expression patterns have not yet been convincingly interpreted. It seems that Dll does not determine a telopodite but is more generally the prerequisite for an out-pouching. The expression patterns of Dll can nevertheless support accepted or hypothesized homologies. Two examples shall be referred to.

In a comparative study on leg development in two onychuran branchiopods, *Cyclestheria hislopi* with foliaceous limbs and the raptorial water flea *Leptodora kindtii* with rod-like articulated limbs, Olesen et al. (2001) showed that the early vestiges are very much alike. There is a comparable pattern of *Dll* expression so that the transformation of a phyllopodous limb to a stenopodous one becomes cognizable.

Another field of application is the question whether mandibles in crustaceans on one hand, and insects and myriapods on the other hand, are completely different and have evolved independently from each other as Manton (1964, 1977) has postulated in her Uniramia concept, or whether the gnathal processes of all these groups are comparable. The investigations into various groups have shown that the mandibles are in principle gnathobasic and that Manton's concept of a whole-limb mandible in Uniramia is not supported (Popadic et al. 1998; Scholtz et al. 1998; Scholtz 2001, Richter 2002).

ACKNOWLEDGEMENTS

We are indebted to Ralf Schnabel (Braunschweig) for providing access and facilities to 4D microscopy. Roy Swain (Hobart) was of great help for collecting *Anaspides tasmaniae* in the wild. Honor Cooper-Kovacs helped improving the English style of the manuscript. André Reimann helped formatting the figures. To these persons we offer our sincere thanks. We thank the Tasmanian Parks and Wildlife Service and the Hobart City Council for permits to collect *Anaspides tasmaniae*. Our work has been continuously supported by the Deutsche Forschungsgemeinschaft (DFG) (Do 619/1-1 and Scho 442/5-1,2,3).

REFERENCES

Abzhanov, A. & Kaufman, T.C. 1999. Homeotic genes and the arthropod head: expression patterns of the *labial, proboscipedia,* and *Deformed* genes in crustaceans and insects. *Proc. Natl. Acad. Sci. USA* 96: 10224-10229.

Abzhanov, A. & Kaufman, T. C. 2000a. Evolution of distinct expression patterns for *engrailed* paralogues in higher crustaceans (Malacostraca). *Dev. Genes Evol.* 210: 493-506

Abzhanov, A. & Kaufman, T. C. 2000b. Embryonic expression patterns of Hox genes in the crayfish, *Procambarus clarkii* (Decapoda, Crustacea). *Evol. Dev.* 2: 271-283.

Abzhanov, A. & Kaufman, T. C. 2000c. Crustacean (malacostracan) Hox genes and the evolution of the arthropod trunk. *Development* 127: 2239-2249.

Abzhanov, A. & Kaufman, T.C. 2000d. Homologs of *Drosophila* appendage genes in the patterning of arthropod limbs. *Dev. Biol.* 227: 673-689.

Abzhanov, A. & Kaufman, T.C. 2003. HOX genes and tagmatization of the higher Crustacea (Malacostraca). In: Scholtz, G. (ed), *Crustacean Issues 15, Evolutionary Developmental Biology of Crustacea*: 43-74. Lisse: Balkema.

Anderson, D. T. 1965. Embryonic and larval development and segment formation in *Ibla quadrivalvis* Cuv. (Cirripedia). *Aust. J. Zool.* 13: 1-15. Anderson, D. T. 1967. Larval development and segment formation in the branchiopod crustaceans *Limnadia stanleyana* King (Conchostraca) and *Artemia salina* (L.) (Anostraca). *Aust. J. Zool.* 15: 47-91.

Anderson, D. T. 1969. On the embryology of the cirripede crustaceans *Tetraclita rosea* (Krauss), *Tetraclita purpurascens* (Wood), *Chthamalus antennatus* (Darwin) and *Chamaesipho columna* (Spengler) and some considerations of crustacean phylogenetic relationships. *Phil. Trans. R. Soc. Lond. B* 256: 183-235.

Anderson, D. T. 1973. *Embryology and phylogeny in annelids and arthropods.* Oxford: Pergamon Press.

Averof, M. & Akam, M. 1995. Insect-crustacean relationships: insights from comparative developmental and molecular studies. *Phil. Trans. R. Soc. Lond. B* 347: 293-303.

Averof, M. & Cohen, S. M. 1997. Evolutionary origin of insect wings from ancestral gills. *Nature* 385: 627-630.

Averof, M. & Patel, N.H. 1997. Crustacean appendage evolution associated with changes in *Hox* gene expression. *Nature* 388: 682-686.

Ax, P. 1999. *Das System der Metazoa II. Ein Lehrbuch der phylogenetischen Systematik.* Stuttgart: Gustav Fischer Verlag.

Bate, C. M. 1976. Embryogenesis of an insect nervous system I. A map of the thoracic and abdominal neuroblasts in *Locusta migratoria. J. Embryol. exp. Morph.* 35: 107-123.

Bate, C. M. & Grunewald, E. B. 1981. Embryogenesis of an insect nervous system II: A second class of neuron precursor cells and the origin of the intersegmental connectives. *J. Embryol. exp. Morph.* 61: 317-330.

Benesch, R. 1969. Zur Ontogenie und Morphologie von *Artemia salina* L. *Zool. Jb. Anat.* 86: 307-458.

Benton, J.L. & Beltz, B.S. 2002. Patterns of neurogenesis in the midbrain of embryonic lobsters differ from proliferation in the insect and the crustacean ventral nerve cord. *J. Neurobiol.* 53: 57-67.

Bergh, R. S. 1893. Beiträge zur Embryologie der Crustaceen. I. Zur Bildungsgeschichte des Keimstreifens von *Mysis. Zool. Jb. Anat.* 6: 491-528.

Bergh, R. S. 1894. Beiträge zur Embryologie der Crustaceen. II. Die Drehung des Keimstreifens und die Stellung des Dorsalorgans bei *Gammarus pulex. Zool. Jb. Anat* 7: 235-248.

Bock, E. 1939. Bildung und Differenzierung der Keimblätter bei *Chrysopa perla* (L.). *Z. Morph. Ökol. Tiere* 35: 615-702.

Bossing, T. & Technau, G. M. 1994. The fate of the CNS midline progenitors in *Drosophila* as revealed by a new method for single cell labelling. *Development* 120: 1895-1906.

Boyan, G. & Williams, L. 2001. A single cell analysis of *engrailed* expression in the early embryonic brain of the grasshopper *Schistocerca gregaria*: ontogeny and identity of the secondary headspot cells. *Arthr. Struct. Dev.* 30: 207-218.

Broadus, J., Skeath, J. B., Spana, E. P., Bossing, T., Technau, G. & Doe, C. Q. 1995. New neuroblast markers and the origin of the aCC/pCC neurons in the *Drosophila* central nervous system. *Mech. Dev.* 53: 393-402.

Conklin, E.G. 1896. Cleavage and differentiation. In: Maienschein, J. (ed), *1986 Reprint in Defining Biology, Lectures from the 1890s*: 151-177. Cambridge: Harvard University Press.

Damen, W. C. M. 2002. Parasegmental organization of the spider embryo implies that the parasegment is an evolutionary conserved entity in arthropod embryogenesis. *Development* 129: 1239-1250.

Davis, G. K. & Patel, N. H. 2002. Short, long, and beyond: molecular and embryological approaches to insect segmentation. *Annu. Rev. Entomol.* 47: 669-699.

Deutsch, J.S., Mouchel-Vielh, E., Quéinnec, É. & Gibert, J.-M. 2003. Genes, segments and tagmata in cirripedes. In: Scholtz, G. (ed), *Crustacean Issues 15, Evolutionary Developmental Biology of Crustacea*: 19-42. Lisse: Balkema.

Doe, C. Q. 1992. Molecular markers for identified neuroblasts and ganglion mother cells in the *Drosophila* nervous system. *Development* 116: 855-863.

Dohle, W. 1970. Die Bildung und Differenzierung des postnauplialen Keimstreifs von *Diastylis rathkei* (Crustacea, Cumacea) I. Die Bildung der Teloblasten und ihrer Derivate. *Z. Morph. Tiere* 67: 307-392.

Dohle , W. 1972. Über die Bildung und Differenzierung des postnauplialen Keimstreifs von *Leptochelia spec.* (Crustacea, Tanaidacea). *Zool. Jb. Anat.* 89: 503-566.

Dohle, W. 1974. The segmentation of the germ band of Diplopoda compared with other classes of arthropods. *Symp. zool. Soc. Lond.* 32: 143-161.

Dohle, W. 1976. Die Bildung und Differenzierung des postnauplialen Keimstreifs von *Diastylis rathkei* (Crustacea, Cumacea) II. Die Differenzierung und Musterbildung des Ektoderms. *Zoomorphologie* 84: 235-277.

Dohle, W. 1989. Zur Frage der Homologie ontogenetischer Muster. *Zool. Beitr.* N. F. 32: 355-389.

Dohle, W. 1999. The ancestral cleavage pattern of the clitellates and its phylogenetic deviations. *Hydrobiologia* 402: 267-283.

Dohle, W. 2001. Are the insects terrestrial crustaceans? *Ann. Soc. Entomol. Fr.* (N.S.) 37: 85-103.

Dohle, W. & Scholtz, G. 1988. Clonal analysis of the crustacean segment: the discordance between genealogical and segmental borders. *Development* 104, *Suppl.*: 147-160.

Dohle, W. & Scholtz, G. 1997. How far does cell lineage influence cell fate specification in crustacean embryos? *Semin. Cell Dev. Biol.* 8: 379-390

Duman-Scheel, M. & Patel, N. H. 1999. Analysis of molecular marker expression reveals neuronal homology in distantly related arthropods. *Development* 126: 2327-2334.

Duman-Scheel, M., Pirkl, N. & Patel, N. H. 2002. Analysis of the expression pattern of *Mysidium columbiae wingless* provides evidence for conserved mesodermal and retinal patterning processes among insects and crustaceans *Dev. Genes Evol.* 212: 114-123.

Emerson, M. J. & Schram, F. R. 1990. The origin of crustacean biramous appendages and the evolution of Arthropoda. *Science* 250: 667-669.

Fioroni, P. 1987. *Allgemeine und vergleichende Embryologie der Tiere.* Berlin: Springer Verlag.

Freeman, J. A. 1986. Epidermal cell proliferation during thoracic development in larvae of *Artemia. J. Crust. Biol.* 6: 37-48.

Freeman, J. A. 1989. Segment morphogenesis in *Artemia* larvae. In: Warner, A. H., Macrae T. H. & Bagshaw, J. C. (eds), *Cell and molecular biology of* Artemia *development*: 77-90. New York: Plenum Press.

Gerberding, M. 1997. Germ band formation and early neurogenesis of *Leptodora kindti* (Cladocera): first evidence for neuroblasts in the entomostracan crustaceans. *Invert. Repr. Dev.* 32: 63-73.

Gerberding, M. & Scholtz, G. 1999. Cell lineage of the midline cells in the amphipod crustacean *Orchestia cavimana* (Crustacea, Malacostraca) during formation and separation of the germ band. *Dev. Genes Evol.* 209: 91-102.

Gerberding, M. & Scholtz, G. 2001. Neurons and glia in the midline of the higher crustacean *Orchestia cavimana* are generated via an invariant cell lineage that comprises a median neuroblast and glial progenitors. *Dev. Biol.* 235: 397-409.

Gerberding, M., Browne, W.E. & Patel, N.H. 2002. Cell lineage analysis of the amphipod crustacean *Parhyale hawaiensis* reveals an early restriction of cell fate. *Development* 129: 5789-5801.

Gibert, J.-M. 2002. The evolution of *engrailed* genes after duplication and speciation events. *Dev. Genes Evol.* 212: 307-318.

Gibert, J.-M., Mouchel-Vielh, E. & Deutsch, J. S. 1997. *engrailed* duplication events during the evolution of barnacles. *J. Mol. Evol.* 44: 585-594.

Gibert, J.-M., Mouchel-Vielh, E., Quéinnec, E. & Deutsch, J. S. 2000. Barnacle duplicate *engrailed* genes: divergent expression patterns and evidence for a vestigial abdomen. *Evol. Dev.* 2: 194-202.

González-Crespo, S. & Morata, G. 1996. Genetic evidence for the subdivision of the arthropod limb into coxopodite and telopodite. *Development* 122: 3921-3928.

Gruner, H.-E. 1993. *Crustacea*. In: Gruner, H.-E. (ed), *Lehrbuch der speziellen Zoologie, Band I, 4. Teil*: 448-1030. Jena: Gustav Fischer Verlag.

Harzsch, S. 2001. Neurogenesis in the crustacean ventral nerve cord: homology of neuronal stem cells in Malacostraca and Branchiopoda? *Evol. Dev.* 3: 154-169.

Harzsch, S. & Dawirs, R. R. 1994. Neurogenesis in larval stages of the spider crab *Hyas araneus* (Decapoda, Brachyura): proliferation of neuroblasts in the ventral nerve cord. *Roux's Arch. Dev. Biol.* 204: 93-100.

Hejnol, A. 2002. Der postnaupliale Keimstreif von *Porcellio scaber* und *Orchestia cavimana* (Crustacea, Peracarida): Zelllinie, Genexpression und Beginn der Morphogenese. Dissertation. Humboldt-Universität Berlin. 101 pp.

Hertzler, P. L. 2002. Development of the mesendoderm in the dendrobranchiate shrimp *Sicyonia ingentis*. *Arthr. Struct. Dev.* 31: 33-49.

Hertzler, P.L., and Clark, W.H. Jr. (1992). Cleavage and gastrulation in the shrimp *Sicyonia ingentis*. *Development* 116: 127-140.

Hessler, R. R. 1983. A defense of the caridoid facies: wherein the early evolution of the Eumalacostraca is discussed. In: Schram, F.R. (ed), *Crustacean Issues 1, Crustacean Phylogeny*: 145-164. Rotterdam: Balkema.

Heymons, R. 1901. Die Entwicklungsgeschichte der Scolopender. *Zoologica* 13 (33): 1-244.

Hidalgo, A. 1998. Growth and patterning from the *engrailed* interface. *Int. J. Dev. Biol.* 42: 317-324.

Hughes, C. L. & Kaufman, T. C. 2002a. Exploring myriapod segmentation: the expression patterns of *even-skipped*, *engrailed*, and *wingless* in a centipede. *Dev. Biol.* 246: 47-61.

Hughes, C. L. & Kaufman, T. C. 2002b. Hox genes and the evolution of the arthropod body plan. *Evol. Dev.* 4: 459-499.

Hughes, S. C. & Krause, H. M. 2001: Establishment and maintenance of parasegmental compartments. *Development* 128: 1109-1118.

Krause, G. 1939. Die Eitypen der Insekten. *Biol. Zbl.* 59: 495-536.

Kühn, A. 1912. Die Sonderung der Keimesbezirke in der Entwicklung der Sommereier von *Polyphemus pediculus* de Geer. *Zool. Jb. Anat.* 35: 243-340.

Kühn, A. 1965. *Vorlesungen über Entwicklungsphysiologie*. 2. Aufl. Berlin: Springer-Verlag.

Lawrence, P. A. 1988. The present status of the parasegment. *Development* 104 Suppl.: 61-65.

Lawrence, P. A. & Martinez-Arias, A. 1985. The cell lineage of segments and parasegments in *Drosophila*. *Phil. Trans. R. Soc. Lond.* B 312: 83-90.

Manton, S. M. 1928. On the embryology of a mysid crustacean, *Hemimysis lamornae*. *Phil. Trans. R. Soc. Lond. B* 216: 363-463.

Manton, S. M. 1964. Mandibular mechanisms and the evolution of arthropods. *Phil. Trans. R. Soc. Lond. B* 247: 1-183.

Manton, S. M. 1977. *The Arthropoda. Habits, functional morphology, and evolution*. Oxford: Clarendon Press.

Manzanares, M., Marco, R. & Garesse, R. 1993. Genomic organization and developmental pattern of expression of the *engrailed* gene from the brine shrimp *Artemia*. *Development* 118: 1209-1219.

Manzanares, M., Williams, T. A., Marco, R. & Garesse, R. 1996. Segmentation in the crustacean *Artemia*: engrailed staining studied with an antibody raised against the *Artemia* protein. *Roux's Arch. Dev. Biol.* 205: 424-431.

Martindale, M.Q. & Henry, J. 1997. Reassessing embryogenesis in the Ctenophora: the inductive role of e_1 micromeres in organizing ctene row formation in the 'mosaic' embryo, *Mnemiopsis leidyi*. *Development* 124: 1999-2006.

Meschenmoser, M. 1996. *Dorsal- und Lateralorgane in der Embryonalentwicklung von Peracariden (Crustacea, Malacostraca)*. Göttingen: Cuvillier Verlag.

Mitchell, B. & Crews, S.T. 2002. Expression of the *Artemia trachealess* gene in the salt gland and epipod. *Evol. Dev.* 4: 344-353.

Mittmann, B. & Scholtz, G. 2001. *Distal-less* expression in embryos of *Limulus polyphemus* (Chelicerata, Xiphosura) and *Lepisma saccharina* (Insecta, Zygentoma) suggests a role in the development of mechanoreceptors, chemoreceptors, and the CNS. *Dev. Genes Evol.* 211: 232-243.

Mouchel-Vielh, E., Blin, M., Rigolot, C. & Deutsch, J. S. 2002. Expression of a homologue of the *fushi tarazu (ftz)* gene in a cirripede crustacean. *Evolution & Development* 4: 76-85.

Mouchel-Vielh, E., Rigolot, C., Gibert, J.-M. & Deutsch, J. S. 1998. Molecules and the body plan: the Hox genes of cirripedes (Crustacea). *Mol. Phyl. Evol.* 9: 382-389.

Nair, K. B. 1939. The reproduction, oogenesis and development of *Mesopodopsis orientalis* Tatt. *Proc. Ind. Acad. Sci.* 9: 175-223.

Nulsen, C. & Nagy, L. M. 1999. The role of *wingless* in the development of multibranched crustacean limbs. *Dev. Genes Evol.* 209: 340-348.

Nüsslein-Volhard, C. & Wieschaus, E. 1980. Mutations affecting segment number and polarity in *Drosophila*. *Nature* 287: 795-801.

Oishi, S. 1959. Studies on the teloblasts in the decapod embryo. I. Origin of teloblasts in *Heptacarpus rectirostris* (Stimpson). *Embryologia* 4: 283-309.

Oishi, S. 1960. Studies on the teloblasts in the decapod embryo. II. Origin of teloblasts in *Pagurus samuelis* (Stimpson) and *Hemigrapsus sanguineus* (de Haan). *Embryologia* 5: 270-282.

Olesen, J., Richter, S. & Scholtz, G. 2001. The evolutionary transformation of phyllopodous to stenopodous limbs in the Branchiopoda (Crustacea) – Is there a common mechanism for early limb development in arthropods? *Int. J. Dev. Biol.* 45: 869-876.

Olsen, C. S. & Clegg, J. S. 1978. Cell division during the development of *Artemia salina*. *Roux's Arch. Dev. Biol.* 184: 1-13.

Panganiban, G., Sebring, A., Nagy, L. & Carroll, S. 1995. The development of crustacean limbs and the evolution of arthropods. *Science* 270: 1363-1366.

Patel, N.H. 1994. The evolution of arthropod segmentation: insights from comparisons of gene expression patterns. *Development* Suppl. 201-207.

Patel, N. H., Kornberg, T. B. & Goodman, C. S. 1989a. Expression of *engrailed* during segmentation in grasshopper and crayfish. *Development* 107: 201-212.

Patel, N. H., Martin-Blanco., Coleman, K. G., Poole, S. J., Ellis, M. C., Kornberg, T. B. & Goodman, C. S. 1989b. Expression of *engrailed* proteins in arthropods, annelids, and chordates. *Cell* 58: 955-968.

Patel, N.H., Schafer, B., Goodman, C.S. & Holgren, R. 1989c. The role of segment polarity genes during *Drosophila* neurogenesis. *Genes Dev.* 3: 890-904.

Peterson, M. D., Popadic, A. & Kaufman, T. C. 1998. The expression of two *engrailed*-related genes in an apterygote insect and a phylogenetic analysis of insect *engrailed*-related genes. *Dev. Genes Evol.* 208: 547-557.

Popadic, A. Panganiban, G., Rusch, D., Shear, W. A. & Kaufman, T. C. 1998. Molecular evidence for the gnathobasic derivation of arthropod mandibles and for the appendicular origin of the labrum and other structures. *Dev. Genes Evol.* 208: 142-150.

Quéinnec, E., Mouchel-Vielh, E., Guimonneau, Gibert, J. M., Turquier, Y. & Deutsch, J. S. 1999. Cloning and expression of the *engrailed.a* gene of the barnacle *Sacculina carcini*. *Dev. Genes Evol.* 209: 180-185.

Rabet, N., Gibert, J.-M., Quéinnec, É., Deutsch, J. S. & Mouchel-Vielh, E. 2001. The *caudal* gene of the barnacle *Sacculina carcini* is not expressed in its vestigial abdomen. *Dev. Genes Evol.* 211: 172-178.

Remane, A. 1952. *Die Grundlagen des natürlichen Systems der vergleichenden Anatomie und der Phylogenetik.* Leipzig: Geest & Portig.

Richter, S. 2002. The Tetraconata concept: hexapod-crustacean relationships and the phylogeny of Crustacea. *Org. Divers. Evol.* 2: 217-237.

Richter, S. & Scholtz, G. 2001. Phylogenetic analysis of the Malacostraca (Crustacea). *J. Zool. Syst. Evol. Research* 39: 113-136.

Sander, K. 1983. The evolution of patterning mechanisms: gleanings from insect embryogenesis and spermatogenesis. In: Goodwin, B. C., Holder, N. & Wylie, C. G. (eds), *Development and Evolution*: 137-158. Cambridge: Cambridge University Press.

Schierenberg, E. 1997. Spezifikation des Zellschicksals im frühen Embryo von *Caenorhabditis elegans. Naturwissenschaften* 84: 55-64.

Schmidt, M. 2001. Neuronal differentiation and long-term survival of newly generated cells in the olfactory midbrain of the adult spiny lobster, *Panulirus argus. J. Neurobiol.* 48: 181-203.

Schmidt-Ott, U. & Technau, G. M. 1992. Expression of *en* and *wg* in the embryonic head and brain of *Drosophila* indicates a refolded band of seven segment remnants. *Development* 116: 111-125.

Schnabel, R. 1997. Why does a nematode have an invariant lineage? *Semin. Cell Dev. Biol.* 8: 341-349.

Schnabel, R., Hutter, H., Moerman, D. G. & Schnabel, H. 1997. Assessing normal embryogenesis in *Caenorhabditis elegans* using a 4D microscope: variability of development and regional specification. *Dev. Biol.* 184: 234-265.

Scholtz, G. 1984. Untersuchungen zur Bildung und Differenzierung des postnauplialen Keimstreifs von *Neomysis integer* Leach (Crustacea, Malacostraca, Peracarida). *Zool. Jb. Anat.* 112: 295-349.

Scholtz, G. 1990. The formation, differentiation and segmentation of the post-naupliar germ band of the amphipod *Gammarus pulex* L. (Crustacea, Malacostraca, Peracarida). *Proc. R. Soc. Lond. B* 239: 163-211.

Scholtz, G. 1992. Cell lineage studies in the crayfish *Cherax destructor* (Crustacea, Decapoda): germ band formation, segmentation, and early neurogenesis. *Roux's Arch. Dev. Biol.* 202: 36-48.

Scholtz, G. 1993. Teloblasts in decapod embryos: an embryonic character reveals the monophyletic origin of freshwater crayfishes (Crustacea, Decapoda). *Zool. Anz.* 230: 45-54

Scholtz, G. 1995. Head segmentation in Crustacea – an immunocytochemical study. *Zoology* 98: 104-114.

Scholtz, G. 1997. Cleavage, germ band formation and head segmentation: the ground pattern of the Euarthropoda. In: Fortey R.A. & Thomas R.H. (eds), *Arthropod Relationships*: 317-332. London: Chapman and Hall.

Scholtz, G. 2000. Evolution of the nauplius stage in malacostracan crustaceans. *J. Zool. Syst. Evol. Research* 38: 175-187.

Scholtz, G. 2001. Evolution of developmental patterns in arthropods – the analysis of gene expression and its bearing on morphology and phylogenetics. *Zoology* 103: 99-111.

Scholtz, G. 2002a. Evolution and biogeography of freshwater crayfish. In: Holdich D.M. (ed), *Biology of Freshwater Crayfish*: 30-52, plates 10-16. Oxford: Blackwell Science.

Scholtz, G. 2002b. The Articulata hypothesis - or what is a segment? *Organ. Divers. Evol.* 2: 197-215.

Scholtz, G. & Dohle, W. 1996. Cell lineage and cell fate in crustacean embryos - a comparative approach. *Int. J. Dev. Biol.* 40: 211-220.

Scholtz, G. & Gerberding, M. 1997. Das frühe Expressionsmuster des Gens *Distal-less* bei Amphipoden und Isopoden. *Verh. Dtsch. Zool. Ges.* 90.1: 78.

Scholtz, G. & Gerberding, M. 2002. Cell lineage of crustacean neuroblasts. In: Wiese, K. (ed), *The crustacean nervous system*: 406-416. Berlin: Springer Verlag.

Scholtz, G. & Richter, S. 1995. Phylogenetic systematics of the reptantian Decapoda. (Crustacea, Malacostraca). *Zool. J. Linn. Soc.* 113: 289-328.

Scholtz, G. & Wolff, C. 2002. Cleavage, gastrulation, and germ band formation of the amphipod *Orchestia cavimana* (Crustacea, Malacostraca, Peracarida). *Contrib. Zool.* 71: 9-28.

Scholtz, G., Dohle, W., Sandeman, R. S. & Richter, S. 1993. Expression of *engrailed* can be lost and regained in cells of one clone in crustacean embryos. *Int. J. Dev. Biol.* 37: 299-304.

Scholtz, G., Mittmann, B. & Gerberding, M. 1998. The pattern of *Distal-less* expression in the mouthparts of crustaceans, myriapods and insects: new evidence for a gnathobasic mandible and the common origin of Mandibulata. *Int. J. Dev. Biol.* 42: 801-810.

Scholtz, G., Patel, N. H. & Dohle, W. 1994. Serially homologous *engrailed* stripes are generated via different cell lineages in the germ band of amphipod crustaceans (Malacostraca, Peracarida). *Int. J. Dev. Biol.* 38: 471-478.

Schram, F.R. 1986. *Crustacea.* Oxford: Oxford University Press.

Schram, F.R. & Koenemann, S. 2001. Developmental genetics and arthropod evolution, I, on legs. *Evol. Dev.* 3: 343-354.

Schram, F.R. & Koenemann, S. 2003. Developmental genetics and arthropod evolution: on body regions of Crustacea. In: Scholtz, G. (ed), *Crustacean Issues 15, Evolutionary Developmental Biology of Crustacea*: 75-92. Lisse: Balkema.

Seidel, F., Bock, E. & Krause, G. 1940. Die Organisation des Insekteneies. *Naturwissenschaften* 28.

Shankland, M. 1999. Anteroposterior pattern formation in the leech embryo. In: Moody, S.A. (ed), *Cell lineage and fate determination*: 209-224. San Diego: Academic Press.

Shiga, Y., Yasumoto, R., Yamagata, H. & Hayashi, S. 2002. Evolving role of Antennapedia protein in arthropod limb patterning. *Development* 129: 3555-3561.

Shimizu, T. & Nakamoto, A. 2001. Segmentation in annelids: cellular and molecular basis for metameric body plan. *Zool. Sci.* 18: 285-298.

Siewing, R. 1969. *Lehrbuch der vergleichenden Entwicklungsgeschichte der Tiere.* Hamburg: Verlag Paul Parey.

Stent, G.S. 1985. The role of cell lineage in development. *Phil Trans. R. Soc. Lond.* B 312: 3-19.

Stent, G.S. 1998. Developmental cell lineage. *Int. J. Dev. Biol.* 42: 237-241.

Strömberg, J.-O. 1965. On the embryology of the isopod Idotea. *Ark. Zool.* 17: 421-473.

Strömberg, J.-O. 1967. Segmentation and organogenesis in *Limnoria lignorum* (Rathke) (Isopoda). *Ark. Zool.* 20: 91-139.

Strömberg, J.-O. 1971. Contribution to the embryology of bopyrid isopods with special reference to *Bopyroides*, *Hemiarthrus*, and *Pseudione* (Isopoda, Epicaridea). *Sarsia* 47: 1-46.

Tamarelle, M., Haget, A. & Ressouches, A. 1985. Segregation, division, and early patterning of lateral thoracic neuroblasts in the embryos of *Carausius morosus* Br. (Phasmida: Lonchodidae). *Int. J. Insect Morphol. Embryol.* 14: 307-317.

Tautz, D., Friedrich, M. & Schröder, R. 1994. Insect embryogenesis – what is ancestral and what is derived? *Development* Suppl.: 193-199.

Tiegs, O. W. 1941. The embryology and affinities of the Symphyla, based on a study of *Hanseniella agilis. Quart. J. micr. Sci.* 82: 1-225.

Truman, J. W. & Ball, E. E. 1998. Patterns of embryonic neurogenesis in a primitive wingless insect, the silverfish, *Ctenolepisma longicaudata*: comparison with those seen in flying insects. *Dev. Genes Evol.* 208: 357-368.

van den Biggelaar, J.A.M., Dictus, W.J.A.G. & van Loon, A.E. 1997. Cleavage patterns, cell-lineages and cell specification are clues to phyletic lineages in Spiralia. *Semin. Cell Dev. Biol.* 8: 367-378.

Vehling, D. 1994. Die Entwicklung des postnauplialen Keimstreifs von *Porcellio scaber*. Eine zellgenealogische Studie. Diplomarbeit, Freie Universität Berlin.

Wagner, G.P. & Misof, B.Y. 1993. How can a character be developmentally constrained despite variation in developmental pathways? *J. Evol. Biol.* 6: 449-455.

Walley, L. J. 1969. Studies on the larval structure and metamorphosis of *Balanus balanoides* (L.). *Phil. Trans. R. Soc. Lond. B* 256: 237-280.

Weygoldt, P. 1958. Die Embryonalentwicklung des Amphipoden *Gammarus pulex pulex* (L.). *Zool. Jb. Anat.*77: 51-110.

Wheeler, M. W. 1893. A contribution to insect embryology. *J. Morph.* 8: 1-160.

Williams, T. A. 1998. Distalless expression in crustaceans and the patterning of branched limbs. *Dev. Genes Evol.* 207: 427-434.

Whitington, P.M. 2003. The development of the crustacean nervous system. In: Scholtz, G. (ed), *Crustacean Issues 15, Evolutionary Developmental Biology of Crustacea*: 135-167. Lisse: Balkema.

Williams, T.A. 2003. The evolution and development of crustacean limbs: an analysis of limb homologies. In: Scholtz, G. (ed), *Crustacean Issues 15, Evolutionary Developmental Biology of Crustacea*: 169-173. Lisse: Balkema.

Williams, T. A. & Müller, G. B. 1996. Limb development in a primitive crustacean, *Triops longicaudatus*: subdivision of the early limb bud gives rise to multibranched limbs. *Dev. Genes Evol.* 206: 161-168.

Williams, T.A. & Nagy, L.M. 1996. Comparative limb development in insects and crustaceans. *Semin. Cell Dev. Biol.* 7: 615-628.

Williams, T. A., Nulsen, C. & Nagy, L. M. 2001. A complex role for Distal-less in crustacean appendage development. *Dev. Biol.* 241: 302-312.

Wolff, C. & Scholtz, G. 2002. Cell lineage, axis formation, and the origin of germ layers in the amphipod crustacean *Orchestia cavimana. Dev. Biol.* 250: 44-58.

Zacharias, D., Williams, J. L. D., Meier, T. & Reichert, H. 1993. Neurogenesis in the insect brain: cellular identification and molecular characterization of brain neuroblasts in the grasshopper embryo. *Development* 118: 941-955.

Zilch, R. 1974. Die Embryonalentwicklung von *Thermosbaena mirabilis* Monod. (Crustacea, Malacostraca, Pancarida). *Zool. Jb. Anat.* 93: 462-576.

Zilch, R. 1978. Embryologische Untersuchungen an der holoblastischen Ontogenese von *Penaeus trisulcatus* Leach (Crustacea, Decapoda). *Zoomorphologie* 90: 67-100.

Zilch, R. 1979. Cell lineage in arthropods? *Fortschr. zool. Syst. Evolut.-forsch.* 1: 19-41.

The development of the crustacean nervous system

PAUL M. WHITINGTON

Department of Anatomy and Cell Biology, University of Melbourne, Victoria, Australia 3010

1 INTRODUCTION: THE CONTRIBUTION OF STUDIES OF NEURAL DEVELOPMENT TO ISSUES IN CRUSTACEAN EVOLUTION

In recent years, there has been a strong resurgence of interest in using a comparative approach to developmental biology to address fundamental problems in evolution. New approaches and techniques have yielded insights into long-standing questions such as the phylogenetic relationships between different animal groups and the genetic basis for morphological diversity. Nervous systems represent favorable material for such an approach, as they possess a convenient balance of evolutionary conservatism versus diversity. Recent advances in our understanding of cellular and molecular mechanisms of neural development in a variety of animal groups have underpinned this comparative analysis. While our knowledge of mechanisms for neural development within the Arthropoda is dominated by the Insecta, and particularly the fruit-fly *Drosophila melanogaster*, there is an increasing amount of information from other arthropod groups, including the Crustacea. In this article, I will review our knowledge of neural development in Crustacea from the standpoint of insights that can be gained into evolutionary processes. Emphasis will therefore be given to those aspects of neural development that provide the opportunity for comparison between different crustacean groups or with other arthropods.

2 DEVELOPMENT OF THE CENTRAL NERVOUS SYSTEM (CNS)

2.1 *Gross morphology of the CNS*

To provide a framework for consideration of neural development, I will begin with a brief overview of the gross morphology of the crustacean CNS. Crustaceans, like all other arthropods, possess a dorsally located brain connected to a ventral nerve cord (VNC) by paired circumesophageal connectives. Both the brain and the VNC are segmented structures: the segmental units, neuromeres, are joined by paired longitudinal connectives and the bilaterally symmetrical hemi-neuromeres are connected by transverse commissures. The segmental nature of the CNS is most clearly evident at embryonic stages and is generally obscured by condensation of the neuromeres into ganglia during late embryonic or larval development. In decapod embryos, the VNC consists of eight thoracic neuromeres and eight abdominal neuromeres (Bullock & Horridge 1965; Scholtz 1995b; see Fig. 1). The presence of a vestigial ninth abdominal neuromere is suggested by the embryonic expression pattern of the segmentation gene *engrailed* in the crayfish *Cherax destructor* (Scholtz 1995b). The segmental composition of the brain has been a subject of long-standing debate. There is widespread agreement that three neuromeres (mandibular, maxillary 1 and maxillary 2) exist in the brain posterior to the esophagus. (Indeed, some workers consider the three sub esophageal neuromeres to be part of the VNC, rather than the brain). The argument centers on whether the three morphological brain regions anterior to the esophagus (the protocerebrum, deutocerebrum and tritocerebrum) represent neuromeres. On the basis of *engrailed* expression pattern in the head, Scholtz (1995a, 2001) argues in favor of a segmental identity of the deutocerebrum (which innervates the first antennae) and tritocerebrum (which innervates the second antennae) while concluding that the status of the protocerebrum is unresolved. Paired nerves run from each neuromere into the periphery, where they innervate muscles in the body wall and segmental appendages. Two nerves, the segmental nerve and the intersegmental nerve, arise from each hemi-neuromere posterior to the mandibular neuromere, while the mandibular and more anterior neuromeres possess only single nerves. Based on the relatively conserved pattern of body segmentation and gross structure of the adult CNS amongst the Malacostraca (Gruner 1993), the decapod pattern of neuromeres described above is probably representative of that group as a whole, but is subject to considerable modification in non-malacostracans (Entomostraca). For example, many entomostracans, such as the branchiopod *Triops* possess a much higher number of trunk neuromeres than malacostracans (Gruner 1993; Harzsch 2001).

2.2 *Patterns of development*

A broad range of developmental patterns is found within the Crustacea, ranging from direct developing species that have a full number of body segments at hatching (epimeric de-velopment), to species that hatch as nauplius larvae, which possess only the segments of the antenna 1, antenna 2 and the mandibles (anameric development). In anameric developing species, additional segments with their corresponding neuromeres are generated by budding from a growth zone at the posterior end of the nauplius larva. In some cases, new appen-dages and VNC neuromeres are continually generated throughout adult life (Fryer, 1988; Harzsch 2001). The body region posterior to the naupliar segments is referred to as the post-naupliar region.

Indirect developing species can undergo a substantial change in body form during larval life. New appendages may develop on body segments that previously lacked them or existing appendages can undergo dramatic changes in form and function. This metamor-phosis often involves the adoption of different modes of feeding and locomotion. Clearly, one would expect these changes in external body form and function to be associated with significant developmental changes in the nervous system during larval life.

Finally, even in those species that show little change in body form after hatching, certain regions of the nervous system, such as the olfactory system, can undergo developmental changes that extend through larval development into adult life (see section 2.4.2).

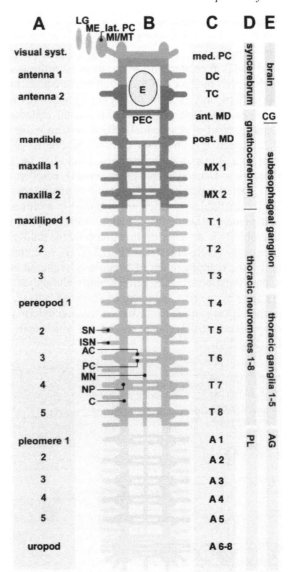

Figure 1. [See Plate 15 in color insert.] Schematic representation of the pattern of neuromeres in the CNS of the decapod embryo (after Harzsch et al. 1998 and Sullivan & Macmillan 2001). (A-E) Column A indicates the appendages associated with each neuromere, B shows the arrangement of the neuromeres, C gives the nomenclature used in this study for the neuromeres, D indicates the 4 major tagmata of the CNS, while E indicates the gross morphological subdivisions usually recognized in the CNS (Bullock & Horridge 1965). Abbreviations: A = abdominal; AC = anterior commissure; AG = abdominal ganglion; ant.MD = anterior mandibular; C = connective; CG = commissural ganglion; DC = deutocerebrum; E = esophagus; LG = lamina ganglionaris; ME = medulla externa; ISN = intersegmental nerve; lat.PC = lateral protocerebrum; med.PC = medial protocerebrum; MI/MT = medulla interna/medulla terminalis; MN = median nerve; MX = maxillary; NP = neuropil; PC = posterior commissure; PL = pleon neuromeres; post. MD = posterior mandibular; SN = segmental nerve; T = thoracic neuromeres; TC = tritocerebrum.

2.3   *Neurogenesis in the ventral nerve cord*

2.3.1   *Formation of the ectoderm*

The ventral nerve cord (VNC) in crustaceans arises, as in all arthropods, from cells within the ventral ectoderm. However, the mode of formation of the ectoderm differs substantially in different crustacean groups and even in different segments (Scholtz & Dohle 1996; Dohle & Scholtz 1997; see also Dohle et al. 2003). In all malacostracans examined a stereotypic, grid-like arrangement of ectodermal cells is evident in the post-naupliar region of the early embryo, although this invariant cellular pattern is generated by different mechanisms in different taxa. In most malacostracans, (including the Phyllocarida, Stomatopoda, Syncarida, Peracarida, Decapoda and Pleocyemata), ectodermal rows in the post-naupliar region from the posterior part of the 1st thoracic segment onwards are generated by a set of large stem cells, ectoteloblasts, which are arranged in a crescent or transverse row at the posterior end of the embryo (Dohle 1976; Scholtz 1992; see Fig. 2B). Iterative rounds of division of the ectoteloblasts in an anterior direction produce multiple transverse rows of cells, parallel to the generative ectoteloblast row (Fig. 2 A, B). In contrast, the most anterior ectodermal rows form by the rearrangement of irregularly scattered ectodermal cells. One malacostracan group, the Amphipoda, lacks ectoteloblasts altogether; in these embryos all of the ectodermal rows arise by the rearrangement of scattered cells (reviewed in Scholtz & Dohle 1996). Irrespective of their mode of formation, each ectodermal row undergoes two rounds of division to produce four descendent rows from which cells of the VNC and the ventral epidermis arise (Fig. 2A).

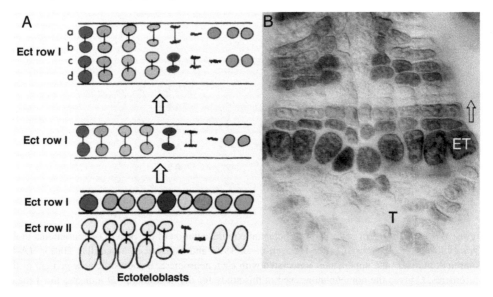

Figure 2. [See Plate 16 in color insert.] (A) Diagram (after Scholtz & Dohle 1996) to show how repeated divisions of the row of ectoteloblasts generate transverse rows of ectodermal (Ect) cells, each of which subsequently undergoes two rounds of divisions to produce four daughter rows, a, b, c, d. The most medial cells in a row (to the left in the figure) begin mitosis before more lateral cells. (B) Caudal papilla of a 30% *Palaemonetes argentinus* embryo showing the ectoteloblasts (ET) located in front of the telson (T) budding off rows of ectodermal cells in an anterior direction (indicated by the arrow) (from Harzsch 2001).

Entomostracan crustaceans do not exhibit the same ordered pattern of ectodermal cells as malacostracans and, with the exception of the Cirripedia (Anderson 1965), lack ectoteloblasts. For example, in the cladoceran *Leptodora kindtii*, randomly oriented and distributed cell divisions generate an ectoderm in which there is no regular cellular arrangement (Gerberding 1997).

2.3.2 *Segregation of neural precursors (neuroblasts) from ventral ectoderm*

Soon after its formation, two major cell types appear within the ventral ectoderm – (i) epidermoblasts, which give rise to epidermal cells and (ii) neuroblasts, which produce neurons and/or, in some cases, glial cells. In this article, I define a neuroblast as a neural precursor with stem cell properties i.e., a cell that undergoes repeated rounds of unequal (in size) mitotic divisions in a single direction to produce neurons/glia or neural/glial precursors. This is a relatively broad use of the term, but one that is useful when seeking to recognize common patterns of neurogenesis in different taxa. A more restrictive set of features (including basal nuclear migration, cell enlargement, see section 2.3.2.1) defines neuroblasts in insects, the arthropod group in which neurogenesis has been most intensively studied to date.

Even this relatively broad definition of a neuroblast may be too restrictive. It is possible that the products of division of a neural stem cell may be unequal in **developmental potential**, i.e., one of the daughter cells retains a stem-cell character, the other is committed to a neural fate, but equal in cell size. Neuroblasts that divide in this fashion would fail to be identified when solely applying the operational definition of a difference in daughter cell size.

2.3.2.1 *Malacostraca*.

In malacostracan crustaceans, neuroblasts arise within the ventral ectoderm after each of the descendent ectoteloblast rows has generated four transverse ectodermal rows. Many of the ensuing divisions are unequal and/or are oriented in a variety of directions other than longitudinal (Dohle 1976; Dohle & Scholtz 1988; Scholtz 1990, 1992; Scholtz et al. 1994). The spindles of the "differential" divisions of a subset of the ectodermal cells are oriented radially such that their smaller progeny, ganglion mother cells, come to lie internally (Fig. 3). Ganglion mother cells subsequently divide equally to produce neurons, although it remains an open question as to whether in crustaceans, as in insects, some ganglion mother cell progeny develop into glial cells.

Malacostracan neuroblasts have a stem-cell character, undergoing multiple unequal divisions to produce a chain of internally oriented progeny. This pattern of neuroblast division is reminiscent of that seen in the Insecta (Hartenstein & Campos-Ortega 1984; Doe & Goodman 1985a; Hartenstein et al. 1994). However, a number of differences exist between the two groups. Unlike the insects, the nuclei of malacostracan neuroblasts do not migrate towards the basal side of the ectoderm (Fig. 3A). Early arising malacostracan neuroblasts do not exhibit the cell enlargement that insect neuroblasts undergo soon after their formation (Fig. 4B), although increased neuroblast size is evident at later embryonic stages (Harzsch et al. 1998). At least some crustacean neuroblasts (Dohle 1976) can revert to an equal mode of division after having generated a number of ganglion mother cells (Fig. 4A, B). This phenomenon does not appear to occur in insect neuroblasts (although Tamarelle et al. 1985 report that neuroblasts in the phasmid *Carausius morosus* can arise by equal divisions of precursor cells). Finally, unlike the malacostracan situation, insect neuroblasts are not generated by invariant cell lineages. Rather, a process of lateral inhibition results in one cell from amongst a group of developmentally equivalent cells being selected for a neuroblast fate in a stochastic manner (Doe & Goodman 1985b).

2.3.2.2 *Entomostraca.* Neurogenesis has been examined carefully in only a few non-malacostracan taxa. Earlier studies using classical histological techniques have in general reported that neurogenesis proceeds by the "inward proliferation of numerous small cells from the ectoderm" (Benesch 1969; see Anderson 1973:344).

However, cells with some of the characteristics of malacostracan neuroblasts have been reported in the cladoceran *Leptodora kindti* (Gerberding 1997). While there is no stereotypic division pattern within the ectoderm of this species, a bilaterally symmetric set of enlarged cells is found within the ventral ectoderm lateral to the midline. These cells divide unequally, their spindles being oriented perpendicular to the body surface, in a stem cell manner, giving rise to chains of internally located progeny. However it has not been shown definitively that these progeny divide to produce neurons or differentiate into neurons themselves.

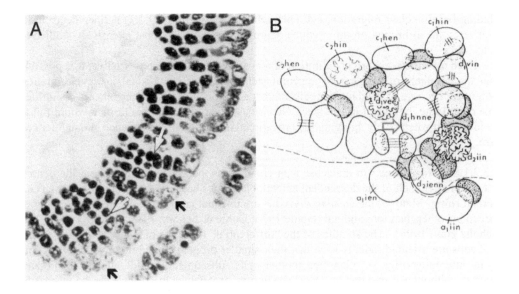

Figure 3. (A) Sagittal section through an advanced embryo of *Diastylis* showing neuroblasts located on external surface of the neuroectoderm (dark arrows) and rows of ganglion mother cell progeny oriented internally (from Dohle & Scholtz 1988). White arrows show divisions of ganglion mother cells. The two patches of cells in the bottom right hand corner of the figure belong to limb buds. (B) Diagram (from Dohle & Scholtz 1988) showing the stereotypic pattern of early neuroblasts generated within ectodermal row II of the *Diastylis* embryo. The right hemisegment is shown viewed from a ventral aspect (midline to right). Clear nuclei belong to neuroblasts, while stippled nuclei are ganglion mother cells, connected to their mother neuroblast by lines. Three lines indicate that the ganglion mother cell is a product of the third differential division of the neuroblast, four lines represent the fourth differential division.

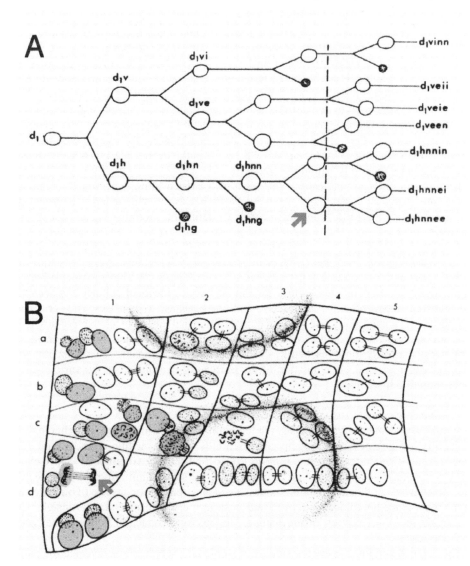

Figure 4. (A) The early lineage of the most posterior and medial cell, d_1, within ectodermal row II of the embryo of *Diastylis* (from Dohle & Scholtz 1988). One of the progeny of the first division of d_1 is a neuroblast that produces ganglion mother cells ($d_1$hg, $d_1$hng) in its next two rounds of divisions. Its next division is equal; one of the progeny reverts to a neuroblast fate, the other continues to divide equally (red arrow). Cell dhnne is also shown in Fig. 3 B. (B) Drawing (from Dohle 1976) of the descendents of cells 1-5 in row (3) of *Diastylis*. The left hemi-segment is shown with the posterior part of the second maxilla and the anterior part of the first thoracic limb indicated with shading. Nuclei of sister cells are connected by thin straight lines, with the number of lines denoting whether they are the product of the first, second, third or fourth differential division. Neuroblasts which have either produced ganglion mother cells or which will do so in the following divisions are colored green. Note that neuroblast nuclei are not consistently larger than other, presumed epidermoblast nuclei. The cell indicated with a red arrow is the progeny of $d_1$hnn that subsequently divides equally.

Duman-Scheel and Patel (1999) reported a pattern of regularly arranged rows of large cells in the ventral ectoderm of *Artemia franciscana* and *Triops longicaudatus* and speculated that these large cells were neuroblasts. However, no direct evidence was put forward to support this claim. Harzsch (2001) used BrdU labeling to reveal a population of relatively large, mitotically active cells, frequently associated with smaller labeled nuclei, in the ventral neurogenic region of the notostracan crustacean, *Triops cancriformis*. These large cells were suggested to be neuroblasts, although definitive evidence that their progeny generated neurons was not advanced. A similar population of enlarged cells was seen in the anostracan *Artemia salina*, although the size difference between the two labeled cell types was not as marked as in *Triops* (Harzsch 2001). In both species, the larger cells were arranged apparently randomly within the ectoderm.

In summary, there are indications that stem cells exist within the neurogenic regions of several entomostracan crustaceans. However, whether these cells generate neuronal progeny and therefore meet the "minimal" neuroblast definition given above, has not yet been definitively shown.

2.3.2.3 *Homology of insect and crustacean neuroblasts?* It remains an open question as to whether insect and malacostracan neuroblasts are homologous. We have to decide between two alternatives: (i) neural stem cells with similar but not identical cellular characteristics have evolved independently in insects and crustaceans; (ii) insect and crustacean neuroblasts are homologous but have acquired somewhat different characteristics in the course of the separate evolution of these two groups. A decision between these two alternatives is unlikely to be reached by examination of the comparative cellular properties of neuroblasts in general. Examination of the **pattern** of neuroblasts in the various groups may be more illuminating.

2.3.3 *Pattern of neuroblasts in the Malacostraca*
The spatial arrangement of neuroblasts, the temporal order in which they are generated, and the identities of their neuronal progeny are essentially invariant within an individual insect species (Bate 1976; Hartenstein & Campos-Ortega 1984; Doe & Goodman 1985a; Schmidt et al. 1997; Schmid et al. 1999). Furthermore, there are strong similarities in these features across diverse hexapod taxa, including both winged and apterygotic insects (Truman & Ball 1998). What is the neuroblast pattern in malacostracan crustaceans and what insights does this provide into conservation of mechanisms for neuron production between the Insecta and Crustacea?

The detailed cell lineage tracing studies of Dohle and Scholtz have revealed the pattern of early neuroblasts in a number of peracaridan species. These workers used the criterion of unequal mitotic division, with spindles oriented internally, to identify neuroblasts. The neuroblast pattern, like the ectodermal cell division pattern in general, is stereotypic in space and time and shows a high degree of conservation both between species and between segments in any one species. By the end of the second differential division, 13 neuroblasts per hemi-segment are evident (Figs. 3B, 4B): all except four of these are progeny of cells from the ectodermal column closest to the midline and within this medial region, neuroblasts greatly outnumber epidermoblasts (Fig. 4B). The method of lineage tracing used in these studies (mapping positions of dividing nuclei in a developmental series of fixed embryos) did not allow determination of the neuroblast pattern beyond the fourth differential division.

Scholtz (1992) found that the decapod *Cherax destructor* has a very similar pattern of differential ectodermal divisions to the peracaridan crustaceans. While cell lineages could be traced beyond the first differential division in only a few cases, the first two neuroblasts were identified, and found to be generated by a homologous lineage to the peracaridans. Neuroblast number subsequently increases to "a total of 25-30 per hemisegment arranged in 4-5 irregular files containing 6-7 neuroblasts" (Scholtz 1992). A single unpaired median neuroblast was found at the posterior margin of the segment. However, as the criteria for identification of these later neuroblasts were not explicitly stated, it is possible that some epidermoblasts were included in the count. As noted above, epidermoblasts lie in the same cellular layer as neuroblasts and the two cell types are similar in size at early stages of neurogenesis.

A number of workers have used mitotic incorporation of the thymidine analogue BrdU to map the positions of neuroblasts in malacostracan embryos. Labeled cells in the most ventral ectodermal layer are classified as neuroblasts if smaller, labeled cells (assumed to be ganglion mother cell progeny) are found in close association. However, there is a potential ambiguity in this method. If labeled ventral cells are closely spaced, one cannot be confident that they have all contributed to the production of nearby labeled ganglion mother cells, since neuroblast progeny do not invariably lie in strict vertical columns above their mother cell. Thus, some of the labeled ventral cells may be dividing epidermoblasts. Nonetheless, using this approach in the decapod, *Palaemonetes argentinus*, Harzsch (2001) has found five rows of neuroblasts with 3-4 neuroblasts in a row in each hemineuromere, together with a single median neuroblast at 35% of embryogenesis (Fig. 5A). This pattern is repeated in the second maxillary to the second thoracic segment and is similar but not identical to the "insect" thoracic/abdominal pattern of around 30 neuroblasts per hemisegment arranged in 6-7 rows (Truman & Ball 1998; Doe & Goodman 1985a; Hartenstein & Campos-Ortega 1984).

BrdU incorporation has also been used to determine the pattern of later dividing neuroblasts in three species of decapod crustaceans. Neuroblasts can be identified with more certainty at later stages using this method because they have enlarged considerably and are spaced further apart. However, it is possible that some early dividing neuroblasts may fail to undergo enlargement and will therefore not be represented in maps made at the later stages. Sullivan and Macmillan (2001) report 7 paired and 2 unpaired mitotically active neuroblasts in thoracic neuromeres 4-8 of the 85% *Cherax destructor* embryo. In each hemi-neuromere four neuroblasts are located superficially in the ventral cortex of the ganglion, two are found laterally and one lies on the dorsal surface close to the midline at the posterior segmental boundary (Fig. 5B). All except the dorsal neuroblast are present and mitotically active in the first larval stage (POI). By the second larval stage (POII), three of the neuroblasts have ceased proliferation. Abdominal neuromeres show a greatly reduced array of neuroblasts compared to the thoracic neuromeres, with a maximum of 2 neuroblasts per hemi-neuromere in A1-5.

Harzsch et al. (1998) examined the pattern of neuroblasts in embryos and larvae of the lobster *Homarus americanus* and the crab *Hyas araneus* and found a very similar pattern to that reported in *Cherax*. At a mid stage of neurogenesis (neuropil formation was well advanced), some dividing neuroblasts could be identified individually due to their stereotypic positions. Typically, four neuroblasts are found at the ventral ganglionic surface in each hemi-neuromere, a single neuroblast lies close to the midline at the posterior margin of the segment and a single ventrally located unpaired neuroblast is present. The pattern is similar in *Hyas araneus* (Fig. 5C), except that this species possesses an additional pair of lateral neuroblasts.

While neurogenesis has ceased in *Homarus* by the end of embryogenesis, it continues in *Hyas* after hatching as would be expected given the substantial differentiation of thoracic and pleon segments that takes place post-embryonically in the latter species. Harzsch and Dawirs (1994) have suggested that these post-embryonically generated neurons may be incorporated into the existing larval nervous system as new neural circuits are formed during metamorphosis. The larval (zoea 1 and zoea 2) population of neuroblasts in thoracic neuromeres appears to be somewhat larger than in the embryo, consisting of a dorsal neuroblast, a ventral-median neuroblast, 3-6 ventral neuroblasts and 1-3 lateral neuroblasts (Harzsch & Dawirs 1994). Neuroblast divisions have ceased by the megalopa stage.

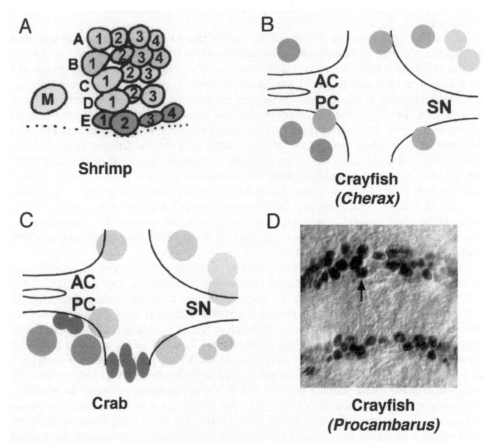

Figure 5. [See Plate 17 in color insert.] (A) Diagram (from Harzsch 2001) showing the pattern of BrdU labeled neuroblast nuclei in the first maxillary hemi-neuromere (left hand side) of *Palaemonetes argentinus* at 35% of embryogenesis. A stereotypic pattern of 5 rows of nuclei is evident, with a single median unpaired neuroblast (M). (B) Diagram (adapted from Sullivan & Macmillan 2001) showing the pattern of mitotically active neuroblasts in thoracic neuromeres of late stage *Cherax destructor* embryos. (C) Diagram (from Harzsch et al. 1998) showing the pattern of mitotically active neuroblasts in thoracic neuromeres of late stage *Hyas araneus* embryos. In both B and C, blue neuroblasts lie on the ventral surface of the ganglion, pink neuroblasts on the lateral side, the purple neuroblast on the dorsal surface, while the mauve cells are dorsal, unpaired, medial neuroblasts. The smaller cells in C are *engrailed*-positive neurons and/or glial cells. (D) Ventral view of the neuroectodermal region of the crayfish *Procambarus clarkii* (from Duman-Scheel & Patel 1999), showing a repeating pattern of two rows of *engrailed*-positive nuclei with a single *engrailed*-positive nucleus in the next most posterior row (arrow).

Duman-Scheel and Patel (1999) used large size with respect to smaller epidermal cells to identify neuroblasts in the embryos of a range of crustaceans, including the malacostracans *Procambarus clarkii, Porcellio laevis, Mysidium columbiae* and the branchiopods *Triops longicaudatus* and *Artemia franciscana*. They state that in *Procambarus* the neuroblasts are arranged in seven transverse rows and that the gene *engrailed* is expressed in neuroblasts in the most posterior two rows. Their figures suggest that there are up to 7 *engrailed*-positive cells in each hemisegmental row (Fig. 5D). This description is at variance with reports of 6-7 neuroblasts per hemisegment in other decapod species (Harzsch et al. 1998; Sullivan & Macmillan 2001) and the claim that none of the neuroblasts in mid-stage embryos of *Homarus americanus* express *engrailed* (Harzsch et al. 1998). Furthermore, while it has been reported that *engrailed* is expressed in progeny of two of the four early ectodermal rows in each segment of the amphipods *Orchestia* and *Gammarus* (Scholtz et al. 1994) and the decapods *Cherax* and *Neomysis* (Scholtz & Dohle 1996), lineage data (Scholtz & Dohle 1996) suggests that only two of these *engrailed* expressing cells develop into neuroblasts. Whether the discrepancies between these various reports can be accounted for by species differences, inclusion of ectodermal cells or epidermoblasts in the cells identified as neuroblasts or changes in neuroblast pattern or *engrailed* expression with embryonic stage remains to be determined.

In summary, cells with some of the characteristics of insect neuroblasts have been identified in malacostracans. The data to hand suggests that the pattern of neuroblasts is probably conserved within the Eumalacostraca, and that the early pattern of neuroblasts bears some similarities to the insect pattern. However, a definitive description of the pattern of neuroblasts throughout embryogenesis has not yet been obtained for any single crustacean species. Such data are essential if we are to make valid comparisons of the pattern of neuroblasts in different groups of crustaceans or between the Crustacea and the Insecta. The application of more direct and reliable lineage tracing methods, such as the DiI-labeling method previously used in *Drosophila* (Bossing & Technau 1994; Schmidt et al. 1997; Schmid et al. 1999) is required to fill this gap in our knowledge. Such methods could also reveal the identities of neurons generated by individual neuroblasts, information which will be invaluable in establishing homology between individual neuroblasts across different crustacean taxa or in determining whether crustacean neuroblasts are homologous to those found in other arthropod groups. The feasibility of application of the DiI lineage tracing method to selected crustacean species is evidenced by the studies of Gerberding and Scholtz (1999, 2001).

2.3.4 *Midline cells*

In both malacostracan (Gerberding & Scholtz (1999, 2001) and non-malacostracan embryos (Gerberding 1997), a distinct set of unpaired cells can be recognized at the ventral midline of the post-naupliar neuroectoderm at early stages of neurogenesis. In the amphipod *Orchestia cavimana*, it has been shown that these cells arise from a midline cell belonging to the original transverse cell rows and mimicking their first two rounds of division (Fig. 6). These divisions generate four progeny, a_0, b_0, c_0, d_0, aligned in the anterior-posterior axis (Gerberding & Scholtz 1999). The most anterior cell, a_0, and its immediate progeny, express *engrailed*, like the more lateral cells in row a. However, the median cells subsequently display a number of significant differences to the lateral ectodermal cells (Fig. 6). The next divisions of the median cells take place only after divisions have begun in the most medial lateral cells of the same row. The spindles of the third median cell divisions are all longitudinally oriented, whereas the lateral cells divide in a number of different directions. Each of the three most anterior median cells undergoes a single, equal division:

these progeny all differentiate into glial cells, some of which enwrap the anterior or posterior commissures (Gerberding & Scholtz 2001). In contrast, the most posterior median cell, d_0, undergoes several unequal divisions to produce approximately 10 progeny that differentiate into neurons (Fig. 6B). These appear to be unpaired neurons, and include both motorneurons and interneurons, with distinctive bifurcating axon morphologies similar to the progeny of the insect median neuroblast and midline progenitors (Thompson & Siegler 1991; Bossing & Technau 1994; Condron & Zinn 1994; Schmid et al. 1999). Thus, in malacostracans as in insects, the unpaired midline cells generate a specific population of commissural glia and neurons with distinctive, bifurcating axon morphologies.

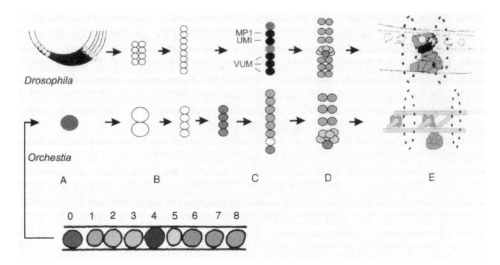

Figure 6. [See Plate 18 in color insert.] (Adapted from Gerberding & Scholtz 2001). Diagram to show the fates of the progeny of median ectodermal cells in *Drosophila* and the amphipod *Orchestia*. Median cells in *Orchestia* (cell 0 in each ectodermal row) (A) undergoes two rounds of division (B). The three most anterior progeny (green) each undergo one equal division (C) and differentiate into glial cells associated with the commissures (E). The fourth, most posterior progeny (red) undergoes several unequal divisions to produce approximately 10 progeny (D, pink), which differentiate into median neurons with distinctive bifurcating axon morphologies (E). Divisions of midline cells in *Drosophila* generate between 5 to 9 progeny (B). Some of these differentiate into the identified VUM and UM1 neurons (C), while another (red) undergoes further divisions to produce a cluster of neurons (D) which like the VUM/UM1 neurons and *Orchestia* median neurons have bifurcating axon morphologies. One of the midline cell progeny (C, green) divides (D) to produce midline glial cells (E).

2.3.5 *Significance of differences in mode of neurogenesis*
While the process of neurogenesis has been incompletely documented in crustaceans, available data suggests that malacostracans may utilize somewhat different cellular mechanisms for neuron production to entomostracans. Certainly, the mode of formation of the ectoderm from which the neuron precursors arise differs substantially between these groups. It has been suggested that ectoteloblasts (Dohle & Scholtz 1988, 1997) and an invariant arrangement of neuroblasts in a grid-like pattern (Harzsch 2001) are an autapo-

morphic character of the malacostracan crustaceans, not present in the entomostracan crustaceans. Do divergent modes of cell division underlying ectoderm and neuroblast formation in different crustacean taxa reflect the operation of fundamentally different molecular/genetic mechanisms in these groups? This seems unlikely given that these different modes of cell division are found in relatively closely related malacostracan taxa, e.g., Amphipoda vs. Isopoda, and in different segments of individual malacostracan species, viz., naupliar vs. post-naupliar segments. However, a definitive answer to this question and an explanation for why these different developmental modes have evolved will have to await future molecular/genetic studies of neurogenesis in the Crustacea. It will be interesting to see whether, for example, the well-documented Notch/Delta system of lateral inhibition uncovered in insects (Campos-Ortega 1994) plays any role in those crustaceans that generate neuroblasts by invariant cell divisions of ectodermal cells.

2.3.6 *Neurogenesis in other arthropods*

For a sound evolutionary interpretation of the apparent similarities in neurogenesis between the malacostracans and insects as well as the differences between malacostracans and entomostracans, we require information from the other arthropod groups – the Myriapoda and Chelicerata. For the most part, we have to rely on descriptions using classical histological staining techniques in these groups. Scattered reports of the presence of neuroblasts in chilopods (Knoll 1974; Hertzel 1984), pycnogonids (Winter 1979), and arachnids (reviewed in Weygoldt 1985) have emerged from such studies. However, while the figures that accompany these reports do in some cases show relatively large cells in the ventral neuroectoderm, no direct evidence is provided that these cells divide in a stem-cell fashion to produce neural progeny. Two more recent studies that have used either BrdU incorporation or anti-Phospho-Histone 3 immuno-histochemistry to label dividing cells in the neuroectoderm, have failed to find cells with the characteristics of insect/malacostracan neuroblasts in the centipede *Ethmostigmus rubripes* (Whitington et al. 1991) or the spider *Cupiennius salei* (Stollewerk et al. 2001). It thus seems unlikely that the possession of neuroblasts, as defined in this review, is a symplesiomorphic character of the Arthropoda.

2.4 *Neurogenesis in the brain*

2.4.1 *Overall pattern of neurogenesis*

Neurogenesis in the brain of malacostracan crustaceans proceeds by similar cellular mechanisms to the VNC. Harzsch and Dawirs (1996a) report the presence of large mitotically active neuroblasts in the brain of the spider crab *Hyas araneus* in the newly hatched zoea 1 larva. These are found in a relatively stereotypic arrangement. Most are located in the ventral cortical region, but each of the major brain regions possesses some dividing neuroblasts. The timing of neurogenesis in the brain of this species is similar to that seen in the VNC. While the full set of neuroblasts has been formed by the zoea 1 stage, they continue dividing into the zoea 2 stage, although at a reduced rate. Neuroblast divisions have ceased by the megalopa stage, but divisions of ganglion mother cells can still be detected in the brain of the megalopa and in the case of the olfactory lobe, into the adult crab. Neurogenesis in the olfactory system of decapod crustaceans has received special attention and shows some unique features (see section 2.4.2).

2.4.2 *Neurogenesis and neuron turnover in the olfactory system*

A distinctive feature of neurogenesis in the olfactory system of decapods is its prolonged time-course. While cell proliferation has ceased in most regions of the brain by the end of embryogenesis (in the case of the shore crab *Carcinus maenas* and the lobster *Homarus americanus*; Schmidt 1997, Harzsch et al. 1999b) or by late larval stages (in the case of the spider crab *Hyas araneus*; Harzsch & Dawirs, 1996a), neurons continue to be produced in the olfactory lobe throughout adult life. This phenomenon has been documented in a wide range of decapod species using both direct neuronal counts and BrdU labeling (Schmidt 1997, 2001; Sandeman et al. 1998; Harzsch et al. 1999b; Schmidt & Harzsch 1999).

In all species examined to date, ongoing proliferation occurs in the lateral soma cluster of the olfactory lobe, which houses olfactory projection neuron (OPN) cell bodies. In *Panulirus argus, Cherax destructor* and *Homarus americanus* (Sandeman et al. 1998; Harzsch et al. 1999a; Schmidt & Harzsch 1999; Schmidt 2001) proliferation is also observed in the paired medial soma clusters of the olfactory lobes, which contains local interneurons, while in *Carcinus maenas, Cancer pagurus* and *Pagurus bernhardus* (Schmidt 1997; Schmidt & Harzsch 1999) a cluster of cells associated with the hemiellipsoid bodies, which contains local interneurons that are targets of the OPNs, undergoes continued proliferation. In all cases, the BrdU labeled cells have a neural, rather than a glial morphology and move away from the proliferative zones towards the periphery of the soma clusters, probably as a result of passive displacement by newly generated cells.

Schmidt (2001) has recently used double-labeling with antibodies to BrdU and the neuropeptides FMRFamide and substance P to provide direct evidence that cells in the medial cluster of the olfactory lobe of *Panulirus argus* that are generated during adult life differentiate into neurons. His results show that these neurons can survive for at least 14 months. In addition, they suggest that the BrdU labeled neurons are the products of symmetrical divisions of precursor cells that are equivalent to ganglion mother cells, rather than neuroblasts.

Harzsch et al. (1999b) find that in *Homarus americanus* a substantial degree of apoptosis occurs in the same olfactory brain regions in parallel with neural proliferation, leading to the suggestion that there is continual turnover of interneurons in the olfactory system. However, the observation that pyknotic nuclei in the olfactory lobe of *Panulirus argus* are scarce and only observed in the immediate vicinity of the zones of cell proliferation (Schmidt 2001) is at odds with this conclusion.

In any event, the significance of neuronal cell death in the olfactory lobes is unclear. It may be related to the normal increase in number and turnover of olfactory receptor neurons in the antennules during adult growth (Sandeman & Sandeman 1996; Steullet et al. 2000). That olfactory interneurons are sensitive to their afferent input is shown by the loss of projection and local neurons after antennule amputation (Sandeman et al. 1998). A recent antennule ablation study in *Carcinus maenas* (Hansen & Schmidt 2001) has provided evidence that neuron generation in the olfactory lobes and hemiellipsoid bodies, as well as survival of post-mitotic neurons, is influenced by olfactory sensory input.

Another explanation for continued neural proliferation and apoptosis in the adult is that it may provide the anatomical substrate for adaptation to changing olfactory environments (Schmidt 1997). Consistent with this idea, Sandeman and Sandeman (2000) have shown that proliferation of interneurons in the olfactory and accessory lobes of adult crayfish (the latter receiving input from higher order visual and tactile receptor neurons) is sensitive to conditions in which the animals are reared.

2.5   *Neural differentiation*

Neurogenesis is the first step in a complex series of developmental events involved in the construction of the nervous system. The later events, which can be collectively termed neural differentiation, include such processes as axon outgrowth, innervation of post-synaptic targets, formation of dendritic branches, acquisition of neural excitability and neurotransmitter synthesis. Complex, specific sequences of gene expression underlie and regulate these events. Great advances have been made in our understanding of the molecular and cellular basis for neuron differentiation in *Drosophila* and a few other select insect model species. This provides an excellent base from which to explore other arthropods, where our knowledge in these areas is relatively rudimentary. In this section, I will review the current state of our understanding of neural differentiation in crustaceans. I will start with the question of whether the processes of neural differentiation in different crustacean taxa result in the formation of a common set of neurons, and whether this set of neurons represents an evolutionary "ground plan" (*Grundmuster* Ax 1987) for the Malacostraca, Entomostraca, or their common ancestor.

2.5.1   *Early pattern of neurons in the VNC*

Several lines of evidence argue for the existence of sets of homologous neurons in the VNC of a wide range of crustacean taxa.

2.5.1.1   *Pattern of serotonin-immunoreactive neuron.*   A number of workers have described the pattern of serotonin-immunoreactive neurons in the adult VNCs of crustaceans. On reviewing the described patterns of serotonergic neurons in three branchiopods, *Artemia salina*, *Triops cancriformis*, *Leptestheria dahalacensis* (Harzsch & Waloszek 2001b), the cephalocarid *Hutchinsoniella macracantha* (Elofsson 1992) and three barnacles *Balanus nubilis*, *Semibalanus cariosus*, *Pollicipes pollicipes* (Callaway & Stuart 1999), Harzsch and Waloszek (2001b) conclude that the entomostracan ground plan consists of a bilateral pair of anterior neurons, which project neurites into the anterior commissure and a bilateral pair of posterior neurons, which project neurites into the posterior commissure. The Cirripedia show a segment-specific reduction in this set, ranging from the loss of the posterior bilateral pair in the first, third, fifth and sixth thoracic ganglia to its complete absence in the fourth thoracic ganglion.

The pattern of serotonergic neurons has been described for representatives of most of the major malacostracan taxa, including the Phyllocarida (Harzsch & Waloszek 2001b), Decapoda (Beltz & Kravitz 1983; Real & Czternasty 1990; Harrison et al. 1995), Syncarida (Harrison et al. 1995) and Peracarida (Breidbach et al. 1990; Thompson et al. 1994). Amongst these groups, the pattern of serotonergic neurons shows considerable inter-specific and inter-segmental variability, making it difficult to reconstruct the ground plan. In any event, the pattern in all malacostracan species examined is substantially different to the proposed entomostracan ground plan. Whether this reflects a different ground plan between these groups or the relative plasticity of serotonin expression during evolution remains unclear at this stage.

2.5.1.2   *Pattern of even-skipped and engrailed expressing neurons.*   Duman-Scheel and Patel (1999) have reported some striking similarities between diverse crustacean taxa in the pattern of embryonic VNC neurons that express the genes *even-skipped* and *engrailed*. The three malacostracans *Porcellio laevis* (Isopoda), *Mysidium columbiae* (Mysidacea) and

Procambarus clarkii (Decapoda) as well as the branchiopod *Triops longicaudatus* all show *even-skipped* expression in a set of 3 bilateral dorsal neurons, two bilateral clusters of more ventral neurons and an unpaired midline neuron (Fig. 7). Remarkably, a very similar pattern of *even-skipped*-expressing neurons can be recognized in the Hexapoda (*Drosophila melanogaster* - Doe et al. 1988; grasshopper *Schistocerca americana* – Patel et al. 1992; springtail *Folsomia candida* – Duman-Scheel & Patel 1999). The authors argue for homology of these crustacean and hexapod neurons on the basis of similarity in cell body position. This homology claim is supported by earlier findings (Thomas et al. 1984; Whitington et al. 1993) that the 3 dorsal *even-skipped*-expressing neurons in *Porcellio scaber* and the decapods *Cherax destructor* and *Procambarus clarkii* have matching axon morphologies to the three even-skipped expressing insect neurons aCC, pCC and RP2, which are located in corresponding neuromere positions to their putative crustacean homologues (Whitington 1996).

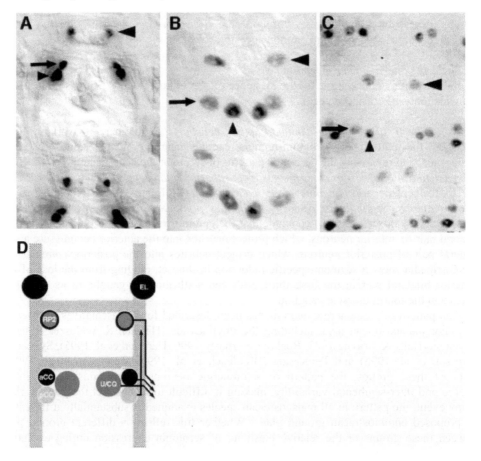

Figure 7. (From Duman-Scheel & Patel 1999). A conserved pattern of *engrailed*-expressing dorsal neurons in the CNS of the grasshopper *Schistocerca americana* (A), the isopod *Porcellio laevis* (B) and the branchiopod *Triops longicauditus* (C). The positions of these neurons suggest they may be homologues of the insect neurons RP2 (large arrowheads), aCC (arrow) and pCC (small arrowhead). D shows the position of these neurons within a neuromere together with the locations of other *even-skipped*-expressing neurons that are conserved across these species.

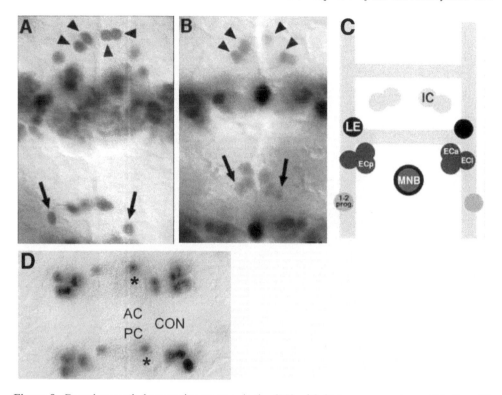

Figure 8. Dorsal *engrailed*-expressing neurons in the CNS of *Schistocerca americana* (A), *Porcellio laevis* (B) (from Duman-Scheel & Patel 1999) and *Homarus americanus* (D, from Harzsch et al. 1998). The IC (black arrowheads) and LE (black arrows) are found in both *Schistocerca* and *Porcellio*. Their positions within the neuromere are shown in C, together with other *engrailed*-expressing neurons that are conserved between these two species. Harzsch et al. (1998) claim that the lateral clusters of *engrailed*-positive cells in *Homarus* are likely to be glial cells, while they identify the cell marked with an asterisk behind the posterior commissure as a neuron. This may be homologous to one of the EC cluster.

Duman-Scheel and Patel (1999) also report that a range of crustacean and hexapod species shows strong similarities in the pattern of *engrailed*-expressing neurons (Fig. 8A-C). They identified 4 individual neurons and 3 clusters of *engrailed*-expressing neurons on the basis of cell body position in *Schistocerca americana*. From similarity in cell body position, they claim that homologues to all of the individual neurons and 2 of the 3 clusters can be found in *Folsomia candida*, *Drosophila melanogaster,* and the apterygotic insect *Thermobia domestica*. Putative homologues to all of the neural clusters and all except one of the individual neurons could be identified in the crustaceans *Mysidium columbiae, Porcellio laevis* and *Procambarus clarkii*. Homologues to only one of the clusters, the progeny of the median neuroblast, could be found in *Triops longicaudatus*. As no axon morphology data is available for these *engrailed*-expressing neurons, the homology assignments must be considered to be more tentative than for the *even-skipped*-expressing neurons.

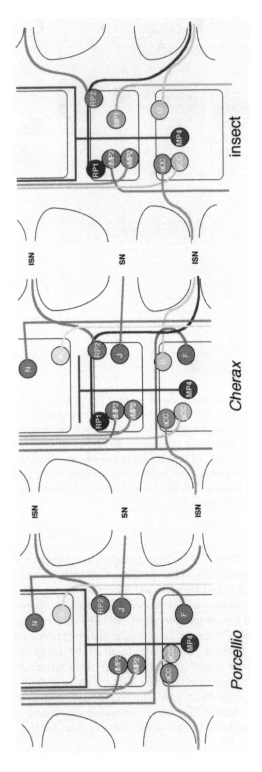

Harzsch et al. (1998) have examined the pattern of *engrailed*-positive cells in the embryonic VNC of the decapods *Homarus americanus* and *Hyas araneus*. They report a similar pattern in both species of dorsal bilateral pairs of neurons just behind the posterior commissure and at the posterior segmental border, clusters of putative glial cells located dorsally within the connectives at the posterior segmental border, and pairs of weakly labeled lateral cells in the ventral cortex (Fig. 8D). This appears to be a substantially different pattern to that described by Duman-Scheel and Patel (1999) in *Mysidium columbiae, Porcellio laevis,* and *Procambarus clarkii.* The reasons for these differences are unclear, but are unlikely to be due to species differences, given that each group examined both decapod and peracaridan species and reported a high degree of similarity between those species.

2.5.1.3 *Comparative axon morphology in embryos.* Further evidence for neural homologies within the Eumalacostraca and between the Eumalacostraca and Insecta comes from comparative studies of axon morphology in the embryonic VNC. Whitington et al. (1993) found that at least 10 embryonic central neurons in *Cherax destructor* and *Porcellio scaber* have similar cell body positions and axon morphologies in the two species and that 6 of these neurons initiate axon growth at the same relative time (Fig. 9). Three of these putative homologues are the dorsal *even-skipped* expressing neurons, which, as discussed in the previous section, are probably homologous to the insect neurons aCC, pCC and RP2. At least one other conserved crustacean neuron may have an insect homologue (vMP2), one *Porcellio* neuron (not found in *Cherax*) is likely to be homologous to the insect U neuron and one *Cherax* neuron (not found in Porcellio) may be homologous to the insect RP1/3/4 neurons (Whitington et al. 1993).

2.5.1.4 *Conclusion.* The data presented above provides a persuasive argument for homology of subsets of embryonic neurons both within and between the Malacostraca and Entomostraca, and even between the Crustacea and Hexapoda. However, much additional work needs to be done on comparative gene expression, neurotransmitter expression, and axon morphology to determine the extent of similarity between these groups in their neuronal populations and to define the neural ground plan for each. This is a worthwhile goal as determining the ground plan of a taxon will provide the necessary framework for understanding how developmental changes have brought about changes in neural organization during the evolution of that group.

2.5.2 *Axon growth and the establishment of major axon tracts*

2.5.2.1 *Brain.* Axon growth in the embryonic malacostracan CNS begins with the formation of a nerve ring around the stomodeum (Elofsson 1969; Helluy et al. 1993; Harzsch et al. 1997). This structure is the primordium of the brain. The post-esophageal commissure, which connects the ganglion *Anlagen* of the tritocerebrum, represents the posterior portion of the ring. Two closely spaced commissural tracts make up the anterior portion of the ring; the deutocerebral commissure and just anterior to it, the protocerebral commissure. The protocerebral tract coursing anteriorly from the anterior-lateral corners of

Figure 9. [See Plate 20 in color insert.] Diagram of axon morphologies of putatively homologous neurons in the embryos of *Porcellio scaber*, *Cherax destructor* and insects.

the ring connects the *Anlagen* of the medial protocerebrum to the eyestalk *Anlagen* (Harzsch et al. 1997). Axons descend posteriorly from this ring into the VNC along the path of the future longitudinal connectives. In Cherax destructor, some of these axons have been shown to originate from cell bodies anterior to the stomodeum, just lateral to the midline, a region that appears to correspond to the medial protocerebrum (Whitington et al. 1993).

2.5.2.2 *Ventral nerve cord.* There is a clear rostro-caudal gradient in the timing of axon growth in the crustacean VNC: axonogenesis in anterior segments precedes posterior segments. This phenomenon has been reported for both malacostracans (Harzsch et al. 1997; Harzsch et al. 1998; Whitington et al. 1993) and entomostracans (Harzsch 2001) and mirrors rostro-caudal gradients in other developmental processes, such as segment formation, e.g., Scholtz 1992, and neurotransmitter expression (Cournil et al. 1995; Harzsch & Dawirs 1995/6; Foa & Cooke 1998). In some cases, e.g., Cherax destructor (Whitington et al. 1993), the rostro-caudal gradient is so pronounced that longitudinal axons descending from the brain have reached thoracic segments before neurons in those segments have begun axogenesis.

The sequence of axon growth underlying the establishment of the major axon pathways in VNC ganglia – longitudinal connectives, transverse commissures and peripheral nerves – has been examined at the level of single, identified neurons for two species; the peracaridan *Porcellio scaber* and the decapod *Cherax destructor* (Whitington et al. 1993). The pattern of pioneering axon growth is very similar in both species. Furthermore, these crustaceans display many similarities to insects in the mode of establishment of major axon tracts. However, comparable descriptions of axon growth need to be obtained for other crustacean groups in order to properly assess the phylogenetic significance of observed similarities and differences. For example, we presently have no information on the pattern of central axon growth for any entomostracan.

The first axon tract to appear in the VNC is the median nerve, an unpaired nerve that originates in the mandibular segment and runs medially along the dorsal side of the VNC, linking adjacent ganglia. In *Porcellio*, this nerve is pioneered by a posteriorly-coursing axon that originates from a large neuron in the midline of the mandibular segment. This neuron has additional, paired, anteriorly-directed processes, which extend to the brain.

In both *Porcellio* and *Cherax*, the paired longitudinal connectives are pioneered by an anteriorly-directed axon from a neuron that is a likely homologue of the insect pCC neuron. Intersegmental axonal connections are established by contact between the axon of this neuron and its homologue in the next most anterior segment (Fig. 10). This axon is closely followed by a pair of anteriorly directed axons from neurons that are likely homologues of the insect vMP2/dMP2 or MP1 neurons (Goodman et al. 1984; see Fig. 10). The pattern in the winged insects differs slightly from this in that axons from the dMP2/MP1 neurons grow posteriorly between adjacent neuromeres either at the same time or immediately after the anteriorly directed vMP2/pCC axons (Goodman et al. 1984; Jacobs & Goodman 1989).

Interestingly, in the apterygotic insect *Ctenolepisma longicaudata*, only the MP1 pioneering axon grows posteriorly while both the dMP2 and vMP2 axons grow anteriorly (Whitington et al. 1996). Posteriorly directed growth of the dMP2 axon in the winged insects may therefore represent an autapomorphy of that group.

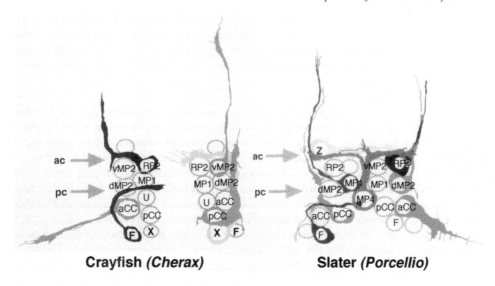

Crayfish *(Cherax)*   Slater *(Porcellio)*

Figure 10. [See Plate 19 in color insert.] Axon morphologies of neurons that pioneer the longitudinal connectives, commissure and intersegmental nerve in *Cherax destructor* and *Porcellio scaber* (from Whitington et al. 1993).

The *Porcellio* anterior commissure is pioneered at around the same time as the longitudinal connectives by a neuron that may be a homologue to the insect SP1 neuron (Goodman & Doe, 1993). The medially directed axon from this anteriorly-sited neuron (neuron Z in Fig. 10) meets the laterally directed, bifurcating axon branches from a dorsal unpaired median neuron, thereby completing the anterior commissural connection. The anterior commissure in *Cherax* is pioneered in a different fashion, by medially-directed axon growth from a neuron (neuron X in Fig. 10) that lies just posterior to pCC.

Formation of the posterior commissure is similar in both crustacean species: it is pioneered just after the anterior commissure by medially-directed axon growth from a neuron (neuron F in Fig. 10) adjacent to the pCC homologue. Harzsch et al. (1997) confirm that in the shrimp *Palaemonetes argentinus*, the anterior commissure starts to form before the posterior commissure. A definitive comparison between the crustacean and insect modes of commissure formation is not possible at this time as the neurons that pioneer these pathways in winged insects have not yet been identified with certainty (Goodman & Doe 1993). However, it is clear that there are substantial differences between the apterygotic and winged insects in the mode of commissure formation (Whitington et al. 1996).

The intersegmental peripheral nerve in both *Cherax* and *Porcellio* is founded, as in insects, by the laterally-directed axonal growth from the likely homologue of the insect aCC neuron (Fig. 10). This is an early event, taking place around the same time as the pioneering longitudinal axon growth from the pCC neuron.

In summary, all of the major central axonal tracts in these two eumalacostracan embryos, except for the anterior commissure, are pioneered in a similar mode and sequence by homologous neurons. Furthermore, many elements of this pattern are conserved between malacostracans and insects.

The pattern of axon growth described above is repeated essentially unchanged in all thoracic ganglia (Whitington et al. 1993). Abdominal and maxillary ganglia may show some variations on this thoracic pattern, although this has not been investigated. The mandibular ganglion must be considerably modified, as it possesses only a single commissure (Harzsch et al. 1997).

2.5.2.3 *Mechanisms of axon guidance.* Several conclusions regarding cellular mechanisms of axon growth and guidance in malacostracan embryos can be drawn from the study of Whitington et al. (1993). Firstly, axons from individual neurons follow stereotypic pathways from the outset. In some cases, axon branches are seen in inappropriate pathways at early stages of growth, e.g., the pCC homologue usually sends a short branch posteriorly at the point where it turns anteriorly, but these are invariably corrected later in development. Secondly, axons show specific and stereotypic patterns of fasciculation with other axons. Thirdly, some neurons that form collateral branches off the main axon, e.g., the pCC homologue sends collaterals from the anterior axon into the intersegmental nerves in every segment) do so only after the main axonal pathway has been established. Fourthly, growth cones of these early differentiating central neurons are often broad and complex, with filopodia oriented in many different directions. Finally, growth cones tend to grow between the regions of contact of adjacent cells, rather than over their dorsal or ventral surfaces.

All of these observations concur with those made in numerous studies of axon growth in insect embryos and suggest that common cellular mechanisms for axon growth and guidance are employed in these two arthropod groups. At present, we have virtually no information about the molecular mechanisms that underlie these cellular processes in crustacean embryos. An obvious approach to this problem is to clone crustacean orthologs of some of the many genes that have been implicated in axon guidance in *Drosophila* and other insects and determine whether they play similar roles in the Crustacea. Knowledge of the molecular basis for axon guidance in different species will ultimately lead to an understanding of what types of genetic change underlie changes in neural connectivity during evolution.

2.5.3 *Development of neuropil and neurotransmitter expression*
After sending axons to their synaptic targets, neurons form axonal arborizations over those targets and soon thereafter begin to elaborate dendritic branches. Within the CNS, these events lead to the expansion of the neuropil. Little is known about these developmental events in crustaceans. Harzsch et al. (1997, 1998) have found that synapsin-like immunoreactivity is apparent at early embryonic stages in the CNS of the decapods *Palaemonetes argentinus*, *Homarus americanus*, and *Hyas araneus*. At these stages, anti-*Drosophila* synapsin antibody stains axons in the connectives, commissures and peripheral nerve roots in addition to the neuropil, whereas at later embryonic stages staining is confined to the ganglionic neuropil. It is unclear whether this change in expression reflects a change in synapsin function from early to later embryonic stages.

While the first signs of neuropil development are apparent at early embryonic stages, much brain development, including that of the optic and olfactory neuropils begins only in the latter half of embryogenesis and continues through into larval stages (Helluy et al. 1993; Helluy et al. 1996). This is consistent with the protracted period of neurogenesis in these brain regions (see section 2.4).

Several studies have been carried out on developmental changes in neurotransmitter expression in crustaceans. Some of the conclusions that emerge from these studies are: a) different neurotransmitters show different timetables of expression; b) in some cases, the time of onset of neurotransmitter expression correlates with the functional requirement of that neurotransmitter for neurotransmission; c) in other cases, the neurotransmitter appears long before it is required for neurotransmission, suggesting that it may carry out other functions.

2.5.3.1 *Dopamine.* In the lobster *Homarus gammarus*, dopamine first appears in mid-embryonic life in neurons in ganglia of the eye stalks, brain and subesophageal ganglia (Cournil et al. 1995). Approximately, 30% of the complement of dopaminergic neurons present in the brain and VNC of juvenile lobsters is detected by the end of embryogenesis. This number increases to 60% by the early post-larval stage. This protracted period of acquisition of dopamine expression is similar to that seen for other neurotransmitters (Beltz & Kravitz 1987; Beltz et al. 1990).

2.5.3.2 *Serotonin and proctolin.* Some 50% of the adult complement of serotonergic neurons is seen in 50% embryos, whereas only 10% of proctolin-staining neurons are seen at the same stage (Beltz et al. 1990). The number of proctolin-staining neurons increases gradually through larval life, but even in late stage larvae only 50% of the final number of neurons have developed. Therefore the developmental appearance of proctolin is relatively late and protracted compared to the appearance of serotonin. This is due, at least in part, to a delay of proctolin expression relative to serotonin expression rather than a delay in the production of proctolinergic neurons, as some identified neurons in the VNC of the lobster express only serotonin at the 50% embryonic stage, but express both neurotransmitters at late embryonic and early larval stages (Beltz et al. 1990). The early onset of serotonergic expression, in some cases well before the appearance of the post-synaptic targets for this neurotransmitter, has led to the suggestion that it may serve non-neurotransmitter functions during development (Beltz et al. 1990). Consistent with this hypothesis is the finding that depletion of serotonin levels in the lobster brain following 5,7-dihydroxytryptamine injection, leads to a marked reduction in the size of olfactory projection neurons in the deutocerebrum (Benton et al. 1997; Sullivan et al. 2000).

2.5.3.3 *FMRFamide.* Harzsch and Dawirs (1996b) report a two-fold increase in the number of FMRFamide-like immunoreactive neurons in the brain and VNC of the crab Hyas araneus between the late zoea 1 and zoea 2 larval stages. In addition, there is a dramatic growth of glomerular-like FMRFamide immunoreactive neuropil structures in the thoracic ganglia over this period. This correlates with the delay in development of thoracic appendages until after metamorphosis, suggesting that in this case the onset of expression of this neuromodulatory substance is regulated to coincide with the appearance of its target tissues.

Schmidt (2001) has presented evidence that the development of FMRFamide and substance P expression is a relatively protracted process for interneurons in the olfactory lobe of adult *Panulirus argus*. Neurons do not begin to express FMRFamide neuropeptides until 3 months after birth, while substance P expression is delayed until 6-14 months. This prolonged differentiation period corresponds to the extended time needed for the functional maturation of olfactory receptor neurons in this species (2 to 8 months; Steullet et al. 2000).

2.5.3.4 *GABA and glutamate.* Foa and Cooke (1998) have compared the developmental pattern of expression of GABA and glutamate in embryos of the crayfish, *Cherax destructor*. They find that GABA-like immunoreactive neurons and fibers are some of the earliest neural elements to appear in the CNS and that they become extensively distributed by 70% of development. A rostro-caudal gradient in onset of GABA immunoreactivity is evident. In contrast, glutamate-like immunoreactivity first appears considerably later, at around 60-65% and is much more restricted in its distribution. The precocious appearance of GABA-like immunoreactive neurons is consistent with a role for GABA in trophic regulation of developmental processes, independent of its role as a neurotransmitter. This might include the modulation of activity-dependent processes involved in the establishment and stabilization of the excitatory innervation of musculature.

3 DEVELOPMENT OF THE VISUAL SYSTEM

Comparative studies of visual system development have taken on a special significance in light of the recent finding that genes that play key regulatory roles in eye formation are conserved across the entire animal kingdom (Callearts et al. 1997). Given that the animal that has been pivotal in advancing our understanding of the molecular/genetic basis for eye development – the fruit fly *Drosophila melanogaster* – is an arthropod it would be highly desirable to know more about visual system development in other arthropod groups, such as crustaceans. In this context, I will review recent work at the cellular level on the development of the crustacean visual system. These studies have pointed to conclusions regarding malacostracan, entomostracan and insect affinities that largely parallel those derived from research on CNS development.

3.1 *Malacostraca*

3.1.1 *Morphogenesis and pattern formation in the retina*
Several studies have examined the embryonic development of the retina in decapod crustaceans, including the crayfish, *Procambarus clarkii,* and *Homarus americanus*, the prawn, *Palaemonetes argentinus*, and the crab, *Hyas araneus* (Hafner et al. 1982; Hafner & Tokarski 1998; Harzsch et al. 1999a). In all of these species, the optic primordia appear as round elevations on the ventral surface of the embryo that are lifted into a dorso-ventral orientation as the head grows. The retina develops from the surface epithelial layer covering these optic primordia. A wave of mitotic activity, oriented parallel to the antero-posterior axis of the eye, moves across the epithelium from lateral to medial, generating the cells from which the ommatidia will be constructed (Hafner & Tokarski 1998; Harzsch et al. 1999a).

Hafner and Tokarski (1998) have identified a transition zone in the *Procambarus* retina, which lies just in the wake of this mitotic wave. This zone defines the border between medial undifferentiated retina and the lateral retina with its parallel rows of differentiated ommatidia. The youngest ommatidial cell clusters, which lie immediately lateral to the transition zone, contain a near mature complement of cell types, including eight retinula cells (R1-R8), four cone cells, distal pigment cells and corneagenous cells: no signs of emergent ommatidia are seen in the transition zone itself. Further laterally, ommatidia show a progressive change in the spatial arrangement of their constituent cell types. Cone cells send processes towards the surface of the retina, covering the retinula cells while the corneagenous and distal pigment cells gradually enclose the cone and retinula cells. R8 moves from a central to a peripheral position within the retinula cell array.

Neurogenesis in the retina does not cease with the end of embryonic development, at least in some species. The crabs *Hyas araneus* and *Carcinus maenas* show a zone of proliferating cells at the medial rim of the retina in all larval and post-larval stages (Harzsch & Dawirs 1995/6), suggesting that eye formation in these species extends into adult life. However, it is unclear whether this protracted proliferative activity is related to specific structural changes that take place in the eye during larval and adult life, such as addition of facets (Meyer-Rochow 1975).

Many of these features of retinal development in malacostracan crustaceans are strikingly similar to what is seen in insects of both the hemi- and holometabolous type (Meinertzhagen 1973; Wolff & Ready 1993; Anderson 1978). In particular, the transition zone shows many characteristics in common with the morphogenetic furrow found in the *Drosophila* eye disc. Differences between these structures include the absence of a surface depression in the crustacean transition zone and the progressive appearance of different cell types within differentiating *Drosophila* ommatidia vs. their apparent simultaneous origin in *Procambarus*.

3.1.2 Development of optic ganglia

In their BrdU study of cell proliferation in the decapod eye, Harzsch et al. (1999a) identified proliferative zones in other optic regions. A zone of mitotically active cells, PZ2, was found internal to the retinal proliferative zone (PZ1) and closely associated with the second optic neuropil, the lamina ganglionaris. These cells in this zone are arranged in two rows that lie parallel to PZ1 and are several cells deep. They have large ovoid nuclei and are closely associated with smaller BrdU-labeled nuclei, suggesting that they are neuroblast-like stem cells. This pattern of cell proliferation closely resembles that found in the outer optic *Anlage* in insects that generates neurons of the lamina and medulla externa (reviewed in Meinertzhagen 1973; Meinertzhagen & Hanson 1993).

A third proliferation zone, PZ3, is located at the surface of the brain and spans the third optic neuropil, the medulla externa. It was not shown whether the progeny of these cells contribute to the formation of the medulla externa or the more internal optic neuropils, the medulla interna and medulla terminalis.

Each of these three proliferation zones, PZ1, PZ2 and PZ3, is flanked by zones of apoptosis (Harzsch et al. 1999a). Cell death has been similarly reported during the development of the retina and each of the optic neuropil regions in insects (Nordlander & Edwards 1968; Anderson 1978; Wolff & Ready 1991; Monsma & Booker 1996).

In summary, the development of both the retina and the optic ganglia in malacostracan crustaceans shows many similarities to the insects. The extent of these similarities would seem to justify the claim that the eyes in these two groups are homologous structures, as has previously been suggested by comparison of the morphology of the adult compound eyes (Osorio & Bacon 1994; Melzer et al. 1997).

3.1.3 Axon growth and neuropil formation

No detailed study has been made of the cellular events underlying the formation of axonal projections from the retina to the lamina in a malacostracan embryo. This is somewhat surprising given that a detailed study of the development of retinula-lamina connectivity was carried out some time ago in an entomostracan embryo (see section 3.2.2). Axon growth from the retina to the lamina was briefly examined in *Palaemonetes argentinus* embryos using acetylated α-tubulin immunohistochemistry (Harzsch et al. 1999a). This method reveals single fibers projecting from the retina to the lamina in 50% embryos, while by 60%, bundles of fibers, each of which may represent the retinula axons from a single proto-ommatidium are evident.

Immunohistochemistry with the anti-*Drosophila* synapsin antibody reveals the developing neuropils of the embryonic visual system in the decapods *Palaemonetes argentinus*, *Homarus americanus* and *Hyas araneus* (Harzsch et al. 1997; Harzsch et al. 1999a). In all three species, the first neuropil to stain is a common *Anlage* of the medulla terminalis and medulla interna, followed by the medulla externa and the lamina. There is a massive growth of these visual neuropils during larval development in the crabs *Hyas araneus* and *Carcinus maenas* (Harzsch & Dawirs 1995/6).

3.2 Entomostraca

3.2.1 Cell proliferation and pattern formation in the eye

Eye development, as with most other aspects of neural development, has been neglected in the entomostracan crustaceans. However, two recent studies in the branchiopod *Triops* have begun to redress this lack of data. Melzer et al. (2000) have examined the arrangement of cells within ommatidial pre-clusters of *Triops cancriformis*. Unlike previous studies in malacostracans, they find that the various cell types within an ommatidium appear progressively as that structure matures. The earliest ommatidia consist of six cells, composed of a vertex cell, R8, two symmetrical cell pairs, R2/5 and R3/4, and a central mystery cell. Older pre-clusters contain an additional symmetrical pair of cells, R1/6, and a new single cell, R7, lying close to the former vertex cell, which has by now been displaced into the center of the cluster. The most mature ommatidia contain an additional pair of cells, the cone cells. This sequence of cell differentiation and rearrangement of cell positions during ommatidial morphogenesis accords closely with that seen in insects (Tomlinson 1985; Tomlinson & Ready 1987; Wolff & Ready 1993; Friedrich et al. 1996).

A recent BrdU study (Harzsch & Waloszek 2001a) in the compound eye of the meta-nauplius of *Triops longicaudatus* has revealed three band-shaped zones of proliferating cells, PZ1, 2 and 3, arranged parallel to each other. PZ1, the most dorsal zone, lies close to the midline on the medial side of the eye and is separated from the lateral differentiated ommatidia by a transition zone of largely undifferentiated tissue. Pre-ommatidial cell clusters can be seen in the transition zone. PZ2 is located deeper and more lateral within the eye and consists of a 2-3 cell wide band of large labeled nuclei, some of which divide asymmetrically. PZ3 is the deepest proliferative zone and courses around the medulla externa in a semicircle. This pattern of cell proliferation is very similar to that described for the decapods (see section 3.1.2).

In summary, the pattern of cell proliferation and ommatidial pattern formation within the eye of this entomostracan crustacean is very similar to the malacostracan/insect pattern. This finding of similarity in developmental mechanisms for eye formation contrasts with basic differences seen between the adult entomostracan and malacostracan eye in the organization of optic neuropils: non-malacostracans have only two optic ganglia, a lamina and a medulla, and lack the lamina/medulla and medulla/lobula chiasmata (reviewed in Nilsson & Osorio, 1998). Depending on the weight that one attaches to similarities in developmental mechanisms versus adult structure as a criterion for homology, this may suggest that the common ancestor of entomostracans, malacostracans and insects had a compound eye. An alternative view is that these three groups of arthropods evolved from an ancestor that possessed only ocelli and acquired their compound eyes independently. Yet another possibility is that insects are descendents of a malacostracan crustacean that possessed compound eyes, while entomostracans are the sister group to an insect/ malacostracan clade (see Nilsson & Osorio, 1998).

3.2.2 *Axon growth from retina to lamina*

The spatio-temporal pattern of retinula axon growth to the lamina in the branchiopod, *Daphnia magna,* has been documented in detail by serial electron microscopy (LoPestri et al. 1973, 1974; Macagno 1978; Flaster & Macagno 1984). This entomostracan has a single, reduced compound eye containing only 22 ommatidia, each of which possesses 8 photoreceptors or retinula neurons. The axons from retinula neurons in a single ommatidium associate with a cartridge of 5 neurons in the lamina. Retinula axons within an ommatidium grow out in succession, the later axons fasciculating closely with earlier axons. The leading retinula axon makes contact with each of the five undifferentiated cells in a lamina cartridge in turn, wrapping around them and forming transient gap junctions with them (Lopresti et al. 1973, 1974). Laminar cells become post-mitotic just prior to or during this initial contact with the retinula axon (Flaster & Macagno 1984). Shortly after being contacted by the leading retinula axon, each of the lamina neurons sends out an axon that grows internally, following the overtaking retinula axon. A similar close association is seen between retinula axons and lamina neurons in the insect compound eye (reviewed in Meinertzhagen & Hanson 1993). Experimental evidence exists to show that contact between the retinula axon and the lamina neurons in necessary for survival and differentiation of the lamina neurons. If an ommatidium with its eight photoreceptor neurons is laser ablated, the lamina neurons that would have been contacted by these retinula axons degenerate (Macagno 1978, 1979). Delaying in-growth of retinula axons leads to a delay in differentiation of the laminar neurons contacted by them (Macagno 1981). Similarly, an inductive interaction between retinula axons and lamina neurons has been well documented in the insects (Meinertzhagen & Hanson 1993; Huang & Kunes 1996; Huang et al. 1998).

The more lateral ommatidia begin to mature first, the more medial later. All retinula axons grow towards and along the midline of the eye as they advance to the lamina, perhaps guided by glial cells located at the midline. The retinula axons then contact the lamina neurons that happen to lie closest to the midline at that time. Following contact by retinula axons, this group of five lamina neurons migrates laterally, allowing a new set of lamina neurons to move to the midline. These "virgin" lamina neurons are contacted by retinula axons from the adjacent medial cartridge and the process is repeated (Lopresti et al. 1973, 1974; Macagno 1979).

These observations suggest a model in which retinula axon-lamina neuron connectivity is determined by some simple cellular rules: retinula axons grow along the midline following the leading axon in the ommatidium; retinula axons innervate which ever set of lamina neurons happens to lie closest to the midline; lamina neurons move laterally following contact with the lead retinula axon. This contrasts with alternative models involving an active molecular matching between retinula and lamina neurons. The results of experiments in which ommatidia have been ablated (Macagno 1978) speak against a rigid "one ommatidium-one lamina cartridge" matching model. However, they do not exclude a possible role for active interactions between retinula growth cones and lamina cells in determining some aspect of the pattern of termination of photoreceptor axons. Studies on the formation of retina-lamina connections in *Drosophila* point to an active signaling between glial cells in the lamina and retinula growth cones in causing photoreceptor axons to terminate in the appropriate optic layer (Perez & Steller 1996). It would not be surprising to find that a similar phenomenon exists in *Daphnia*.

4 SUMMARY – THE STATE OF THE FIELD

Research in the field of crustacean neural development is currently at an exciting, albeit somewhat frustrating point. There can be no doubting the importance of extending the detailed studies on neural development that have been carried out in *Drosophila* and other insects to other arthropods. A comparison between the insects and their closest evolutionary relatives will provide significant insights into the ancestral roles of developmental regulatory genes and how those functions change in evolution to bring about changes in morphology. Our current knowledge of the development of the crustacean nervous system supports the view that insects and crustaceans (or at least malacostracans) share a common ground plan for construction of the CNS and eye. This will make the task of understanding how developmental mechanisms have changed during the evolution of the insects and crustaceans much easier than if these groups shared little in common.

Despite the significance and promise of studies of neural development in Crustacea, progress in this field is being impeded by a lack of probes to analyze gene function during development. An urgent priority for the future must lie in cloning crustacean homologues of key *Drosophila* neural development genes. This will allow the development of reagents for determining and manipulating the expression of genes whose action underlies the cellular events detailed in this review. In parallel with these molecular studies, continued effort should be given to widening the scope of cellular descriptions of neural development to include poorly represented taxa.

ACKNOWLEDGEMENTS

I am indebted to Gerhard Scholtz and Wolfgang Dohle for many illuminating discussions on arthropod evolution. I thank Kerri-Lee Harris for assistance in the preparation of figures.

REFERENCES

Anderson, D.T. 1965. Embryonic and larval development and segment formation in *Ibla quadrivalvis* Cuv. (Cirripedia). *Austr. J. Zool.* 13: 1-15.

Anderson, D.T. 1973. *Embryology and phylogeny of annelids and arthropods.* Oxford: Pergamon Press.

Anderson, H. 1978. Postembryonic development of the visual system of the locust, *Schistocerca gregaria*: patterns of growth and developmental interactions in the retina and optic lobe. *J. Embryol. Exp. Morphol.* 45: 55-83.

Ax, P. 1987. *The Phylogenetic System. The Systematization of Organisms on the Basis of their Phylogenetics.* New York: J.Wiley & Sons.

Bate, C.M. 1976. Embryogenesis of an insect nervous system I. A map of the thoracic and abdominal neuroblasts in *Locusta migratoria. J. Embryol. Exp. Morph.* 35: 107-123.

Beltz, B.S. & Kravitz, E.A. 1983. Mapping of serotonin-like immunoreactivity in the lobster nervous system. *J. Neurosci.* 3: 585-602.

Beltz, B.S.& Kravitz, E.A. 1987. Physiological identification, morphological analysis, and development of identified serotonin-proctolin containing neurons in the lobster ventral nerve cord. *J. Neurosci.* 7: 533-546.

Beltz, B.S., Pontes, M.S., Helluy, S.M. & Kravitz, E.A. 1990. Patterns of appearance of serotonin and proctolin immunoreactivities in the developing nervous system of the American lobster. *J. Neurobiol.* 21: 521-542.

Benesch, R. 1969. Zur Ontogenie und Morphologie von *Artemia salina*. *Zool. Jb. Anat.* 86: 307-458.

Benton, J., Huber, R., Ruckhoeft, M., Helluy, S. & Beltz, B. 1997. Serotonin depletion by 5,7-dihyroxytryptamine alters deutocerebral development in the lobster, *Homarus americanus*. *J. Neurobiol.* 33: 357-373.

Bossing, T. & Technau, G. M. 1994. The fate of the CNS midline progenitors in Drosophila as revealed by a new method for single cell labelling. *Development* 120: 1895-1906.

Breidbach, O., Wegerhoff, R. & Dennis, R. 1990. Patterns of serotonin-like immunoreactivity in the ventral nerve cord of the wood louse, *Oniscus asellus*. *Zool. Beitr.* (N.F.) 33: 311-329.

Bullock, T.H. & Horridge, G.A. 1965. *Structure and Function in the Nervous System of Invertebrates*. San Francisco: Freeman & Company.

Callaway, J.C. & Stuart, A.E. 1999. The distribution of histamine and serotonin in the barnacle's nervous system. *Micros. Res. Tech.* 44: 94-104.

Callearts, P., Halder, G. & Gehring, W.J. 1997. *Pax-6* in development and evolution. *Annu. Rev. Neurosci.* 20: 483-532.

Campos-Ortega, J. A. 1994. Genetic mechanisms of early neurogenesis in *Drosophila melanogaster*. *Advances Insect Physiol.* 25: 75-103.

Condron, B.G. & Zinn, K. 1994. The grasshopper median neuroblast is a multipotent progenitor cell that generates glia and neurons in distinct temporal phases. *J. Neurosci.* 14: 5766-5777.

Cournil, I., Casasnovas, B., Helluy, S.M. & Beltz. B.S. 1995. Dopamine in the lobster *Homarus gammarus*: II. Dopamine immunoreactive neurons and development of the nervous system. *J. Comp. Neurol.* 362: 1-16.

Doe, C.Q. & Goodman, C.S. 1985a. Early events in insect neurogenesis I. Development and segmental differences in the pattern of neuronal precursor cells. *Dev. Biol.* 111: 193-205.

Doe, C.Q. & Goodman, C.S. 1985b. Early events in insect neurogenesis II The role of cell interactions and cell lineage in the determination of neuronal precursor cells. *Dev. Biol.* 111: 206-219.

Doe, C., Smouse, D. & Goodman, C.S. 1988. Control of neuronal fate by the Drosophila segmentation gene *even-skipped*. *Nature* 333: 376-378.

Dohle, W. 1976. Die Bildung und Differenzierung des postnauplialen Keimstreifs von *Diastylis rathkei* (Crustacea, Cumacea) II Die Differenzierung und Musterbildung des Ektoderms. *Zoomorphologie* 84: 235-277.

Dohle, W. & Scholtz, G. 1988. Clonal analysis of the crustacean segment: the discordance between genealogical and segmental boundaries. *Development* 104 (Suppl.): 147-160.

Dohle, W. & Scholtz, G. 1997. How far does cell lineage influence cell fate specification in crustacean embryos? *Sem. Cell Dev. Biol.* 8: 379-390.

Dohle, W., Gerberding, M., Hejnol, A. & Scholtz, G. 2003. Cell lineage, segment differentiation, and gene expression in crustaceans. In: Scholtz, G. (ed), *Crustacean Issues 15, Evolutionary Developmental Biology of Crustacea*: 95-133. Lisse: Balkema.

Duman-Scheel, M. & Patel, N. H. 1999. Analysis of molecular marker expression reveals neuronal homology in distantly related arthropods. *Development* 126: 2327-2334.

Elofsson, R. 1969. The development of the compound eyes of *Penaeus duodarum* (Crustacea: Decapoda) with remarks on the nervous system. *Z. Zellforsch.* 97: 323-350.

Elofsson, R. 1992. Monoaminergic and peptidergic neurons in the nervous system of *Hutchinsoniella macracantha* (Cephalocarida). *J. Crust. Biol.* 12: 531-536.

Flaster, M.S. & Macagno, E.R. 1984. Cellular interactions and pattern formation in the visual system of the branchiopod crustacean, *Daphnia magna*. III. The relationship between cell birthdates and cell fates in the optic lamina. *J. Neurosci.* 4: 1486-1498.

Foa, L. C. & Cooke, I. R. C. 1998. The ontogeny of GABA- and glutamate-like immunoreactivity in the embryonic Australian freshwater crayfish, *Cherax destructor*. *Dev. Brain Res*. 107: 33-42.

Friedrich, M., Rambold, I. & Melzer, R.R. 1996. The early stages of ommatidial development in the flour beetle *Tribolium confusum* (Coleoptera; Tenebrionidae). *Dev. Genes Evol*. 206: 147-152.

Fryer, G. 1988. Studies on the functional morphology and biology of the Notostraca (Crustacea: Branchiopoda). *Phil. Trans. R. Soc. Lond. B* 321: 27-124.

Gerberding, M. 1997. Germ band formation and early neurogenesis of *Leptodora kindti* (Cladocera): first evidence for neuroblasts in the entomostracan crustaceans. *Invert. Repr. Dev*. 32: 63-73.

Gerberding, M. & G. Scholtz 1999. Cell lineage of the midline cells in the amphipod crustacean *Orchestia cavimana* (Crustacea, Malacostraca) during formation and separation of the germ band. *Dev. Genes. Evol*. 209: 91-102.

Gerberding, M. & G. Scholtz 2001. Neurons and glia in the midline of the higher crustacean *Orchestia cavimana* are generated via an invariant cell lineage that comprises a median neuroblast and glial progenitors. *Dev. Biol*. 235: 397-409.

Goodman, C. S., Bastiani, M. J., Doe, C.Q., duLac, S., Helfand, S.L., Kuwada, J.Y. & Thomas, J.B. 1984. Cell recognition during neuronal development. *Science* 225: 1271-1279.

Goodman, C.S & Doe, C.Q. 1993. Embryonic development of the Drosophila central nervous system. In: Bate, M. & Martinez Arias, A. (eds), *The Development of Drosophila melanogaster*. Vol. II: 1131-1206. Cold Spring Harbor: Cold Spring Harbor Press.

Gruner, H-E. 1993. 1. Klasse Crustacea. In: Gruner, H.-E. (ed), *Lehrbuch der Speziellen Zoologie. Band I: Wirbellose Tiere. 4. Teil: Arthropoda*: 448-1030. Jena: Gustav Fischer Verlag.

Hafner, G. S. & Tokarski, T. R. 1998. Morphogenesis and pattern formation in the retina of the crayfish *Procambarus clarkii*. *Cell Tissue Res*. 293: 535-550.

Hafner, G.S., Tokarski, T.R. & Hammond-Soltis, G. 1982. Development of the crayfish retina: a light and electron microscopic study. *J. Morphol*. 173: 101-118.

Hansen, A. & Schmidt, M. 2001. Neurogenesis in the central olfactory pathway of the adult shore crab *Carcinus maenas* is controlled by sensory afferents. *J. Comp. Neurol*. 441: 223-233.

Harrison, P.J., Macmillan, D.L. & Young, H.M. 1995. Serotonin immunoreactivity in the ventral nerve cord of the primitive crustacean *Anaspides tasmaniae* closely resembles that of crayfish. *J. Exp. Biol*. 198: 531-535.

Hartenstein, V. & Campos-Ortega, J.A. 1984. Early neurogenesis in wildtype *Drosophila melanogaster*. *Roux's Arch. Dev. Biol*. 193: 308-325.

Hartenstein, V., Younossi-Hartenstein, A. & Lekven, A. 1994. Delamination and division in the *Drosophila* neuroectoderm: Spatiotemporal pattern, cytoskeletal dynamics and common control by neurogenic and segment polarity genes. *Dev. Biol*. 165: 480-499.

Harzsch, S. 2001. Neurogenesis in the crustacean ventral nerve cord: homology of neuronal stem cells in Malacostraca and Branchiopoda? *Evol. .Dev*. 3: 154-169.

Harzsch, S. & Dawirs, R. R. 1994. Neurogenesis in larval stages of the spider crab *Hyas araneus* (Decapoda, Brachyura) - proliferation of neuroblasts in the ventral nerve cord. *Roux's Arch. Dev. Biol*. 204: 93-100.

Harzsch, S. & Dawirs, R. R. 1995/6. Maturation of the compound eyes and eyestalk ganglia during larval development of the brachyuran crustaceans *Hyas araneus* L. (Decapoda, Majidae) and *Carcinus maenas* L. (Decapoda, Portunidae). *Zoology* 99: 189-204.

Harzsch, S. & Dawirs, R.R. 1996a. Neurogenesis in the developing crab brain: postembryonic generation of neurons persists beyond metamorphosis. *J. Neurobiol*. 29: 384-398.

Harzsch, S. & Dawirs, R. R. 1996b. Development of neurons exhibiting FMFRamide related immunoreactivity in the central nervous system of larvae of the spider crab *Hyas araneus* L. (Decapoda: Majidae). *J. Crust. Biol*. 16: 10-19.

Harzsch, S., Anger, K. & Dawirs, R.R. 1997. Immunocytochemical detection of acetylated alpha-tubulin and *Drosophila* synapsin in the embryonic crustacean nervous system. *Int. J. Dev. Biol*. 41: 477-484.

Harzsch, S., Benton, J., Dawirs, R.R. & Beltz, B. 1999a. A new look at embryonic development of the visual system in decapod crustaceans: Neuropil formation, neurogenesis, and apoptotic cell death. *J. Neurobiol*. 39: 294-306.

Harzsch, S., Miller, J., Benton, J. & Beltz, B. 1999b. From embryo to adult: Persistent neurogenesis and apoptotic cell death shape the lobster deutocerebrum. *J. Neurosci.* 19: 3472-3485.

Harzsch, S., Miller, J. , Benton, J., Dawirs, R. R. & Beltz, B. 1998. Neurogenesis in the thoracic neuromeres of two crustaceans with different types of metamorphic development. *J. Exp. Biol.* 201: 2465-2479.

Harzsch, S. & Waloszek, D. 2001a. Neurogenesis in the developing visual system of the branchiopod crustacean *Triops longicaudatus* (LeConte, 1846): corresponding patterns of compound-eye formation in Crustacea and Insecta? *Dev. Genes. Evol.* 211: 37-43.

Harzsch, S. & Waloszek, D. 2001b. Serotonin-immunoreactive neurons in the ventral nerve cord of Crustacea: a character to study aspects of arthropod phylogeny. *Arthr. Struct. Dev.* 29: 307-322.

Helluy, S. M., Benton, J. L., Langworthy, K.A., Ruchhoeft, M.L. & Beltz, B.S. 1996. Glomerular organization in developing olfactory and accessory lobes of American lobsters: Stabilization of numbers and increase in size after metamorphosis. *J. Neurobiol.* 29: 459-472.

Helluy, S., Sandeman, R., Beltz, B. & Sandeman, D. 1993. Comparative brain ontogeny of the crayfish and clawed lobster: implications of direct and larval development. *J. Comp. Neurol.* 335: 343-354.

Hertzel, V. 1984. Die Segmentation des Keimstreifens von *Lithobius forficatus* (L.). *Zool. Jb. Anat.* 112: 369-386.

Huang, Z. & Kunes, S. 1996. Hedgehog, transmitted along retinal axons, triggers neurogenesis in the developing visual centers of the *Drosophila* brain. *Cell* 86: 411-422.

Huang, Z., Shilo, B.Z. & Kunes, S. 1998. A retinal axon fascicle uses spitz, an EGF receptor ligand, to construct a synaptic cartridge in the brain of *Drosophila*. *Cell* 95: 693-703.

Jacobs, J. R. & Goodman, C. S. 1989. Embryonic development of axon pathways in the *Drosophila* CNS. 2. Behavior of pioneer growth cones. *J. Neurosci.* 9: 2412-2422.

Knoll, H.J. 1974. Untersuchungen zur Entwicklungsgeschichte von *Scutigera coleoptrata* L. (Chilopoda). *Zool. Jb. Anat.* 92: 47-132.

LoPresti, V., Macagno, E.R. & Levinthal, C. 1973. Structure and development of neuronal connections in isogenic organisms: cellular interactions in the development of the optic lamina of *Daphnia. Proc. Natl. Acad. Sci. USA* 70: 433-437.

LoPresti, V., Macagno, E.R. & Levinthal, C. 1974. Structure and development of neuronal connections in isogenic organisms: transient gap junctions between growing optic axons and lamina neuroblasts. *Proc. Natl. Acad. Sci. USA* 71: 1098-1102.

Macagno, E.R. 1978. Mechanism for the formation of synaptic projections in the arthropod visual system. *Nature* 275: 318-320.

Macagno, E. R. 1979. Cellular interactions and pattern formation in the development of the visual system of *Daphnia magna* (Crustacea, Branchiopoda). I. Interactions between embryonic retinular fibers and laminar neurons. *Dev. Biol.* 73: 206-238.

Macagno, E. R. 1981. Cellular Interactions and Pattern Formation in the Development of the Visual System of *Daphnia magna* (Crustacea, Branchiopoda) II. Induced Retardation of Optic Axon Ingrowth Results in a Delay in Laminar Neuron Differentiation. *J. Neurosci.* 1: 945-955.

Meinertzhagen, I.A. 1973. Development of the compound eye and optic lobes of insects. In: Young, D. (ed), *Developmental Neurobiology of the Arthropods*: 51-104. London: Cambridge University Press.

Meinertzhagen, I.A. & Hanson, T.E. 1993. The development of the optic lobe. In: Bate, M., & Martinez Arias, A. (eds), *The Development of* Drosophila melanogaster. Vol. II: 1363-1491. Cold Spring Harbor: Cold Spring Harbor Press.

Melzer, R.R., Diersch, R., Dicastro, D. & Smola, U. 1997. Compound eye evolution: highly conserved retinula and cone cell patterns indicated a common origin of the insect and crustacean ommatidium. *Naturwissenschaften* 84: 542-444.

Melzer, R.R., Michalke, C. & Smola, U. 2000. Walking on insect paths: early ommatidial development in the compound eye of the ancestral crustacean *Triops cancriformis*. *Naturwissenschaften* 87: 308-311.

Meyer-Rochow, V.B. 1975. Larval and adult eye of the western rock lobster (*Paniluirus longipes*). *Cell Tissue Res.* 162: 439-457.

Monsma, S.A. & Booker, R. 1996. Genesis of the adult retina and outer optic lobes of the moth, *Manduca Sexta*. I. Patterns of proliferation and cell death. *J.Comp.Neurol.* 367: 10-20.

Nilsson, D. & Osorio, D. 1998. Homology and parallelism in arthropod sensory processing. In: Fortey, R.A. & Thomas, R.H. (eds), *Arthropod relationships*: 333-348. London: Chapman & Hall.

Nordlander, R.H. & Edwards, J.S. 1968. Morphological cell death in the post-embryonic development of the insect optic lobes. *Nature* 218: 780-781.

Osorio, D., & Bacon, J.P. 1994. A good eye for arthropod evolution. *BioEssays* 16: 419-424.

Patel, N. H., Ball, E. E. & Goodman, C.S. 1992. Changing role of *even-skipped* during the evolution of insect pattern formation. *Nature* 357: 339-342.

Perez, S.E. & Steller, H. 1996. Migration of glial cells into retinal axon target field in *Drosophila melanogaster*. *J. Neurobiol.* 30: 359-373.

Real, D. & Czternasty, G. 1990. Mapping of serotonin-like immunoreactivity in the ventral nerve cord of crayfish. *Brain Res.* 521: 203-212.

Sandeman, R. E. & Sandeman, D. C. 1996. Pre- and postembryonic development, growth and turnover of olfactory receptor neurones in crayfish antennules. *J. Exp. Biol.* 199: 2409-2418.

Sandeman, R. & Sandeman, D.C. 2000. "Impoverished" and "enriched" living conditions influence the proliferation and survival of neurons in crayfish brain. *J. Neurobiol.* 45: 215-226.

Sandeman, R.E., Clarke, D., Sandeman, D.C. & Manly, M. 1998. Growth-related and antennular amputation-induced changes in the olfactory centers of crayfish brain. *J. Neurosci.* 18: 6195-6206.

Schmid, A., Chiba, A. & Doe, C.Q 1999. Clonal analysis of *Drosophila* embryonic neuroblasts: neural cell types, axon projections and muscle targets. *Development* 126: 4653-4689.

Schmidt, H., Rickert, C., Bossing, T., Vef, O., Urban, J. & Technau, G.M. 1997. The embryonic central nervous system lineages of *Drosophila melanogaster* II. Neuroblast lineages derived from the dorsal part of the neuroectoderm. *Dev. Biol.* 198: 186-204.

Schmidt, M. 1997. Continuous neurogenesis in the olfactory brain of adult shore crabs, *Carcinus maenas*. *Brain Res.*762: 131-143.

Schmidt, M. 2001. Neuronal differentiation and long-term survival of newly generated cells in the olfactory midbrain of the adult spiny lobster, *Panulirus argus*. *J. Neurobiol.* 48: 181-203.

Schmidt, M. & Harzsch, S. 1999. Comparative analysis of neurogenesis in the central olfactory pathway of adult decapod crustaceans by *in vivo* BrdU labeling. *Biol. Bull.* 196: 127-136.

Scholtz, G. 1990. The formation, differentiation and segmentation of the post-naupliar germ band of the amphipod *Gammarus pulex* L. (Crustacea, Malacostraca, Peracarida). *Proc. R. Soc. Lond. B* 239: 163-211.

Scholtz, G. 1992. Cell lineage studies in the crayfish *Cherax destructor* (Crustacea, Decapoda): germ band formation, segmentation, and early neurogenesis. *Roux's Arch. Dev. Biol.* 202: 36-48.

Scholtz, G. 1995a. Head segmentation in Crustacea – An immunocytochemical study. *Zoology* 98: 104-114.

Scholtz, G. 1995b. Expression of the *engrailed* gene reveals nine putative segment-anlagen in the embryonic pleon of the freshwater crayfish *Cherax destructor* (Crustacea, Malacostraca, Decapoda). *Biol. Bull.* 188: 157-165.

Scholtz, G. 2001. Evolution of developmental patterns in arthropods - the analysis of gene expression and its bearing on morphology and phylogenetics. *Zoology* 103: 99-111.

Scholtz, G. & Dohle, W. 1996. Cell lineage and cell fate in crustacean embryos – A comparative approach. *Int. J. Dev. Biol.* 40: 211-220.

Scholtz, G., Patel, N.H. & Dohle, W. 1994. Serially homologous *engrailed* stripes are generated via different cell lineages in the germ band of amphipod crustaceans (Malacostraca, Peracarida). *Int. J. Dev. Biol.* 38: 471-478.

Stollewerk, A., Weller, M. & Tautz, D. 2001. Neurogenesis in the spider *Cupiennius salei*. *Development* 128: 2673-2688.

Steullet, P., Cate, H.S. & Derby, C.D. 2000. A spatiotemporal wave of turnover and functional maturation of olfactory receptor neurons in the spiny lobster *Panulirus argus*. *J. Neurosci.* 20: 3282-3294.

Sullivan, J.M., Benton, J.L. & Beltz, B.S. 2000. Serotonin depletion *in vivo* inhibits the branching of olfactory projection neurons in the lobster deutocerebrum. *J. Neurosci.* 20: 7716-7721.

Sullivan, J. M. & D. L. Macmillan 2001. Embryonic and postembryonic neurogenesis in the ventral nerve cord of the freshwater crayfish *Cherax destructor*. *J. Exp. Zool.* 290: 49-60.

Tamarelle, M., Haget, A. & Ressouches, A. 1985. Segregation, division, and early patterning of lateral thoracic neuroblasts in the embryos of *Carausisus morosus* Br. (Phasmida: Lonchodidae). *Int. J. Insect Morphol. Embryol.* 14: 307-317.

Thomas, J. B., Bastiani, M. J, Bate, C.M. & Goodman, C.S. 1984. From grasshopper to *Drosophila*: a common plan for neuronal development. *Nature* 310: 203-207.

Thompson, K.J. & Siegler, M.V.S. 1991. Anatomy and physiology of spiking local and intersegmental interneurons in the median neuroblast lineage of the grasshopper. *J. Comp. Neurol.* 305: 659-675.

Thompson, K. S. J., Zeidler, M. P., & Bacon, J.P. 1994. Comparative anatomy of serotonin-like immunoreactive neurons in isopods: putative homologues in several species. *J. Comp. Neurol.* 347: 553-569.

Tomlinson, A. 1985. The cellular dynamics of pattern formation in the eye of *Drosophila*. *J. Embryol. Exp. Morphol.* 89: 313-331.

Tomlinson, A. & Ready, D.F. 1987. Neuronal differentiation in the *Drosophila* ommatidium. *Dev. Biol.* 120: 366-376.

Truman, J.W. & Ball, E.E. 1998. Patterns of embryonic neurogenesis in a primitive wingless insect, the silverfish *Ctenolepisma longicaudata*: comparison with those seen in flying insects. *Dev. Genes Evol.* 208: 357-368.

Weygoldt, P. 1985. Ontogeny of the arachnid central nervous system. In: Barth, F.G. (ed), *Neurobiology of Arachnids*: 20-37. Berlin: Springer Verlag.

Whitington, P.M. 1996. Evolution of neural development in the arthropods. *Sem. Cell Dev. Biol.* 7: 605-614.

Whitington, P.M., Harris, K-L & Leach, D. 1996. Early axonogenesis in the embryo of a primitive insect, the silverfish *Ctenolepisma longicaudata*. *Roux's Arch. Dev. Biol.* 205: 272-281.

Whitington, P. M., Leach, D. & Sandeman, R.E. 1993. Evolutionary change in neural development within the arthropods - Axonogenesis in the embryos of two crustaceans. *Development* 118: 449-461.

Whitington, P.M., Meier, T. & King, P. 1991. Segmentation, neurogenesis and formation of early axonal pathways in the centipede, *Ethmostigmus rubripes* (Brandt). *Roux's Arch. Dev. Biol.* 199: 349-363.

Winter, G. 1980. Beiträge zur Morphologie und Embryologie des vorderen Körperabschnitts (Cephalosoma) der Pantopoda Gerstaecker, 1863. I. Entstehung und Struktur des Zentralnervensystems. *Z. zool. Syst. Evolut.-Forsch.* 18: 27-61.

Wolff, T. & Ready, D.F. 1991. Cell death in normal and *rough* eye mutants of *Drosophila*. *Development* 113: 825-839.

Wolff, T. & Ready, D.F. 1993. Pattern formation in the Drosophila retina. In: Bate, M. & Martinez Arias, A. (eds), *The Development of* Drosophila melanogaster. Vol.II: 1277-1325. Cold Spring Harbor: Cold Spring Harbor Press.

The evolution and development of crustacean limbs: an analysis of limb homologies

TERRI A. WILLIAMS

Department of Ecology and Evolutionary Biology, Yale University, New Haven, USA

1 INTRODUCTION

Crustaceans exhibit the most diverse body plans of any arthropod taxon. Coupled to this diversity of body plans is an incredible diversity of limb morphologies. In thinking about the evolution of crustacean limbs, we might like to know how this enormous diversification has taken place and what particular evolutionary transformations have produced specific limb morphologies. Unfortunately, we do not presently have a well-established phylogeny of the major crustacean taxa that could help in the precise formulation of these questions. Such a phylogeny could, for example, point to likely ancestral limb characters or specific transformations to be explored.

Since clarifying limb homologies will also be a necessary part of understanding the evolutionary transformations of limbs, I will discuss some of the problems of drawing homologies between different crustacean limbs. It is interesting that homologies of virtually all parts of the limbs have been called into question, e.g., how the limb stem and its branches are homologous across taxa and whether the main axial branches (endopod and exopod) are fundamentally different from other branches (endites and exites; see Snodgrass 1935; Hessler 1982; McLaughlin 1982; Calman 1909; Walossek 1993). Difficulties in establishing homology are not limited to particular regions of the crustacean limb in the way that, for example, questions of homology in vertebrate limbs center on the elements of the "hand" (autopod). Instead, many regions in crustacean limbs have proved problematic for comparative biologists.

Recently, comparative expression data of genes that pattern the legs in the fly, *Drosophila melanogaster*, have revealed deep-seated commonalties in some aspects of leg patterning both between insects and crustaceans and between crustaceans with different limb morphologies, particularly in the early establishment of a proximal and distal patterning domain (reviewed in Nagy & Williams 2001). In crustaceans, an early distal domain, defined by gene expression in developing limbs, forms the two main branches of biramous limbs (endopod and exopod) and provides a new criterion for identifying these branches in taxa where adult morphology has been ambiguous. However, the proximal patterning domain, contiguous with the body wall, has not provided a similarly discrete and recognizable link with adult structure. Although studies of morphogenesis and comparative gene expression data cannot hope to quickly resolve the many problematic issues of homology, they do provide a large, relatively untapped pool of comparative information for analyzing homology. I will discuss some of the problems of identifying homologies among limbs and point out ways that developmental data address them. I point out that we are still limited by our tendency to analyze the evolution of limb morphology on a limb-by-limb basis. I end the chapter by speculating on whether the analysis of development may help us decompose limbs into a set of characters distinct from adult morphology – a set that could transform at least some of the ways we interpret adult morphology.

2 MORPHOLOGY AND HOMOLOGY

2.1 *Crustacean limb morphology*

Because of the great diversity of limb morphology in crustaceans, the use of adult structures to establish homologies between limbs has had limited success in comparisons of the most disparate limb types. To illustrate this, I begin by describing three distinct types of limb morphologies (Fig. 1): a malacostracan pleopod and stenopod, and a branchiopod phyllo-pod. All three limb types have a stem from which branches (rami) arise. In the stenopod, the stem is composed of two leg segments, coxa and basis, defined by joints and intrinsic musculature (Snodgrass 1935). The main inner branch of the limb (endopod) is a five-segmented continuation of the stem such that the entire walking leg has seven segments. The lateral branch (exopod) arises from the second leg segment (basis) and may be multi-articulate or flap-like. In addition to the two main branches defining the biramous limb, other branches or outgrowths may be present: lateral branches from the coxa/basis (exites) or medial outgrowths from the coxa/basis (endites). In a pleopod, the stem of the limb is not divided into segments and the entire structure is called the protopod. Two unjointed branches, more or less equal in size, arise from the distal margin of the protopod. In a phyllopod, the limb stem is generally unjointed and not distinctly delineated from the branches. The two distal-most branches are termed the endopod and exopod although these designations are problematic (see below). Typically there is a medial series of setose endites and a flap-like exite, called the epipod.

2.2 *Problems of homology between distinct limb types*

Despite the straightforward terminology used above, a number of structural homologies between these limbs are controversial. Adult structure alone has not been able to resolve the homologies. Below, I discuss the difficulties in drawing homologies based on adult limb morphology in both the proximal and distal regions of the limb. Specific questions I raise are: 1) can the endopod and exopod be identified unambiguously and are they homologous across all taxa? 2) what are the homologies amongst limb stems and the outgrowths that they bear?

2.2.1 *The biramous limb: homologies of the endopod and exopod*
Although biramous limbs are widespread in crustaceans, the actual interpretation of the bi-ramous structure can be controversial or unclear. One issue is simply the identification of the endopod and exopod in different limbs. In comparing malacostracan pleopods to a typical biramous stenopod, the position of the inner branch has been the main criterion for defining it as the endopod since it lacks segments that would make it comparable to the en-

dopod in a stenopodous limb. In branchiopod phyllopods, the identification of the endopod is even more tenuous. The limbs lack structural features to distinguish coxa and basis (thus position is not an unambiguous criterion). In addition, the endopod often appears to be a serial homologue of the other medial endites (Fig. 2). Comparison within branchiopods shows that the endopod can be more or less fused to the stem of the limb, as in anostracans, or articulating with the stem, as in notostracans. Some researchers have hypothesized that branchiopod limbs completely lack an endopod (Sanders 1963; Hessler & Newman 1975; Hessler 1982; Martin 1992) or an exopod (Packard 1883; Borradaile 1926).

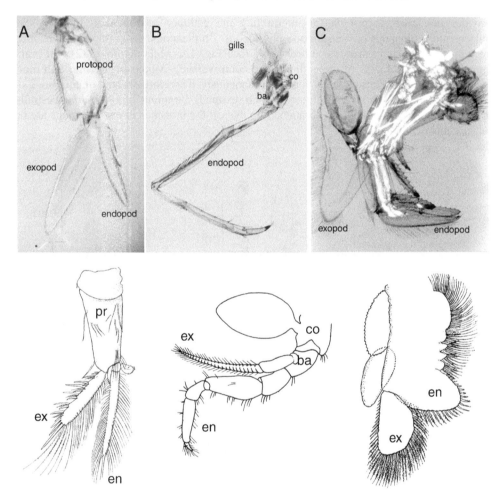

Figure 1. Three types of crustacean limbs used to illustrate some of the difficulties in drawing homologies between limbs with very disparate morphology. (A) pleopod (*Palaemonetes pugio*); (B) stenopod (*Palaemonetes pugio*); (C) phyllopod (*Triops longicaudatus*). Beneath the photographs, are standard outline sketches of the three limb types (A, C after Calman 1909; B after Kozloff 1990). Medial is to the right, lateral, to the left. Abbreviations: en = endopod; ex = exopod; co = coxa; ba = basis; pr = protopod.

Another unresolved issue in the interpretation of the biramous limb is whether the endopod represents the main axis of the limb from which other branches arise in a subsidiary manner. Snodgrass (1935) argued that crustacean legs were derived from uniramous legs and that the exopod was merely a well-developed exite (and that any of the segments of the endopod could bear exites). Similarly, Borradaile (1926), in his description of the ancestral crustacean limb asserted that the endopod is the main axis of the leg, while the exopod is an "appendage of the outer side of the limb". In overviews of limb morphology, a stenopodous limb is often depicted to illustrate the biramous limb morphology, reinforcing an idea that the endopod is the main limb axis. However, adult comparative morphology provides no structural criteria for interpreting the endopod as the main limb axis in malacostracan pleopods or branchiopod phyllopods. In pleopods, the endopod and exopod are typically similar in size and morphology (see Fig. 1A; McLaughlin 1982). In branchiopods, the homologies of the endopod are controversial. Moreover, during limb morphogenesis both branches typically form simultaneously. Therefore, instead of consisting of a main axis (endopod) with a subsidiary branch (exopod), biramous limbs could ancestrally have had a single axis with a two-branched tip even if the branches in extant groups are no longer equivalent.

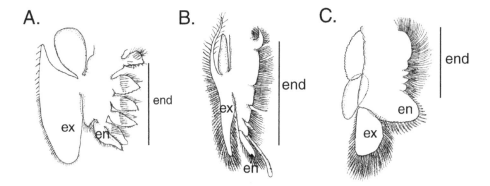

Figure 2. Comparison of limbs from three branchiopod orders. The endopods (en) and endites (end) form a structurally similar series. Note variations in the number of endites and the degree to which endites are distinct from the main axis of the limb. (A) notostracan, (B) conchostracan, (C) anostracan (after Calman 1909). Medial is to the right, lateral, to the left. Ex = exopod.

Part of the problem arises from the controversy over the ancestral state in crustacean limbs. Researchers have variously argued that the ancestral limb is a biramous paddle (Cannon & Manton 1927), a biramous stenopod (Snodgrass 1935), a triramous limb (Hessler & Newman 1975), or a polyramous limb (Fryer 1992). Each of these options gives varied status to the limb branches. This issue has been particularly controversial since two taxa (cephalocarids and branchiopods) that were long believed to be early branches in crustacean phylogeny, do not have typical biramous limbs. The cephalocarid limb has been designated triramous (Hessler & Newman 1975) and the branchiopod phyllopodous or polyramous (Calman 1909; McLaughlin 1982; Schram 1986; Martin 1992; Fryer 1992). Although the plesiomorphic state for crustacean limbs remains unresolved, the discovery from the upper Cambrian of a number of crustacean stem group taxa all with biramous trunk limbs (Walossek & Müller 1997) has made the hypothesis of an ancestral biramous limb more widely accepted. While this adds support for designating the endopod and exopod as homologs throughout crustaceans, it does not resolve whether in the ancestral state the endopod was the main limb axis.

2.2.2 *Some homologies of the proximal limb and its outgrowths remain unresolved*

All crustacean limbs have a stem from which branches or outgrowths arise. The precise delineation of the stem of the limb and the homologies between limb stems vary for different types of limbs. To illustrate the difficulties in comparing the proximal part of the limb, I return to the three limb types described above. The most straightforward comparison is apparently between the protopod of the pleopod and the coxa/basis of a stenopod since both limbs have two branches that could define the distal margin of the limb stem. It is generally assumed that the protopod is homologous to the coxa plus basis. This rests on the fact that the branches arise from the distal margin in both cases. Thus, the distal margin of the protopod must be the basis. However, the branches are often not easily compared between stenopods and pleopods. In addition, there are no studies of the internal musculature of the pleopods showing a sub-division that would suggests a remnant of two segments that subsequently fused. This contrasts, for example, with studies of the biramous swimmerets of remipedes, in which morphological evidence for a coxal and basal segment was found (Hessler & Yager 1998).

Defining the limb stem in phyllopodous limbs and relating it to the coxa/basis or protopod is even more problematic. Because of the relatively thin cuticle and scarcity of joints, the stem of the limb is only partially distinct from the endopod, exopod, and the numerous endites and exites (Figs. 1, 2). No landmarks have been described from muscle patterns that suggest that ancestrally the stem of phyllopodous limbs was divided into coxa and basis.

Because they arise from the stem of the limb, endites and exites can suffer the same ambiguities of identity as the structures that bear them. Are the endites and exites that arise from the limb stem evolutionary novelties within crustaceans and have they evolved once or numerous times? The hypothesis that a biramous limb is plesiomorphic in crustaceans does not in and of itself resolve the status of various proximal limb branches. This question is complicated by the fact that outgrowths on the medial and lateral margins of limbs are highly variable both within and among taxa. Exites are perhaps most variable and often lack unambiguous positional criteria for drawing homologies. In some groups, exites are simple flaps, e.g., branchiopods; in other groups, they are elaborated into complex structures, e.g., decapod gills. Even within the stenopodous limbs typical of malacostracans, both the number and position of exites varies (Fig. 3). Although the definition of an exite is that it arises from the coxa, this is only useful when a coxal segment can be identified. So, for example, the status of the exites of the branchiopods with regard to exites of malacostracans is unresolved. In addition, some taxa have lateral outgrowths above the limb boundary, e.g. pleurobranchs in decapods, dorsolateral lobes in *Branchinecta*, (Lynch 1958). Although some of these structures originate embryonically as outgrowths of the limb and "migrate" (Hong 1988), others remain undescribed.

2.3 *Summary of homology issues and some general questions about limb evolution*

Perhaps not surprisingly given the great diversity of limb morphologies among crustaceans, some homologies between limbs remain unresolved. Based on adult morphology, it has not been possible to unambiguously identify the endopod and exopod between stenopods and phyllopods. In addition, ambiguity remains in drawing homologies between the coxa/basis of thoracic limbs, protopods of pleopods, and the limb stem in phyllopods. In addition to these specific questions about homology between limbs, we would generally like to know certain things. What limb structures are evolutionary novelties within the crustaceans? Given the wide diversity of limb morphology throughout crustaceans, what constrains limb form, as, for example, in the constancy of stenopodal limbs on the thorax of eumala-costracans (Richter & Scholtz 2001)? None of these questions regarding the evolution of limbs in crustaceans will be easily resolved. Below I discuss how recent developmental data, specifically comparative gene expression data, bear on some of these unanswered questions.

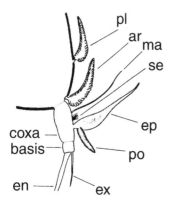

Figure 3. Composite summary sketch of variety and position of exites in decapods (after Hong 1988). En = endopod; ex = exopod; po = podobranch; ep = epipod; se = setobranch; ma = mastigobranch; ar = arthrobranch; pl = pleurobranch.

3   WHAT IS THE ROLE OF DEVELOPMENT IN EXPLAINING LIMB EVOLUTION?

It is a well known fact that developmental characters do not necessarily supercede adult characters as indicators of homology (de Beer 1971; Roth 1988). At the same time, there is a long history demonstrating that developmental characters can be useful. To a limited degree, developmental characters have been used in crustaceans to analyze issues of homology although such studies were never as numerous as vertebrate studies, for example. With the success of molecular biology in providing tools for almost any research paradigm, comparative studies of morphogenesis have been superceded by comparative studies of gene expression. In this section, I review the comparative studies of gene expression in crustaceans and point out the bearing they have in long-standing controversies of homology in crustacean limbs.

Homologies, especially those that have historically proved intractable, are never resolved by recourse to a single character and genes provide no exception. Indeed, if any-thing, we should expect less, not more, of gene expression patterns since the molecules thus deployed are much less complex than the adult structures of which they are a part. In the cases I describe where expression of genes have been informative of homologies, they typically have been so in the context of other comparative morphological data. Therefore, I argue at the end of this section that morphogenesis provides a much needed context for interpreting comparative expression data.

Interpretations of gene function in crustaceans are not based on mutant analyses but on inferences from gene action in *Drosophila* where the genes were first characterized. Only now are RNA interference techniques beginning to provide a tool for assaying function in non-model systems (Hughes & Kaufman 2000; Schoppmeier & Damen 2001). Most crustacean expression data consist of *in situ* hybridization to visualize mRNA of cloned gene products or antibody reactions to visualize proteins using antibodies that cross-react with a large number of taxa. Since the basic model for how arthropod legs are patterned during development is derived from *Drosophila*, I begin with a brief overview of *Drosophila* leg patterning. I then discuss how comparative gene expression data bear on some problematic issues of homology and end with some caveats about the limits to gene expression.

3.1 *Regulation of leg patterning in* Drosophila

Unlike crustaceans and most insects, *Drosophila* does not develop its legs as simple out-pocketings from the body wall. Instead, cells are set aside during embryogenesis in in-pocketings of the epithelium called imaginal discs. Discs are patterned and grow as the worm-like larvae develop in preparation for the emergence of the leg during metamorphosis to the adult. Certain processes of leg patterning are apparently conserved across some arthropod taxa, others are not at all (see Nagy & Williams 2001; Williams & Nagy 2001). One conserved aspect of limb patterning is the establishment of the proximal/distal axis for outgrowth (Fig. 4). In *Drosophila*, the proximal/distal (P/D) patterning of the leg is accomplished by progressive subdivision of the disc into domains of distinct gene expression that provide information to cells within the leg about their position along that axis (reviewed in Held 1995; Brook et al. 1998; Blair 1999). P/D patterning is initiated by the expression of the growth factors *wingless* (*wg*) and *decapentaplegic* (*dpp*) that, in a concentration dependent manner, cooperatively activate target genes that are expressed in discrete domains along the P/D axis of the leg. First, the disc is divided into two expression domains that are mutually exclusive: *extradenticle* (*exd*) in the region of the leg fated to be proximal and *Distal-less* (*Dll*) in the region of the leg fated to be distal (Gonzalez-Crespo & Morata 1996). As development proceeds, the leg becomes further subdivided by the expression of new genes, e.g., *dachshund (dac)*, or regions of overlap between genes. Thus, position within the leg becomes defined by the ordered expression of genes along the P/D axis.

3.2 *How comparative gene expression data bear on some problematic issues of homologies between limbs*

When the genes known to pattern the leg in *Drosophila* have been examined in crustaceans, some conservation has been found. However, it is important to note both that very few species and very few genes have been sampled at all (see Table 1) (see also Dohle et al. 2003). The most widely surveyed gene, *Dll*, is expressed in the distal region of the leg and, although less often analyzed, *exd* in the proximal region (see references, Table 1). Thus, an early step in *Drosophila* P/D patterning, the subdivision of the leg into two domains, appears widely conserved in crustaceans (Abzhanov & Kaufman 2000a; Willams et al. 2002) as well as other arthropods (Abzhanov & Kaufman 2000a). Comparisons between arthropod taxa show that the P/D patterning genes cannot be mapped in a simple way onto discrete adult structures, e.g., by corresponding directly to segment boundaries in uniramous legs (Abzhanov & Kaufman 2000a). This may also be the case within crustaceans, although comparisons are complicated by the disparate leg morphologies described above.

It does appear that within malacostracans with biramous thoracopods, *Americamysis* (Panganiban et al. 1995) and *Nebalia* (Williams 1998), *Dll* is expressed in both the endopod and exopod (although careful comparisons of the proximal boundary of *Dll* expression have not been made).

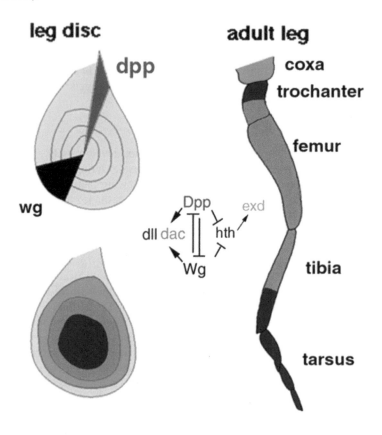

Figure 4. [See Plate 21 in color insert.] Diagram of a larval limb disc (left) and adult leg (right) of *Drosophila*. The disc is patterned from periphery to center; this translates into proximal to distal of the adult leg when the disc everts during metamorphosis. The genes involved in specifying the P/D axis are indicated by color on the bottom disc and the adult leg (*Dll*, blue; *dac*, green; *exd*, red). The growth factors *wg* and *dpp* are expressed in ventral and dorsal sectors respectively within the leg disc (top disc). Different levels of WG and DPP directly activate and maintain cell fates: cells fated to form the distal tip of the leg receive the highest levels of DPP and WG and activate the expression of the transcription factor *Dll*. Slightly more proximal cells express the transcription factor *dachshund* (*dac*), which is repressed in more distal cells by the higher levels of DPP and WG (Lecuit & Cohen 1997). While the coxal leg segment is outside the domain affected by WG and DPP, it is nonetheless patterned indirectly by these two growth factors. High levels of WG and DPP activate *Dll* and *dac*, and repress the otherwise ubiquitous expression of *homothorax (hth)*, in the distal domain of the leg. *hth* is necessary for the function of the transcription factor *exd*. Restricting the expression of *hth* to the proximal domain allows for the translocation of EXD to the nucleus, where it directs proximal limb development (Mann & Abu-Shaar 1996; Aspland & White 1997; Rieckhof et al. 1997; Wu & Cohen 1999).

Table 1. Trunk leg patterning genes described in crustaceans.

	Dll	*exd*	*dac*	*wg*	*pdm*
Triops longicaudatus	Williams 1998	Williams et al. 2002		Nulsen & Nagy 1999	
Artemia franciscana	Panganiban et al. 1995; Gonzalez-Crespo & Morata 1996	Gonzalez-Crespo & Morata 1996			Averof & Cohen 1997
Thamnocephalus platyurus	Williams, et al. 2002	Williams et al. 2002			
Leptodora kindtii	Olesen et al. 2001				
Cyclestheria hislopi	Olesen et al. 2001				
Nebalia pugettiensis	Williams 1998				
Americamysis bahia	Panganiban et al. 1995				
Porcellio scaber	Abzhanov & Kaufman 2000a	Abzhanov & Kaufman 2000a	Abzhanov & Kaufman 2000a		
crayfish (species not identified)	Browne & Patel 2000				
Pacifastacus leniusculus					Averof & Cohen 1997

Whether the upstream regulators of P/D patterning, *wg* and *dpp*, or the subsequent P/D patterning events, e.g., *dac* expression, are conserved remains basically unknown. *wg*, *dpp*, and *dac* have been more extensively sampled in insects and, based on those data, it appears that *wg* may have a conserved role in other arthropod taxa (Nagy & Carroll 1994; Kraft & Jäckle 1994) but that *dpp* does not (Sanchez-Salazar et al.1996; Jockusch et al. 2000). The gene *dac* is more ambiguous showing a conserved pattern in some cases but not in others (Abzhanov & Kaufman 2000a; Prpic et al. 2001). In spite of the relatively small sample size, comparisons of gene expression have, in some cases, been useful in extending our analyses of homology or limb evolution (Schram & Koenemann 2001). Below I describe three cases that use comparative gene expression to interpret limb evolution: first, the identification of the endopod and exopod in the phyllopodous limbs of branchiopods; second, the gnathobasic nature of the mandible; and, third, the evolution of insect wings from crustacean exites.

3.2.1 *Identification of the endopod and exopod in the phyllopodous limbs of branchiopods*

Despite controversy, most evidence supports the hypothesis that a biramous limb is ancestral for crustaceans. Included in this evidence are patterns of limb morphogenesis in all taxa except branchiopods: the early limb buds are bifurcated and the two branches form the endopod and exopod (Schram 1986). This is true even in cases where adult limb morphology is highly modified from a standard biramous leg, e.g., cephalocarids (Sanders 1963) and leptostracan thoracic limbs (Manton 1934; Williams 1998; Olesen & Walossek 2000).

The one problematic case based on both morphology and morphogenesis has been the branchiopods. As mentioned above, morphological criteria for assigning branch homologies are lacking in adult branchiopod limbs prompting speculation that the endopod could simply be missing in these limbs. Furthermore, studies of morphogenesis have shown that phyllopodous limbs in a basal malacostracan taxon (*Leptostraca*) but not those in branchiopods clearly initiate development as biramous limb buds (Manton 1934; Williams 1998; Williams 1999). In branchiopods, all the endites and exites of the adult limbs arise virtually simultaneously when the initial, unbranched limb bud forms its branches (Williams & Müller 1996; Williams 1998; Williams 1999). However, recent comparative gene expression data from species representing three orders of branchiopods double-labeled for DLL and EXD show early, separate EXD and DLL domains like those seen in *Drosophila* (Williams et al. 2002, unpublished results). The early domain of *Dll* expression extends across the endopod and exopod (Fig. 5, Williams, et al. 2002; Olesen et al. 2001). This corresponds to what has been found in biramous limbs. Double-labels of EXD/DLL show a domain of exclusive *Dll* expression corresponding to the endopod and exopod of the mysid (Panganiban et al. 1995). When, as is most common, only *Dll* expression was examined, the DLL domain includes both the endopod and exopod. These data suggest that the endopod and exopod together form a distal patterning domain homologous to the distal patterning domain of *Drosophila* and other arthropod appendages. The coincidence of this domain with only two branches in branchiopod limbs constitutes a criterion for designating those branches endopod and exopod since the endopod and exopod of biramous crustacean limbs also express *Dll*.

Nonetheless, the result is not wholly unambiguous since it has been argued that the endopod extends much further proximally and corresponds to the whole series of medially reiterated endites (Borradaile 1917, 1926). However, the strength of the gene expression data lies in the double labeling of EXD and DLL. The mutually exclusive expression domains suggest both that a part of a regulatory circuit and its function are conserved. By contrast, the initial analysis of *Dll* expression alone in branchiopod limbs was incon-clusive on this point since *Dll* is expressed in branches other than the endopod and exopod (Williams 1998; Olesen et al. 2001). Therefore, the exclusively *Dll* -expressing distal domain does appear to offer compelling support for the two branches being the endopod and exopod even if their proximal extent remains unresolved. To date, all of the expression data are consistent with the hypothesis that a biramous limb is plesiomorphic for crus-taceans (Table 1) although *exd* and *Dll* co-expression are known for few species.

The same kind of data could help distinguish whether the endopod is the single main axis of the leg or whether ancestrally the main axis had a bifurcated tip, i.e., representing the endopod and exopod. If the endopod and exopod are part of a primary distal region of the limb, the prediction is that the distal DLL domain exclusive of EXD would include the endopod/exopod in all limbs. Only a systematic sampling of the distal DLL domain throughout crustacean taxa can address this issue.

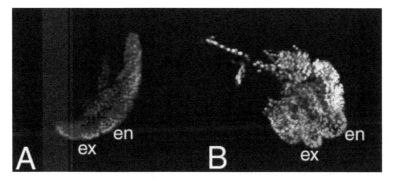

Figure 5. [See Plate 22 in color insert.] The expression of *Dll* and *exd* in the early limb buds of two branchiopods, (A) *Triops longicaudatus*, and (B) *Thamnocephalus platyurus*. Red nuclei express *Dll*; green nuclei express *exd*; yellow nuclei express both. The early expression of *Dll* exclusive of *exd* extends across the endopod/exopod region of the limb bud in both limbs. (Compare A to Fig. 9C for schematic of developing limb branches in *Triops*.) En = endopod; ex = exopod.

3.2.2 *The gnathobasic nature of the mandible*
Given the widespread conservation of *Dll* expression in distal limb patterning in all trunk limbs surveyed, its absence in mandibles and mandibles only is striking and has been used to support the hypothesis that mandibles evolved by the loss of distal limb structures. Crustacean mandibles have long been considered gnathobasic, i.e. composed of only the stem of the appendage (Calman 1909; Snodgrass 1938). However, Manton (1964, 1973, 1977) argued that insects mandibles were derived from a whole limb, i.e., stem plus distal branch. Because their mandibles had different origins, insects and crustaceans (or other groups with gnathobasic mandibles) could not form a monophyletic group.

Because *Dll* is expressed in the distal portion of all limbs, it has been argued that its expression represents the distal portion, telopod, of an ancestral division of the limbs into a proximal and distal part (Gonzalez-Crespo & Morata 1996, Williams et al. 2002). Al-though crustacean adult mandibles are elaborations of the base of the limb, they often have a mandibular palp that represents a reduced limb branch. Within crustaceans mandibles, *Dll* expression is limited to an early transient expression or to the mandibular palp (Popadic et al. 1996; Scholtz et al. 1998; Popadic et al. 1998; Browne & Patel 2000). In all cases, *Dll* is expressed in the earliest *Anlage* of the limb. However, this expression is rapidly lost unless adults have a mandibular palp. If a palp is present, *Dll* continues to be expressed in the palp (Scholtz et al. 1998; Popadic et al. 1998, Browne & Patel 2000). These results differ slightly from insects, which never express *Dll* in the mandible, but nonetheless suggest that the mandibles in both insects and crustaceans are distinct in their patterning from other limbs. Unless a palp is present, both insects and crustaceans lack Dll expression in the adult mandibles suggesting that the distal portion of the limb is lost and arguing against Manton's (1964, 1973, 1977) notion of distinct derivations of insect and crustacean mandibles.

3.2.3 *Crustacean exites and the evolution of wings*

One theory of the evolution of insect wings is that wings arose from leg exites in ancestral species (Wigglesworth 1973; Kukalova-Peck 1983, 1992). Because crustacean limbs have exites, Averof and Cohen (1997) cloned two genes from the branchiopod *Artemia franscicana* known to pattern the wings of *Drosophila*, *pdm* and *apterous*. The gene *pdm* specifies the wing primordium (Ng et al. 1995, 1996) and *apterous* patterns the dorsal side of the wing (Cohen et al. 1992; Carroll et al. 1994). An antibody to these genes revealed protein in the distal exite of *Artemia* and the exites (gills) of the crayfish, *Pacifastacus leniusculus*. Averof and Cohen (1997) interpret these data to support the theory that wings evolved from exites. However, it should be noted that the term "exite" (or the term for respiratory exites "epipod") does not designate a unique homologous structure in crustacean limbs. Any and all lateral branches can be called exites. Thus, the *pdm* expression in the two crustacean species is not in structures already known to be homologous. This high-lights the importance of comparative morphology in establishing the context within which gene expression is interpreted. For example, if the exites are evolutionarily novel structures within the crustaceans then their expression of wing genes would result from convergence based on some other factors, like relative position in the body wall. To help resolve this claim it would be useful to sample more crustacean taxa and see if there are taxa with exites that do not express wing genes or taxa without exites that do. Such data could be used to strengthen or refute the claim that ancestrally (certain) exites expressed *pdm* in crustaceans. It is probably better to view Averof and Cohen's (1997) claim as a very general one, that some lateral region/structure in the ancestor of insects and crustaceans expressed *pdm* and *ap* and became wings in insects and some (but not all) exites in crustaceans.

3.2.4 *Interpretation of gene expression patterns requires a strong morphological context*

Ideally gene expression would be interpreted within the context of limb morphology and morphogenesis since ultimately we would like comparative gene expression to inform us about the evolution of morphology. However, no surveys or synthetic theories of early limb morphogenesis exist for crustaceans. Indeed, it is striking to note how little data exist on comparative limb morphogenesis. Although a number of classical descriptive embryolo-gical studies exist for crustaceans (Kume & Dan 1968; Anderson 1973), few of these focused on limbs in any detail. Similarly, many systematic descriptions of larval stages exist but the morphogenesis driving changes in limb structures at successive developmental stages can not be inferred from outline drawings of limbs intended for systematic purposes. This lack of basic descriptive limb morphogenesis has created a weakness in comparative studies of developmental gene expression. It can be easier to discover RNA or protein expression in a limb than to describe the morphogenesis of that limb. Yet, gene expression can be difficult to interpret in the absence of an understanding of the pattern of limb morphogenesis both in how it produces a particular adult morphology and how it relates to other patterns of morphogenesis.

Therefore, studies of morphogenesis have an important role in aiding our interpretation of patterns of gene expression. For example, the transformation of the first thoracic appendage in the isopod, *Porcellio*, from a stenopod to a flap-like form is hypothesized to occur through the loss of intervening patterning information between the most proximal and most distal regions (Abzhanov & Kaufman 2000a). During early development the first thoracic limb appears rod-shaped and expresses *exd*, *dac*, and *Dll*. However, this leg begins to transform in shape midway through embryonic development by losing its rod like shape and becoming more flap-like. Concurrently, *dac* expression is lost from the limb, suggesting a loss of intermediate patterning values and a truncation of the limb (Abzhanov & Kaufman 2000a). This hypothesis is interesting, but it is difficult to infer growth dynamics from the external morphology of two stages. For example, do patterns of differential growth in this leg start the same as other thoracic limbs then diverge in early development *via* an arrest of proliferation in intermediate cells? Hypothesizing that the loss of intermediate patterning values accounts for the transformation of limb form begs the question of how the specific morphology of these thoracic limbs develop since, with their bulbous, flap-like shape, they are structurally quite distinct from simply a stenopod that has experienced truncated growth. Furthermore, is there even an obvious and general morphogenetic consequence that can be attributed to limb primordia that lose intermediate patterning information? Thus, knowledge of limb morphology and morphogenesis is an invaluable aid in tying our hypotheses based on gene expression to the larger questions of the evolutionary transformations of morphology.

3.2.5 *Interpretation of gene expression patterns confounded by regulatory genes having multiple functions*

It is expedient to focus on the most obvious function for a regulatory gene. However, all regulatory genes serve multiple functions. One example is a potential dual role for *Dll* in crustacean limbs (Mittman & Scholtz 2001; Williams et al. 2002).

The focus of comparative analyses of *Dll* function in arthropods has been its important role in establishing the P/D axis of appendages. However, *Dll*, like many other regulatory genes, performs different roles at different times and places within the developing organism. For example, *Dll* expression is not restricted to developing limbs; it can be detected in both the central and peripheral nervous system (Cohen et al. 1989). In the peripheral nervous system, genetic studies in *Drosophila* suggest a role for *Dll* in bristle development. *Dll⁻* clones in the femur disrupt bristle formation, and *Dll* regulates expression of the pro-neural gene *achaete* in a subset of *Dll*-expressing cells in the wing (Campbell & Tomlinson 1998). *Drosophila* that have null mutations of the *Dll* gene die as embryos and never begin to form certain larval sensory structures (Panganiban 2000). Thus, in *Drosophila*, *Dll* is known to have functions beyond its role in specifying distal limb structures, specifically in the peripheral nervous system. This is interesting given that Mittman & Scholtz (2001) found a correlation between *Dll* expression and mechano/chemoreceptors in legs of two arthropod species: one chelicerate and one insect. They found that patterns of *Dll* expression in the mouthparts, labrum, and trunk limbs correlate with mechano- and chemoreceptors of the peripheral nervous system.

The same correlation was found in two crustacean species, *Triops longicaudatus* and *Thamnocephalus platyurus* (Williams et al. 2002). The endites of both species, which bear numerous setae, strongly express *Dll* and the location of DLL staining is directly correlated with the position of the setae. This is particularly obvious in older *T. platyurus* limbs in which the DLL resolves to the setal bearing margin (Fig. 6). This role of *Dll* for setal outgrowth would also explain *Dll* expression in the malacostracan crustacean *Nebalia*, which shows strong *DLL* expression along the proximal/medial margin of the limb (Williams 1998) as well as in the endopod and exopod. This margin will eventually bear an extensive setal comb for filtering food.

These data suggest that *Dll* may have evolved a dual role of both promoting peripheral sensory structures and appendage outgrowth. Ancestrally, *Dll* may have been functional in neural development, including peripheral sensory structures (Panganiban *et al.* 1997). If there was selection for those structures to extend away from the body wall, *Dll* could have been co-opted for P/D outgrowth since it was already associated with sensory structures (Mittman & Scholtz 2001). This correlation between *Dll* and setal formation, if proved to be functional, demonstrates the need to not assume too quickly that gene expression can be ascribed to a single function.

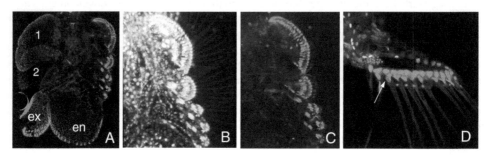

Figure 6. [See Plate 23 in color insert.] Correlation between *Dll* expression and setae within limbs of the branchiopod, *Thamnocephalus platyurus*. (A) *Dll* expression in a juvenile thoracopod. Note how nuclei expressing *Dll* (red) are positioned along the margin of the limb. (B, C) Higher magnification view of medial leg margin. All cells revealed by a general nuclear stain (B, Hoechst), and *Dll*-expressing cells restricted to the marginal edge of the limb that bears setal filtering arrays (C). (D) Larval antenna showing *Dll*-expressing nuclei extending into setae (from Williams et al. 2002).

In addition, this model of a dual role for *Dll* function in arthropod limbs can simplify and unify the *Dll* expression data from different limb morphologies. Three different patterns of *Dll* expression are found in crustacean limbs (Schram & Koenemann 2001): *Dll* expressed at the tip of uniramous or biramous limbs (mysid, Panganiban et al. 1995), *Dll* at the tip of a biramous limb followed by extensive *Dll* expression along the medial margin (leptostracans, Williams 1998), and *Dll* in all or almost all of the branches of the branchiopod limb (Panganiban et al. 1995; Williams 1998; Olesen et al. 2001; Williams et al. 2002). In spite of this diversity, it seems likely that all three are variants of a single pattern that begins with a distal *Dll*-expression domain that is exclusive of *exd*. In both types of phyllopodous limbs, i.e., leptostracan and branchiopod, this is followed by additional *Dll* expression in heavily setose regions of the limb. Thus, one testable prediction of this interpretation is that the secondary *Dll* expression in leptostracans would be co-expressed with *exd*.

3.3 *Summary of role of development*

To date the role of development in interpreting limb evolution in crustaceans has largely been limited to analyzing comparative patterns of gene expression. Perhaps the surprising fact, given how few genes or species have been examined, is that already some of the data seems promising and provides insights not available solely from comparative adult morphology. This in no way undermines the role of comparative morphology in analyzing questions of homology. Indeed, comparative gene expression patterns are most useful when they corroborate or extend a body of comparative morphological data.

4 ONE LIMITATION OF CURRENT STUDIES OF LIMB EVOLUTION IS THAT LIMBS ARE NOT EVALUATED WITHIN THE CONTEXT OF THE WHOLE BODY

All of the preceding analysis is about disembodied limbs, that is, isolated limbs typical of certain groups of tagma. However, understanding the evolution and development of crustacean limbs will ultimately require understanding limbs as serial units along the body axis. This has at least two components (Van Valen 1993): 1) understanding the serial homology among limbs in any particular species, and 2) understanding how different patterns of axial differentiation arise. These questions are important because limbs evolve in series not in isolation. Anyone who has ever watched a crab cracking a gastropod with its highly specialized chelae knows that, although chelae are paradigms of functional specialization for shell cracking, the *in situ* behavior involves other legs for stabilization and support.

4.1 *Understanding serial homologies*

Serially homologous limbs within a single taxon can be quite disparate and the homologies between parts of serial units are not always explicitly resolved. For example, as pointed out above, the homology between the coxa/basis of thoracic stenopods and the protopod of pleopods within eumalacostracans remains unclear. Although assumed to be strictly homologous, these limb stems could have had independent evolutionary origins. This would depend on when the limb stems and limb branches came under independent developmental control allowing them to be independently individualized (see Wagner 1989a). If, for example, the stem and branches of the ancestral biramous limbs did not vary independently prior to tagmatization then both the coxa/basis and the protopod would represent independent evolutionary transformations and, as such, not be strictly homologous to one another (Fig. 7). That is, although the limbs as a whole would be homologous with one another their parts would not have direct homology (for a discussion of degrees of homology, see Roth 1984). This might be the case if the posterior segments came under separate genetic control, became limbless and re-evolved limbs later, or if the limbs on the two tagma differed in size only, i.e., tagmatization did not involve separate specification of stem and branches. If, however, the limb stem and branches did vary independently prior to tagmatization, then the limb stems of the stenopod and protopod would be homologous and we might hope to find some indication in either adult morphology or development. That is, the limb stems could evolve as separate units if they were already under separate genetic control from the branches prior to tagmatization. Resolving this issue will rest on both the phylogeny and the structural organization of limbs at the time of tagmatization.

A   B

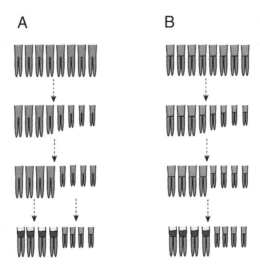

Figure 7. [See Plate 24 in color insert.] The timing of tagmosis with respect to the evolution of the limb stem would determine whether the limb stem of the pleopods and stenopods are strictly homologous. In (A), tagmatization precedes the independent evolution (represented by changes in color) of the limb stems. In this scenario, limbs on different body tagmata would evolve independently and thus the subdivision into limb stem and limb branches would not be strictly equivalent in the two tagmata. Therefore, the limb stems would not be homologous to one another although the limbs as a whole would be homologs. In (B), the limb stems and branches are evolutionarily independent units prior to tagmatization. In spite of later evolutionary modifications, the stems are homologs.

    Certainly developmental data will prove invaluable in our analysis of the evolution of serial homologs. I know of no cases where either patterns of morphogenesis or patterns of gene expression have been used explicitly to compare serially homologous crustacean limbs. However, even the preliminary observations are tantalizing. In one case where I have examined morphogenesis in a taxon with discrete thorax and abdomen, differences in the formation of the proximal part of the limb can arise very early in development (Williams pers. obs.). In the leptostracan, *Nebalia,* the limb buds form a similar bifurcated tip on both thoracic and abdominal segments when they first emerge from the body wall (Manton 1934; Williams 1998; Olesen & Walossek 2000). However, as they grow away from the body wall, thoracic and abdominal limb buds diverge in morphogenesis by forming their proximal part from distinct regions of the body wall (Fig. 8). This distinction in early morphogenesis corresponds exactly to distinct limb morphologies found on thoracic and abdominal segments. In addition, it would be hard to reconcile these differences in the position of the limb field on the body wall in thoracic and abdominal segments with a serially uniform action of developmental genes positioning the limb primordia in each segment (for a review of patterning of primordia, see Nagy & Williams 2001).

    Interestingly, in branchiopods that lack a distinct thorax and abdomen, serially homologous limbs with distinct adult morphologies clearly share an early morphogenetic pathway.

For example, the first thoracic limbs in notostracans are very distinct from more posterior limbs (Fig. 9A, B). Modifications and reduction of branches occur relatively late in development; all limbs pass through the same early development with all branches present and easily compared (Fig. 9C). The degree to which morphogenetic differences like those found in *Nebalia* are simply a modification to early development *versus* representative of systematic differences in the thoracic and abdominal appendages must be resolved through systematic comparisons of limb morphogenesis. These kinds of comparisons coupled with comparisons of expression data could be very revealing both as to the nature of the serial homologs and to the ubiquity of limb patterning mechanisms.

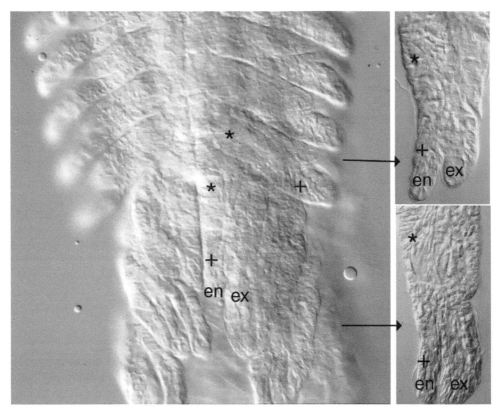

Figure 8. Leg development in the thorax and abdomen of the malacostracan, *Nebalia pugettiensis*. Although early limb buds are similar along the body wall, limb development diverges subsequently. Thoracic limb buds grow away from the body wall in a lateral direction and proliferate cells on the ventral face that will become part of the limb proper. By contrast, abdominal limb buds grow away from the body ventrally (i.e., in an axis perpendicular to the thoracic limb buds; * and + indicate similar positions on the limbs). Because of their position at the posterior part of the body, the ventral body wall near the midline between limb buds is proportionally smaller and does not appear to contribute to the limb bud. Instead the lateral body wall must proliferate disproportionately to cause growth. In as much as in the current view crustacean limbs have a single patterning axis homologous to the one that patterns the fly leg, there is no known model for patterning that could produce medial versus lateral growth; en = endopod; ex = exopod.

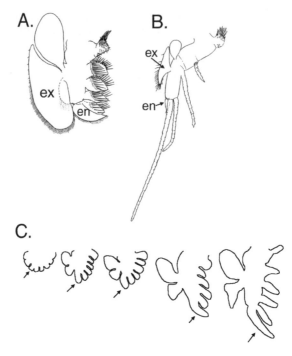

Figure 9. Tenth (A) and first (B) thoracic limb in the notostracan, *Lepidurus apus* (after Fryer 1988). The distinct limb morphologies include a highly reduced endopod on the first thoracic limb. The early morphogenesis in both limbs is nearly identical; divergence in form occurs late in development (C). Outline of a series of limb buds illustrating the early morphogenesis of trunk limbs in notostracans. Limb buds range from approximately 100-300 μm in length. Arrows indicate endopod.

4.2 *Evolution of different patterns of axial differentiation*

Finally, the overall patterns of axial differentiation of limbs can be quite different between taxa. Some taxa have serial homologues with highly discontinuous limb morphologies (termed heteronomy; e.g., decapods) whereas other taxa have serial homologues with only graded differences along the body axis (termed homonomy; e.g., many branchiopods; see discussion in Nagy & Williams 2001). Heteronomous patterns are exemplified by crayfish, which have phyllopodous maxillae, maxillipeds, chelate first walking limbs, stenopodous walking limbs, and pleopods. By contrast, cases of graded changes in limb morphology in homonomous taxa include 1) loss of endopod in cephalocarids (Sanders 1963; Fig. 10) and 2) joints in anterior limbs not present in more posterior limbs (in notostracans, Fryer 1988). Whether these differences in patterns of axial differentiation are qualitative or quantitative remains unexplored for either development or evolution. There is some evidence within heteronomous taxa that changes in the boundaries of *Hox* gene expression are correlated with changes in limb morphology (Averof & Patel 1997; Abzhanov & Kaufman 2000b, 2003). However, no theories of Hox deployment can currently account for the differences in limb morphology in homonomous taxa. The developmental basis for these two patterns may be quite distinct and we still do not understand the evolution of heteronomy from homonomy or *vice versa*.

While analyzing limbs within the natural series of which they are a part adds another layer of complexity to our task of understanding limb evolution, it is a complexity that must be addressed. Fortunately, such studies should have the benefit as well of expanding our knowledge of limb development in any particular limb.

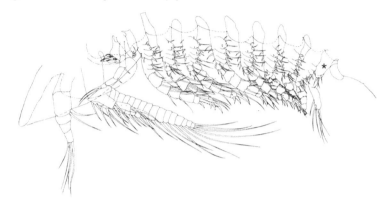

Figure 10. Sagittal view of the cephalocarid *Hutchinsoniella* (after Sanders 1963) illustrating the gradual reduction in medial legs elements and the loss of the endopod (asterisk) in the last limb.

5 CAN THE ANALYSIS OF DEVELOPMENT LEAD US TO NOVEL PERSPECTIVES ON LIMB HOMOLOGY?

As discussed above, crustaceans are both the most diverse arthropod taxa in terms of their body plans and limb morphologies and the least well characterized with respect to their phylogenetic relationships. In addition, comparative adult morphology has not yet solved a number of outstanding issues of homology between adult limbs. Given these facts, perhaps studies of limb development can offer novel insights into limb morphology by providing developmentally based units within limbs for comparison (Wagner 1989b).

To illustrate some outstanding issues of homology among limbs, I initially contrasted three types of limbs: the malacostracan pleopod, thoracic stenopod, and the branchiopod phyllopod. However, this selection is somewhat unorthodox. Typically, the stenopod and phyllopod are contrasted as the two distinct kinds of crustacean limbs whereas the pleopod and stenopod, since they are found in the same species at different axial positions, are assumed to be clearly identified serial homologs. I used the comparisons to point out that the stems of stenopods and pleopods, the coxa/basis and protopod, have never been compared explicitly. Indeed, if we look at the three limbs from a developmental perspective, it is easier to suppose that the phyllopodous and stenopodous limb are transformations of one another and that the pleopod is the most distinct of the three. Figure 11 shows the larval development of maxillae and first maxillipeds from a decapod. Disembodied maxillae look very much like branchiopod phyllopods (Fig. 2): they have a broad stem from which medial lobes (endites) arise and, in at least three stages, the endopod looks very much like a serial homolog of the coxal and basal endites. The maxilliped (the first thoracic limb) has a regular biramous stenopodous form early in larval life. However, this form becomes highly modified during larval development: the endopod is reduced and endites emerge from the coxal and basal segments. The point is not to use this kind of information to draw one-to-one homologies between the endites of phyllopods and the endites of stenopods in malacostracans (an activity which in the past has led to varying attempts to match phyllopod endite number to endopod segment number, Borradaile 1917). Instead, I am suggesting that both types of limbs exhibit a shared developmental potential for creating medially reiterated lobes. By contrast, the medial margin of pleopods never forms endites. Given the great plasticity of limbs in the face of functional specialization, this may represent a true constraint and reflect an evolutionary difference in the limb stem in the thoracic and abdominal limbs in malacostracans.

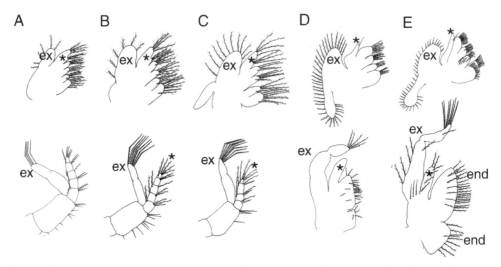

Figure 11. Development of maxillae and maxillipeds in *Pagurus kennerlyi* (after, McLaughlin et al. 1989). Top row = maxillae; second row = maxillipeds. (A) zoea 1; (B) zoea 2; (C) zoea 4; (D) megalopa; (E) first crab. Maxillae have well-developed medial endites. Note the striking transformation of the coxa/basis region of the limb to form large lobed endites in the maxillipeds. * = endopod; ex = exopod; end = endite.

Although endites in stenopods and phyllopods may have no simple one-to-one homology in the adult limbs, they may share homology on the level of development. It is this level of homology that I think can be elucidated by careful studies of limb morphogenesis and then pursued by experimental analysis.

It is interesting in this context that a cladoceran branchiopod, *Leptodora,* which has secondarily evolved a stenopodous limb has a common early morphogenetic pathway with phyllopodous branchiopod (*Cyclestheria hisploi*) limbs (Olesen et al. 2001). Olesen et al. (2001) found that leg segments and endites arise from the same early *Anlagen*. Such morphogenetic data can be coupled with expression data. For example, preliminary data in phyllopodous limbs of the branchiopod *Triops longicaudatus* show that two patterning genes *dac* and *dpp* are expressed in reiterated stripes in the medial endites (Nagy & Williams *pers. obs.*). Looking at the expression of the same genes in malacostracan endites would be interesting indeed.

While this particular reading of limb homologies may well be incorrect, the salient point is that limb development remains largely untapped as a resource for understanding the overall morphology of the limbs. It remains unknown whether endites in malacostracan and branchiopods have a shared morphogenetic and genetic basis. Looking for such commonalties would be analogous to finding the morphogenetic commonalties of segmentation, bifurcation and elongation that are found in tetrapod limbs. This type of morphogenetic data, culled from diverse taxa, provided the basis for insightful and provocative models for the evolutionary transformations seen in vertebrate limbs, even in the absence of precise knowledge of the regulatory molecules controlling particular aspects of morphogenesis (Shubin & Alberch 1986; Oster et al. 1988). While crustacean limbs do not have an equivalent developmental process to cartilage condensation, the careful analysis of temporal patterns of branch formation and mechanisms regulating morphogenesis, such as cell death, cell shape change, and patterns of proliferation, could form a similar database for model building. For example, analyzing comparative patterns of development in copepod limbs, Ferrari and Ivanenko (2001) reject the idea of limb segments as natural units of evolution since setae and arthrodial membranes that define limb segments show independent developmental variation. This search for underlying developmental homologues could provide a new source of homologues for widely disparate limb types, one that could ultimately lead to acceptance of different degrees of homology between distantly related or radically transformed limbs.

ACKNOWLEDGEMENTS

I thank G. Scholtz for the invitation to contribute to this book. This chapter has been improved by discussions with F. Ferrari, R. Hessler, J. Martin, P. McLaughlin, L. Nagy, F. Schram, and G. Wagner. In addition, P. McLaughlin kindly answered numerous detailed questions about comparative limb morphology. K. Dunlap, E. Jockusch, S. Koenemann, H. Larson, L. Nagy, and G. Stopper clarified my thinking by criticizing versions of the manuscript.

REFERENCES

Abzhanov, A. & Kaufman, T. C. 2000a. Homologs of *Drosophila* appendage genes in the patterning of arthropod limbs. *Dev. Biol.* 227: 673-689.

Abzhanov, A. & Kaufman, T. C. 2000b. Crustacean (malacostracan) Hox genes and the evolution of the arthropod trunk. *Development* 127: 2239-2249.

Anderson, D. T. 1973. *Embryology and Phylogeny in Annelids and Arthropods*. Oxford: Pergamon Press.

Aspland, S. E. & White, R. A. 1997. Nucleocytoplasmic localisation of extradenticle protein is spatially regulated throughout development in *Drosophila*. *Development* 124: 741-747.

Averof, M. & Cohen, S. 1997. Evolutionary origin of insect wings from ancestral gills. *Nature* 385: 627-630.

Averof, M. & Patel, N. H. 1997. Crustacean appendage evolution associated with changes in Hox gene expression. *Nature* 388: 682-686.

Blair, S. S. 1999. *Drosophila* imaginal disc development: patterning the adult fly. In: Russo, V.E.A., Cove, D., Edgar L., Jaenisch, R.& Salamini, F. (eds), *Development-Genetics, Epigenetics and Environmental Regulation*. Heidelberg: Springer.

Borradaile, L. A. 1917. On the structure and function of the mouthparts of the palaemonid prawns. *Proc. Zool. Soc. Lond.*: 37-71.

Borradaile, L. A. 1926. Notes upon crustacean limbs. *Ann. Mag. Nat. Hist.* 17: 193-213.

Brook, W. J., Diaz-Benjumea, F. J. & Cohen, S. M. 1998. Organizing spatial pattern in limb development. *Ann. Rev. Cell. Dev. Biol.* 12: 161-80.

Browne, W.E. & Patel, N. H. 2000. Molecular genetics of crustacean feeding appendage development and diversification. *Sem. Cell Dev. Biol.* 11: 427-435.

Calman, W. T. 1909. *Crustacea*. London: Adam and Charles Black.

Campbell, G. & Tomlinson, A. 1998. The roles of the homeobox genes *aristaless* and *Distal-less* in patterning the legs and wings of *Drosophila*. *Development* 125: 4483-4493.

Cannon, H. G. & Manton, S. M. 1927. On the mechanism of a mysid crustacean, *Hemimysis lamornae*. *Trans. R. Soc. Edinburgh* 55: 219-253.

Carroll, S.B., Gates, J., Keys, D.N., Paddock, S.W., Panganiban, G., Selegue, J.E. & Williams, J.A. 1994. Pattern formation and eyespot determination in the butterfly wing. *Science* 265: 109-114.

Cohen, S. M., Bronner, G., Kuttner, F., Jürgens, G. & Jäckle, H. 1989. *Distal-less* encodes a homeodomain protein required for limb development in *Drosophila*. *Nature* 338: 432-434.

Cohen, B., McGuffin, M.E., Pfeifle, C., Segal, D. & Cohen, S. M. 1992. *apterous*, a gene required for imaginal disc development in *Drosophila*, encodes a member of the LIM family of developmental regulatory proteins *Genes Dev.* 6: 715-729.

de Beer, G. 1971. Homology, an Unsolved Problem. In: Head, J.J. & Lowenstein, O.E. (eds), *Oxford Biological Readers*. London: Oxford University Press.

Dohle, W., Gerberding, M., Hejnol, A. & Scholtz, G. 2003. Cell lineage, segment differentiation, and gene expression in crustaceans. In: Scholtz, G. (ed), *Crustacean Issues 15, Evolutionary Developmental Biology of Crustacea*: 95-133. Lisse: Balkema.

Ferrari, F. D. & Ivanenko, V. N. 2001. Interpreting segment homologies of the maxilliped of cyclopoid copepods by comparing stage-specific changes during development. *Org. Divers. Evol.* 1: 113-131.

Fryer, G. 1988. Studies on the functional morphology and biology of the Notostraca (Crustacea: Branchiopoda). *Phil. Trans. R. Soc. Lond. B* 321: 27-124.

Fryer, G. 1992. The Origin of Crustacea. *Acta Zool.* 73: 273-286.

González-Crespo, S. & Morata, G. 1996. Genetic evidence for the subdivision of the arthropod limb into coxopodite and telopodite. *Development* 122: 3921-3928.

Held, L. I. 1995. Axes, boundaries and coordinates: the ABCs of fly leg development. *BioEssays* 17: 721-732.

Hessler, R. H. 1982. Evolution within the Crustacea. In: Abele, L. (ed), *Biology of the Crustacea, Vol. 1, Systematics, the fossil record and biogeography*: 149-185. New York: Academic Press.

Hessler, R. R. & Newman, W. A. 1975. A trilobitomorph origin for the Crustacea. *Foss. Strata* 4: 437-459.

Hessler, R. R. & Yager, J. 1998. Skeletomusculature of trunk segments and their limbs in *Speleonectes tulumensis* (Remipedia). *J. Crust. Biol.* 18: 111-119.

Hong, S. Y. 1988. Development of epipods and gills in some pagurids and brachyurans. *J. Nat. Hist.* 22: 1005-1040.

Hughes, C.L. & T.C. Kaufman 2000. RNAi analysis of *Deformed, proboscipedia* and *Sex combs reduced* in the milkweed bug *Oncopeltus fasciatus*: novel roles for Hox genes in the hemipteran head. *Development* 127: 3683-3694.

Jockusch, E. L., Nulsen, C., Newfeld, S.J. & Nagy, L. M. 2000. Leg development in flies versus grasshoppers: differences in *dpp* expression do not lead to differences in the expression of downstream components of the leg patterning pathway. *Development* 127: 1617-1626.

Kozloff, E.N. 1990. *Invertebrates*. Philadelphia: Saunder College Publisher.

Kraft, R. & Jäckle, H. 1994. *Drosophila* mode of metamerization in the embryogenesis of the lepidopteran insect *Manduca sexta*. *Proc. Natl. Acad. Sci. USA* 91: 6634-6638.

Kukalova-Peck, J. 1983. Origin of the insect wing and wing articulation from the arthropodan leg. *Can. J. Zool.* 61: 1618-1669.

Kukalova-Peck, J. 1992. The "Uniramia" do not exist: the ground plan of the Pterygota as revealed by Permian Diaphanopterodea from Russia (*Insecta: Paleodictyopteroidea*). *Can. J. Zool.* 70: 236-255.

Kume, M. & Dan, K. 1968. *Invertebrate Embryology.* Washington: U.S. Dept. of Health, Education, and Welfare and National Science Foundation.

Lecuit, T. & Cohen, S. M. 1997. Proximal-distal axis formation in the *Drosophila* leg. *Nature* 388: 139-145.

Lynch. J. E. 1958. *Branchinecta cornigera*, a new species of anostracan phyllopod from the state of Washington. *Proc. U.S. Nat'l. Mus.* 108: 25-37.

Mann, R.S. & Abu-Shaar, M. 1996. Nuclear import of the homeo-domain protein Extradenticle in response to Wg and Dpp signaling. *Nature* 383: 630–633.

Manton, S. M. 1934. On the embryology of the crustacean *Nebalia bipes. Phil. Trans. R. Soc. Lond. B* 223: 163-238.

Manton, S. M. 1964. Mandibular mechanisms and the evolution of arthropods. *Phil. Trans. R. Soc. Lond. B* 247: 1-183.

Manton, S. M. 1973. Arthropod phylogeny - a modern synthesis. *J. Zool. London* 171: 11-130.

Manton, S. M. 1977. *The Arthropoda.* Oxford: Clarendon Press.

Martin, J. W. 1992. Branchiopoda. In: Harrison, F.W. & Humes, A.G. (eds), *Microscopic Anatomy of Invertebrates, Vol. 9*: 25-224. New York: Wiley-Liss.

McLaughlin, P. A. 1982. Comparative morphology of crustacean appendages. In: Abele, L.G. (ed), *Embryology, Morphology, and Genetics*: 197-256. New York: Academic Press.

McLaughlin, P. A., Gore, R.H. & W. R. Buce. 1989. Studies on the *Provenzanoi* and other pagurid groups: III. The larval and early juvenile stages of *Pagurus kennerlyi* (Stimpson) (Decapoda: Anomura: Paguridae) reared in the laboratory. *J. Crust. Biol.* 9: 626-644.

Mittmann, B. & Scholtz, G. 2001. Distal-less expression in embryos of *Limulus polyphemus* (Chelicerata, Xiphosura) and *Lepisma saccharina* (Insecta, Zygentoma) suggests a role in the development of mechanoreceptors, chemoreceptors, and the CNS. *Dev. Genes Evol.* 211: 232-243.

Nagy, L. M. & Carroll, S. 1994. Conservation of *wingless* patterning functions in the short-germ embryos of *Tribolium castaneum. Nature* 367: 460-463.

Nagy, L. M. & Williams, T. A. 2001. Comparative limb development as a tool for understanding the evolutionary diversification of limbs in arthropods: challenging the modularity paradigm. In: Wagner, G.P. (ed), *The Character Concept in Evolutionary Biology*: 455-488. San Diego: Academic Press.

Ng, M., Diaz-Benjumea, F.J. & Cohen, S.M. 1995. Nubbin encodes a POU-domain protein required for proximal-distal patterning in the Drosophila wing. *Development* 121: 589-599.

Ng, M., Diaz-Benjumea, F.J., Vincent, J.-P., Wu, J. & Cohen, S.M. 1996. Specification of the wing by localized expression of wingless protein. *Nature* 381: 316-319.

Nulsen, C. & Nagy, L. M. 1999. The role of *wingless* in the development of multi-branched crustacean limbs. *Dev. Genes Evol.* 209: 340-348.

Olesen, J. & Walossek, D. 2000. Limb ontogeny and trunk segmentation in Nebalia species (Crustacea, Malacostraca, Leptostraca). *Zoomorphology* 120: 47-64.

Olesen, J., Richter, S. & Scholtz, G. 2001. The evolutionary transformation of phyllopodous to stenopodous limbs in the Branchiopoda (Crustacea) - Is there a common mechanism for early limb development in arthropods? *Int. J. Dev. Biol.* 45: 869-876.

Oster, G. F., Shubin, N., Murray, J. D. & Alberch, P. 1988. Evolution and morphogenetic rules: the shape of the vertebrate limb in ontogeny and phylogeny. *Evolution* 42: 862-884.

Packard, A. S. 1883. A monograph on the North American phyllopod Crustacea. *Annual Report (1878) U.S. Geol. Geogr. Sur. Territories* 12: 295-592.

Panganiban, G., Sebring, A., Nagy, L.M. & Carroll, S.B. 1995. The development of crustacean limbs and the evolution of arthropods. *Science* 270: 1363-1366.

Panganiban, G., Irvine, S. M., Lowe, C., Roehl, H., Corley, L. S., Sherbon, B., Grenier, J. K., Fallon, J. F., Kimble, J., Walker, M., Wray, G. A., Swalla, B. J., Martindale, M. Q. & Carroll, S.B. 1997. The origin and evolution of animal appendages. *Proc. Natl. Acad. Sci. USA* 94: 5162-5166.

Panganiban, G. 2000. *Distal-less* function during *Drosophila* appendage and sense organ development. *Dev. Dyn.* 218: 554-562.

Popadic, A., Rusch, D., Peterson, M., Rodgers, B.T. & Kaufman, T. C. 1996. Origin of the arthropod mandible *Nature* 380: 395.

Popadic, A., Panganiban, G., Rusch, D., Shear, W. A. & Kaufman, T. C. 1998. Molecular evidence for the gnathobasic derivation of arthropod mandibles and for the appendicular origin of the labrum and other structures. *Dev. Genes Evol.* 208: 142-150.

Prpic, N.-M., Wigand, B., Damen, W. G. M. & Klinger, M. 2001. Expression of *dachshund* in wild-type and *Dll* mutant *Tribolium* corroborates serial homologies in insect appendages. *Dev. Genes Evol.* 211: 467-477.

Richter, S. & Scholtz, G. 2001. Phylogenetic analysis of the Malacostraca (Crustacea). *J. Zool. Syst. Evol. Research* 39: 113-136.

Rieckhof, G. E., Casares, F., Ryoo, H. D., Abu-Shaar, M. & Mann, R. S. 1997. Nuclear translocation of extradenticle requires homothorax, which encodes an extradenticle-related homeodomain protein. *Cell* 91: 171-183.

Roth, V. L. 1984. On homology. *Biol. J. Linn. Soc.* 22: 13-29.

Roth, V. L. 1988. The Biological Basis of Homology. In: Humphries, C.J. (ed), *Ontogeny and Systematics*: 1-26. New York: Columbia University Press.

Sanchez-Salazar, J., Pletcher, M. T., Bennett, R. L., Dandamudi, T. J., Denell, R. E. & Doctor, J.S. 1996. The *Tribolium decapentaplegic* gene is similar in sequence, structure, and expression to the *Drosophila dpp* gene. *Dev. Genes Evol.* 206: 237-246.

Sanders, H.L. 1963. The *Cephalocarida*: Functional morphology, larval development, comparatice external anatomy. *Mem. Connect. Acad. Art. Sci.* 15: 1-80.

Scholtz, G., Mittmann, B. & Gerberding, M. 1998. The pattern of *Distal-less* expression in the mouthparts of crustaceans, myriapods and insects: new evidence for a gnathobasic mandible and the common origin of Mandibulata. *Int. J. Dev. Biol.* 42: 801-810.

Schoppmeier, M. & Damen, W.G.M. 2001. Doublestranded RNA interference in the spider *Cupiennius salei*: the role of *Distal-less* is evolutionarily conserved in arthropod appendage formation. *Dev. Genes Evol.* 211: 76-82.

Schram, F. R. 1986. *Crustacea*. Oxford: Oxford University Press.

Schram, F. R. & Koenemann, S. 2001. Developmental genetics and arthropod evolution; part 1, on legs. *Evo. Dev.* 3: 343-354.

Shubin, N. H. & Alberch, P. 1986. A morphogenetic approach to the origin and basic organization of the tetrapod limb. *Evol. Biol.* 20: 319-387.

Snodgrass, R. E. 1935. *Principles of Insect Morphology*. New York: MacGraw Hill Book Co.

Snodgrass, R. E. 1938. Evolution of the Annelida, Onychophora, and Arthropoda. *Smiths. Misc. Coll.* 97 (6): 1-159.

Van Valen, L. M. 1993. Serial homology: The crests and cusps of mammalian teeth. *Acta Palae. Pol.* 38: 145-158.

Wagner, G. P. 1989a. The origin of morphological characters and the biological basis of homology. *Evolution* 43 (6): 1157-1171.

Wagner, G. P. 1989b. The biological homology concept. *Annu. Rev. Ecol. Syst.* 20: 51-69.

Walossek, D. 1993. The Upper Cambrian *Rehbachiella kinnekullensis* and the phylogeny of Branchiopoda and Crustacea. *Foss. Strata* 32: 1-202.

Walossek, D. & Müller, K. J. 1997. Cambrian 'Orsten' -type arthropods and the phylogeny of Crustacea. In: Fortey, R.A. & Thomas, R.H. (ed), *Arthropod Relationships*: 139-153. London: Chapman and Hall.

Wigglesworth, V. B. 1973. Evolution of insect wings and flight. *Nature* 246: 127-129.

Williams, T. A. & Müller, G. B. 1996. Limb development in a primitive crustacean, *Triops longicaudatus*: subdivision of the early limb bud gives rise to multibranched limbs. *Dev. Genes Evol.* 206: 161-168.

Williams, T. A. 1998. *Distalless* expression in crustaceans and the patterning of branched limbs. *Dev. Genes Evol.* 207: 427-434.

Williams, T.A. 1999. Morphogenesis and homology in arthropod limbs. *Am. Zool.* 39: 664-675.

Williams, T. A. & Nagy, L. M. 2001. Developmental modularity and the evolutionary diversification of arthropod limbs. *J. Exp. Zool. (MDE)* 291: 241-257.

Williams, T. A., Nulsen, C. & Nagy, L. M. 2002. A complex role for *Distal-less* in crustacean limb development. *Dev. Biol.* 241: 302-312.

Wu, J. & Cohen, S. M. 1999. Proximodistal axis formation in the *Drosophila* leg: subdivision into proximal and distal domains by *Homothorax* and *Distal-less*. *Development* 126: 109-117.

III MORPHOLOGY AND PHYLOGENY

The complete cypris larva and its significance in thecostracan phylogeny

JENS T. HØEG, NIKLAS C. LAGERSSON & HENRIK GLENNER

Zoological Institute, University of Copenhagen, Universitetsparken 15, DK-2100, Copenhagen, Denmark

1 INTRODUCTION

The Cirripedia normally hatch as nauplii but the terminal pelagic instar is highly specialized and called cyprid or cypris larva. All Cirripedia are sessile animals and the process in which the larvae settle, attach, and metamorphose has therefore been at the focus of biologists since Darwin's monumental papers on this specialized crustacean taxon (Darwin 1852, 1854). It is the task of the cyprid to complete this process to success. It is therefore often mentioned as one of the unique, derived characteristics (autapomorphies) for the Cirripedia, but unfortunately both the concept "Cirripedia" and "cyprid" are often used in ill-defined ways both in textbooks and professional papers.

Identifying apomorphies in larval morphology characterizing the Cirripedia is very important because the monophyly of this taxon is by no means straightforward. First, one of the three orders of cirripede, the Rhizocephala, is highly specialized for parasitism with adults reduced to a state where they can hardly be recognized as Arthropods. Thus using morphology alone, only larval characters are available. Second, molecular data have failed to support that the three cirripede orders (Acrothoracica, Thoracica and Rhizocephala) form a monophylum. Instead the data, at least when looked at face value, indicated that the Acrothoracica are more closely related to the Ascothoracida (Spears et al. 1994) and the same result was reached by Mizrahi et al. (1998). More recent analyses of basically the same molecular dataset supported cirripede monophyly (Füllen et al. 2001; Perez-Losada et al. 2002), but morphological support for this taxon still remains weak. Evidence for a monophyletic Cirripedia rests basically on the unique possession of frontolateral horns in the nauplius larva, on the presence of a cypris larva and on the possession of a very specialized type of sperm cell (Høeg 1992; Walker 1992). Here we use the available literature, especially Glenner (1999) and later papers (Lagersson in press; Lagersson & Høeg in press) to for the first time break cypris morphology into a list of characters and compare them with the cypris-like larvae found in the putative closest relatives to the Cirripedia (Spears et al. 1994).

2 THE CIRRIPEDIA AND THE THECOSTRACA

The Cirripedia belongs to the Thecostraca which again forms part of the taxon Maxillopoda (Martin & Davis 2001). It is much discussed whether the Maxillopoda is a real monophylum within Crustacea, but the Thecostraca now seems to be a well-established group. The Thecostraca was defined in its present form by Grygier (1987) and comprises the Cirripedia, Ascothoracida, and Facetotecta (see Fig. 6). The Ascothoracida are parasites of echinoderms and gorgonian corals. The Facetotecta are known only from their pelagic larvae (nauplii and cyprids of "type Y"), whereas the adults are unknown (Høeg & Kolbasov 2002), but they are believed to be parasites. The phylogeny of the Thecostraca in Fig. 6 is based on both morphological (principally larval) and molecular datasets (Høeg & Kolbasov 2002; Jensen et al. 1994; Perez-Losada et al. 2002).

The Cirripedia comprises three monophyla: the filter feeding Thoracica (pedunculate and acorn barnacles); the Acrothoracica which are also filter feeders, but lack calcareous shell plates and instead bore into substrates such as gastropod shells inhabited by hermit crabs; and the parasitic Rhizocephala, which infest other Crustacea and where the adults have lost all traits reminiscent of Crustacea. All Thecostraca start out as pelagic nauplii, except in forms with an abbreviated larval development, but since all forms have permanently sessile adults the terminal larval instar is specialized for locating an attachment site and affix itself to this substratum. This instar is often loosely called "cypris".

3 CYPRIDS AND CYPRIDIFORM LARVAE

In the ground pattern, all three subgroups of the Thecostraca start development with 6 naupliar stages. Details of thecostracan naupliar development were reviewed by (Grygier 1994) and Deutsch et al. (2003). The final nauplius molts into a stage that has a morphology specialized for settlement and therefore swims and looks very different from the nauplii (Fig. 1). This stage was first identified in the Cirripedia (Thompson 1830) and called "cypris" (Darwin 1852, 1854). Later authors have tended to use "cyprid" to avoid confusion with the ostracode genus *Cypris*. This is hardly an argument in itself, but cyprid has the advantage of having the easy to use plural form "cyprids". We therefore follow Walker (1992) in using cyprid/cyprids as a noun but cypris as an adjective such as in "cypris larva".

"Cyprid" has been used indiscriminately for the terminal instar in all three orders of the Thecostraca, but in our view and that of Høeg (1992) and Walker (1992) it should only apply to the terminal larva found in the three orders of the Cirripedia. The reason is that morphological studies are now available of cypris larvae from all three cirripede orders and from the homologous terminal larval instar in the Facetotecta and the Ascothoracida, and a comparison shows that the cirripede cyprids share a long suite of unique characteristics that are lacking from the larvae of the other two classes. The purpose of this paper is to highlight these characters. Below we will use the "cyprid" and "cypris" only for the settlement stage of the Cirripedia. We continue to use the well established name "y-cyprid" (or "cypris y") for the settlement stage in the Facetotecta (Grygier 1996b, Høeg & Kolbasov 2002), but we need a new name for the corresponding stage in the Ascothoracida (Grygier 1996a) and therefore suggest the name "a-cyprid" (meaning ascothoracid cyprid) for this larva. By limiting the use of "cypris" or "cypris larva" to the Cirripedia *sensu stricto* we emphasize the uniqueness of this stage. We introduce the new term "cypridiform larva" to comprise all three homologous types of larvae (y-cyprid, a-cyprid, cyprid). The cypridiform larva consists of a cephalon with prehensile antennules, a thorax with six natatory appendages, a limbless abdomen and a telson with furcal rami. A conspicuous head shield furnished with lattice organs (see below) extends laterally and posteriorly and covers most of the body.

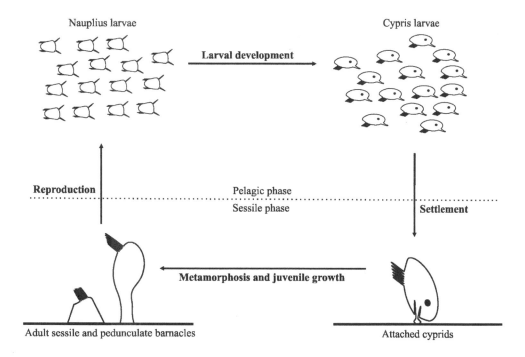

Figure 1. The generalized cirripede life cycle. Pelagic nauplius larvae are released by the sessile adults. Naupliar development encompasses a maximum of 6 instars. The final nauplius molts into the highly specialized cypris larva, unique to the Cirripedia. During the settlement process, the cyprid locates a suitable substrate, cements itself and initiates the metamorphic molt that results in the first juvenile instar.

4 NAUPLIAR DEVELOPMENT

The most distinguishing feature of the crustacean nauplius is that it swims feeds and senses using only the antennules, antennae, and mandibles. In Thecostraca, the separation of the naupliar and post-naupliar (cypridiform) parts of development is very conspicuous. The post-mandibular appendages (maxillules, maxillae, and thoracopods) never develop to a functional state in the nauplii (Walossek 1993; Walossek et al. 1996). Maxillules and maxillae may appear as small buds armed with setae, but the post-cephalic segmentation and appendages are suppressed. At the molt from nauplius to cypridiform larva all six thoracomeres develop in one step and all six pairs of thoracopods appear in a fully functional state. This "condensed" form of naupliar development is an apomorphy for the Thecostraca, or possibly for a slightly more inclusive clade (Walossek et al. 1996).

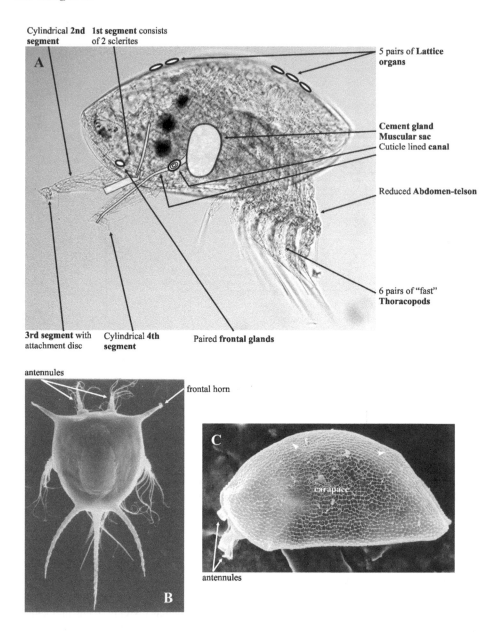

Figure 2. (A) Characters of the cirripede cypris larva (here *Balanus improvisus*). Cirripede cyprids may vary by more than an order of magnitude in size (70 μm – more than 1 mm) but still retain the same very stereotyped morphology adapted to their role in settlement. The torpedo shaped larva is enclosed in a carapace carrying five pairs of chemosensory lattice organs and ventrolaterally the openings of the frontal glands. Using the six pairs of natatory thoracopods operated by "fast" muscles the cyprids can swim very rapidly. Using the very complex, four-segmented antennules operated by "slow" muscles they can walk on the substrate and explore its properties by means of sensillae on the

5 HEAD SHIELD

It follows from the above that the nauplius and cypridiform larvae have distinctly different morphologies (Fig. 2) and the naupliar/cyprid (n/c) molt is therefore often described as a metamorphosis. Yet already Walley (1969) emphasized that many of the internal structures unique to the cypridiform larva develop gradually during the naupliar phase. Walossek et al. (1996) emphasized that this is certainly also true for external morphology as exemplified by the shape of the head shield and the antennules (see below).

Externally, the most distinguishing feature of the cypridiform larva is the presence of a large carapace that in the Ascothoracida and the Cirripedia extends downwards ventrally on both sides so it can completely enclose the body when the appendages are withdrawn (Fig. 2). In contrast, the nauplius is covered dorsally by a cuticular "head shield" (Walossek 1993). In all Cirripedia, the anterolateral corners of the naupliar shield are drawn into a pair of very conspicuous frontolateral horns, while such horns lack in the remaining Thecostraca.

The post cephalic part of the body develops gradually during the naupliar phase and projects anterior-ventrally, hanging completely free of the head shield. The naupliar head shield develops into the cypris carapace during the n/c molt. In the last nauplius, the shield is already very large and extends both laterally and posterior so it virtually overhangs the entire body and foreshadows the cypris carapace. After the n/c molt the shield in the Ascothoracida and in the Cirripedia has flexed ventrally so it becomes U-shaped in cross section and a distinct adductor muscle connects the two sides of the shield, now called "carapace", just behind the mouth. In the Facetotecta, the shield of the y-cyprid maintains more or less its naupliar shape. It overhangs the body laterally but does not bend ventrally. Interestingly, y-cyprids also seem to lack an adductor muscle, since it was not found by Grygier (1987). If true, presence of an adductor muscle becomes a synapomorphy for the Ascothoracida and the Cirripedia in agreement with the most recent evidence from rRNA sequences (Perez-Losada et al. 2002).

6 LATTICE ORGANS

Lattice organs are very specialized chemoreceptors on the head shield of thecostracan cypridiform larvae (Høeg et al. 1998; Høeg & Kolbasov 2002). They are present in all investigated species and always as five pairs situated close to the dorsal midline of the carapace, arranged as a group of two anterior pairs and another group of three posterior pairs. Neither individual lattice organs nor their specific arrangement as 2+3 pairs are from any other Crustacea so they are undoubtedly an apomorphy for the Thecostraca.

3rd and 4th antennulary segments. The cyprid finally attaches by secretions from the multicellular cement gland that exits on the attachment disc located on the 3rd antennulary segment. Most of these characters are apomorphies for the Cirripedia. (B, C) Nauplius (B) and cyprid (C) of *Capitulum mitella*. The morphology of the cyprid is very distinct from the nauplius. In all cirripede nauplii the head shield carries a pair of frontolateral horns on which exits a pair of large unicellular glands. The horns are absent or diminutive in the cyprid, but the glands are retained and exit frontolaterally on each side of the carapace. Note the antennules protruding anteriorly while the thoracopods have been entirely withdrawn into the posterior mantle cavity.

Lattice organs develop ontogenetically from setae on the naupliar head shield and represent one of the most specialized forms of sensillae in all Crustacea (Rybakov & Høeg 2002). In the most plesiomorphic variety the external appearance of the lattice organ is that of a reclined seta (crest) lying prostrate in an elongated depression in the carapace cuticle (Fig. 6A). This type is found in the Facetotecta, the Ascothoracida, and most Cirripedia Acrothoracica. The lattice organs of all Rhizocephala and Thoracica and of some Acrothoracica have lost all external resemblance to a seta and appear in SEM as an elongated area on the cuticle perforated with small, closely spaced pores (pore field in Fig. 6B). Despite this conspicuous difference in the exterior morphology of lattice organ, the two varieties have nearly identical internal structures. TEM shows that they consist of an elongate chamber in the shield cuticle extending under the "crest" or the "pore field". Into this chamber extends branching ciliary extensions from two sensory cells (Høeg et al. 1998), and the other end of the cavity communicates with the exterior through a terminal pore that is very conspicuous in SEM pictures (Fig. 6). In the Ascothoracida and the Facetotecta, the terminal pore lies posteriorly in all five pairs of lattice organs (Fig. 6C). The Cirripedia shares the apomorphy of having an anteriorly situated pore in the second pair of lattice organs and it is an apomorphy for the Thoracica and Rhizocephala to also have the pore of the first pair located anteriorly. The numerous small pores in the pore field type of lattice organ found in the Rhizocephala and Thoracica are in fact deep pits in the cuticular roof of the lattice organ, at the bottom of which a thin layer of epicuticle separates the exterior from the underlying chamber. Homologous, minute pores are found also in all Acrothoracica, although usually only visible with TEM. In the Facetotecta and Ascothoracida, the cuticular roof covering the chamber is a simple cuticle without any pores even at TEM level. The presence of small pores in the roof is therefore an apomorphy of all Cirripedia and may relate to the function of the organ, since they must facilitate the diffusion of substances from the exterior and into the chamber housing the ciliary extensions. Høeg et al. (1998) argued that the location of the lattice organs on the carapace indicated that they operate during the pelagic phase of the larvae rather than during the exploratory phase where the larvae are in contact with the substrate. There is no explanation for the shifts in terminal pore position. It may be a purely serendipitous event in evolution but fortunately one with valid phylogenetic information. The variation in lattice organ morphology within the Acrothoracica does not affect the monophyly, which is based on several apomorphies in the morphology of the adults (Kolbasov & Høeg, 2000).

7 FRONTOLATERAL HORNS AND GLANDS

All cirripede nauplii have a pair of conspicuous frontolateral horns at the anterior end of the head shield (Fig. 2B). Each horn is associated with two large gland cells (frontal horn gland), which exit through a prominent, complexly shaped pore at the tip (Walker 1973). The detailed structure of the horns and especially of the pore may even change from one nauplius instar to the next, and this is of help in groups such as the Rhizocephala, where appendage setation changes but little through development (Walossek et al. 1996). The cyprid retains the two pairs of gland cells, which exit through pores situated on each side of the body near the anterior-ventral edge of the carapace (Fig. 3). Large cyprids, such as those of *Lepas* that may exceed 1 mm in length, can sometimes have the pores situated on small "horns" comparable to the ones in nauplii (Elfimov 1995), but normally the frontal pores lie flush with the surface of the carapace cuticle.

The pore itself can be quite complex and have an outer raised rim within which lies the two pores of the gland cells proper (Fig. 3B). In rhizocephalans, each of the two gland cells exit through separate slit-shaped pores situated very close together (Høeg 1987) and the cells themselves are relatively larger than those in thoracican cyprids (Walley 1969). Some cirripedes have no nauplii but hatch as cyprids. The cyprids of rhizocephalans with such an abbreviated larval development lack the frontal glands, whereas acrothoracicans hatching as cyprids retain the glands and pores (Kolbasov et al. 1999).

The frontal horns and glands are present in all cirripedes with free nauplii but in no other thecostracan or any other crustacean. Some ascothoracid nauplii have a series of smaller spines along the edge of the head shield (Grygier 1990, 1993). Grygier and Ito (1995) observed what could be homologous spines in an unidentified cirripede nauplius thought to be from a lepadid. All this suggests that the nauplius of the ancestor to the Ascothoracida and Cirripedia has such a series of lateral spines and that the cirripede frontolateral horns evolved from an anterior such spine that have become exceedingly large (Grygier & Ito 1995). The spines in ascothoracid nauplii can have up to six separate gland cells and pores while those in the said lepadid nauplius have but one cell and pore, but this does not per se disprove the hypothesis of homology.

The presence of frontal horns and associated glands is probably one of the best apomorphies in all zoology, since it takes no exception among the cirripede species save for a few species with abbreviated development and whose affiliation is not in doubt. It was exactly the presence of nauplii with frontolateral horns that led Thompson (1836) to classify the Rhizocephala as Cirripedia a few years after he himself had used the presence of nauplii and cyprids to classify the Cirripedia as Crustacea (1830, 1835). Adult cirripedes have highly diverse life forms ranging from highly specialized parasites to setose feeders. The omnipresence of frontal horns and glands in cirripede nauplii irrespective of the life form of the adult testifies to their significance. Whatever it is, it must also operate in the cyprid. Secretions that facilitate flotation, antifouling or antipredator secretions are among the several suggestions, but the true function of the frontolateral horns and glands remains an enigma (Walker 1992).

8 CYPRIS APPENDAGES AND LOCOMOTION

In the ground pattern the cyprid body consists of a cephalon, a thorax, a diminutive abdomen and a telson. Anteriorly it carries a pair of prominent antennules (antennae 1), but all other cephalic limbs are either absent or present only as rudiments. Each of the six thoracomeres carries a pair of biramous thoracopods while the telson carries a pair of caudal rami. While the nauplius swims with antennules, antennae, and mandibles, the cypris propels itself through the water using the natatory thoracopods, while using the antennules for walking on the substratum. During the n/c molt the cyprid is still encased in the exoskeleton of the last nauplius. Until ecdysis the cyprid must therefore operate the naupliar set of appendages using its cypris body. In most cases, the cyprid even "lies on its side" within the naupliar cuticle, probably for reasons of limited space. Although it has never been studied, the cyprid must obviously extend parts of its body including muscles into the naupliar appendages prior to ecdysis. The literature contains spurious reports of cyprids with prominent post-antennulary, cephalic appendages and of nauplio-cyprid stages. In all likelihood this covers one or several of the following cases:

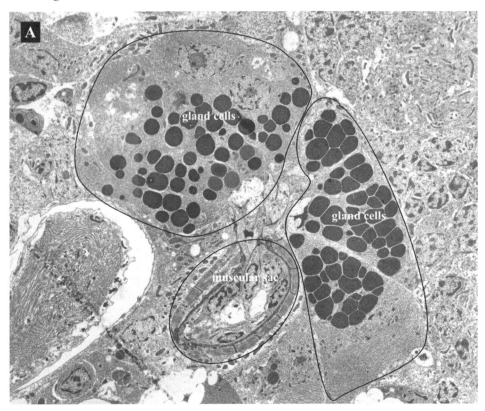

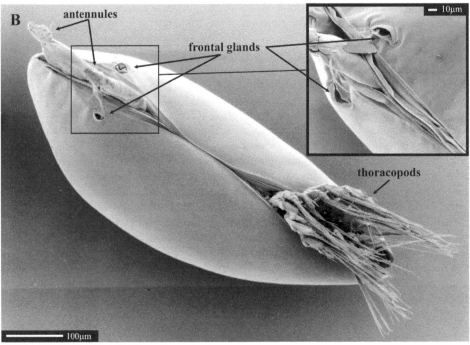

(i) Nauplii that have failed to metamorphose (molt) correctly into cyprids, i.e., metamorphic monster; (ii) cyprids about to appear but still enclosed in the cuticle of the last nauplius and thus still propelling themselves with the naupliar appendages; and (iii) cyprids that are but a few minutes old and therefore still have not retracted the body projections that extended into the naupliar appendages before ecdysis. We have ourselves observed such very prominent buds corresponding to antennae and mandibles in cyprids from cultured larvae and they always consist of soft unsclerotized cuticle that would be indistinguishable from mantle cavity cuticle when retracted. Interestingly, the very detailed studies of cypris musculature have not found any muscles that could operate the "antennae" and mandibles" (Høeg 1985; Lagersson & Høeg 2002; Walley 1969). The "naupliar" muscles operating antennae and mandibles in the young cyprid must therefore degenerate almost immediately after ecdysis.

8.1 *Antennules*

In all Thecostraca, the cypridiform larvae have prehensile antennules used for attaching to the substratum at settlement. Little is known about settlement in the Facetotecta and the Ascothoracida, but from the morphology of the antennules and the scant information on the internal anatomy of the cypridiform it seems certain that attachment is purely mechanical in both taxa. The antennule of the y-cyprid (Facetotecta) consists of only four segments. In contrast, a-cyprid antennule (Ascothoracida) has a total of six segments (Grygier 1987, 1996a), and the distal-most one carries a large movable spine or claw used in attachment (Grygier 1996b; Grygier & Ito 1995; Moyse et al. 1995).

The antennule of the cyprid (Cirripedia) is much more complex than those of the y-cyprids or a-cyprids. It consists of four segments operated by a very specialized musculature and is richly endowed with sensory organs and glands that functions during the exploratory phase and the final attachment. The first segment consists of two individually mobile sclerites and the proximal one articulates in the midline of the cypris body with its counterpart in the other antennule (Glenner 1999; Høeg 1985; Lagersson & Høeg 2002). The second segment is always cylindrical (Fig. 4). The third segment is bell or horseshoe shaped and its ventral (or distal) side forms the attachment organ, which is covered by a carpet of minute hairs that are not setae but rather microcuticular projections (Fig. 4).

Figure 3. (A) Cement gland of cyprid of *Lernaeodiscus porcellenae*. Sagittal TEM section from Høeg (1985). Several gland cells filled with RER and secretion vesicles lie in a half circle around the gland center from which exists a cuticle lined canal (not seen) that exits on the attachment disc of an antennule. The muscular sac surrounds the proximal part of the canal and controls the release of cement. (B) Cyprid of *Scalpellum scalpellum*. SEM. Ventrolateral view. The two antennules protrude from the anterior mantle cavity, the thoracopods from the posterior mantle cavity. Pores of the frontal glands situated anteriorly near the ventral rim of the carapace. Inset shows complex shape of pores of frontal glands; note also how tightly the rim of the carapace fits around the protruding antennules.

Both the cement canal (see below) and numerous unicellular antennulary glands empty onto the attachment disc (Nott & Foster 1969). The fourth segment has a cylindrical shape and projects from the lateral side of the third segment. Altogether these complexly shaped segments, their specialized intrinsic and extrinsic musculature, and their numerous sense organs and glands form an appendage that is very complex compared to either that of the Facetotecta or the Ascothoracida (Lagersson & Høeg 2002). The multisegmented antennule of the ascothoracid a-cyprid is difficult to compare with those in either of the two other thecostracan taxa. In contrast, the antennule of the facetotectan y-cyprid has an overall shape rather similar to that of the cirripede cyprid, but there is no simple homology between the four segments in both these appendages.

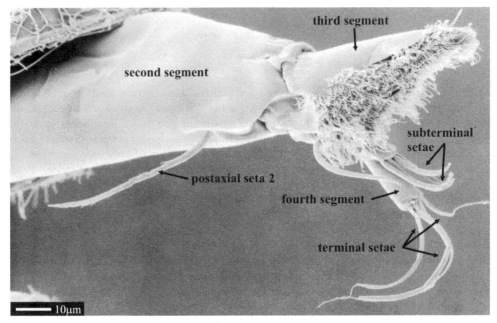

Figure 4. Distal part of antennule in cyprid of *Scalpellum scalpellum*. The attachment disc on the 3rd segment carries a diverse array of sensillae and is covered by a carpet of cuticular projections. Additional sensillae on the cylindrical fourth segment. The only sensilla on the 2nd segment is the ps2 seta situated ventrodistally, it is omnipresent in cirripede cyprids except a few parasitic forms. The shape and sensillar armament of the 3rd and 4th segments vary considerably between the different cirripede taxa.

8.1.1 *Antennulary sense organs*

In all Thecostraca, the antennules of the cypridiform larvae carry a number of sensory organs in the form of modified setae. In the Cirripedia, these sense organs are especially well developed and are almost exclusively situated on the third and fourth antennulary segments.

First and second segment: The first segment is devoid of any setae and the second carries but a single seta ventrodistally, the so called postaxial seta 2 (Fig. 4).

Third segment: Eight peripheral setae are arranged along the perimeter of the attachment organ. More centrally the attachment organ carries an axial sense organ, and a more posteriorly situated postaxial seta. The axial organ varies in morphology between the cirripede taxa, but it appears to be complex and consists of more than one seta. Finally, the third segment carried a postaxial seta 3 posterolaterally at the rim or just outside the attachment disc proper (Glenner et al. 1989; Moyse et al. 1995).

Fourth segment: The sensory armature of the small, fourth segment varies within the Cirripedia, but in the ground pattern it seems to comprise four short and identical subterminally situated setae and five very differently shaped and sized, terminal setae (Fig. 4) (Clare & Nott 1994; Lagersson et al. 2003).

In both the Facetotectan and the Ascothoracida, the antennule of the cypridiform larva carries a prominent aesthetasc (Grygier 1996a, b). These are bag-shaped, thin walled setae used in olfaction. Aesthetascs are also present on the antennule of cyprids of rhizocephalan cirripedes (Glenner et al. 1989) but have not yet been identified in thoracican and acrothoracican cyprids.

8.1.2 *Antennulary glands*
In Cirripedia the attachment is chemical both during the exploratory walking on the substrate and at the final and irreversible attachment, and for this purpose two kinds of glands exit onto the attachment disc, viz., numerous small, unicellular glands and a single, multicellular cement gland. In large-sized cyprids, such as those of the Thoracica, the unicellular glands lie entirely within the antennule and exit by small pores all over the attachment disc. They are normally believed to be responsible for attachment during the exploratory phase (Walker 1992). The two cement glands, one for each antennule, are located anterolaterally in the cypris body and represent some of the largest organs in the larva. The several gland cells are arranged around a central secretory basin, whence leads a cuticular canal that enters the antennules (Fig. 3A). Proximally, the cement canal passes through a muscular sac, which presumably plays a key part in controlling the secretion of cement. Distally, the cement canal proceeds as far as the third antennulary segments where it terminates on the attachment disc by a simple pore or as a complex system of grooves (Høeg 1985).

8.2 *Thoracic appendages, musculature, and setation*

All cypridiform larvae carry six pairs of natatory appendages on the thorax. Relatively little attention has been paid to the morphology and function of these appendages, but the available evidence indicates that those of the Ascothoracida and Facetotecta beat in a metachronal pattern while those of the cirripede cyprid beat comparatively synchronized (Grygier 1987; Lagersson & Høeg 2002). Swimming appears to be rather slow and continuous in the Facetotecta and Ascothoracida, whereas the cyprids of the Cirripedia swim very fast and in bursts separated with quiet periods, where they sink down through the water column (Glenner 1999). Undoubtedly there are numerous details of thoracopod morphology associated with these differences, but even more important may be the thoracic musculature. Lagersson (2002) showed that the muscles associated with the thoracopods of the cirripede cyprid have the ultrastructural characteristics of fast beating muscles, while both the extrinsic and intrinsic antennulary muscles have the characteristic for slow, precise movements. The musculature of the y-cyprid and a-cyprid has not been studied, but we predict from their swimming mode that they are not of the fast beat type seen in the cirripede cyprid. This, however, may have little phylogenetic significance. It seems that wherever a fast beat action is required it also develops and that this has occurred numerous times in Crustacea.

Obviously, we are here faced with the dilemma of highly adaptive characters versus characters whose distributional pattern in taxa reflects historical events (Gould & Lewontin 1979).

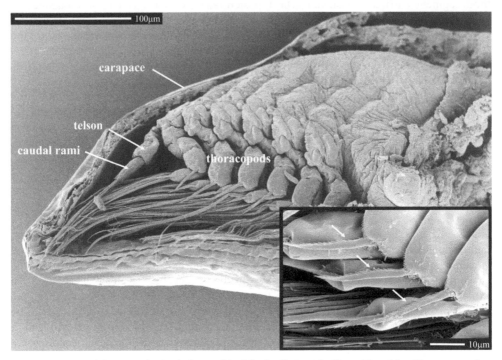

Figure 5. SEM of the posterior body in cyprid of *Scalpellum scalpellum*. The right side carapace has been removed to expose the thorax and telson. Note absence of any conspicuous abdomen between telson and last thoracomere. Insert shows thoracopodal exopods armed with conspicuous, stout and serrated grooming setae (arrows).

All six pairs of thoracopods are biramous. In the ground pattern of the Thecostraca, the protopod consists of a coxa and a basis and carries a three-segmented endopod and two-segmented exopod. This is the pattern generally found in the Facetotecta, the Ascothoracida, and the Cirripedia Acrothoracica. In the Cirripedia Rhizocephala and Thoracica, the endopod is generally only two-segmented. The first or last pairs of thoracopods may sometimes deviate slightly in both segmentation and setation patterns. Basal to the protopod is an exceedingly complex array of cuticular sclerites, somewhat reminiscent in complexity to the wing sclerites in insects, and like these they appear to play a key role in the beating of the appendage (pers. obs.). The detailed morphology of these sclerites has never been worked out, but nothing remotely similar has been observed in y-cyprids or a-cyprids, so they undoubtedly incorporate some apomorphies for the Cirripedia.

Details of thoracopod setation are well described in the y-cyprids and a-cyprids. In most cirripede cyprids, it remains too poorly analyzed to offer a consistent pattern for comparison with the following exception. A short, but stout and conspicuously serrated seta inserts on the exopod in all six pairs of thoracopods of the cirripede cyprid, while the Facetotecta and Ascothoracida lack such a structure (Fig. 5). One function of this omnipresent seta may be grooming of the natatory thoracopods.

8.3 *Abdomen and telson*

In the thecostracan ground pattern, the hind body consists of seven thoracomeres, three abdominal segments and a telson. This, therefore, conforms to the ground pattern of the maxillopodan body plan. However, functionally the cypridiform larvae have a six segment-ed thorax with natatory thoracopods followed by an "urosome" consisting of maximally four segments and a telson. In y-cyprids and a-cyprids, the first urosomal segment, i.e., the seventh thoracomere, may carry the incipient penis corresponding to a fused, seventh pair of thoracopods. Except for the morphologically reduced Rhizocephala, adult Cirripedia also have a penis believed to derive from fused thoracopods, but neither the seventh tho-racomere nor the penis rudiment has ever been found in cypris larvae. One notable exception may be the enigmatic ventral projection situated just anterior to the telson in some rhizocephalan cyprids (Walossek et al. 1996). Both rhizocephalan cyprids and most if not all thoracican cyprids have lost all traces of the abdomen. Acrothoracican cyprids retain a distinct somite with traces of segmentation intercalated between the thorax and the telson. Kolbasov and Høeg (1999) regard this as the abdomen and their interpretation dovetails with recent information from *Hox* gene expression in rhizocephalans (Deutsch et al. 2003).

A telson is always present in cypridiform larvae and carries prominent, unsegmented caudal rami equipped with long setae. In cirripede cyprids, the very small telson seems to insert directly on the sixth thoracomere (but see above).

9   CYPRID MOLTING AND METAMORPHOSIS

Nothing is known about metamorphosis in the Facetotecta, although we expect the adults to be parasites of unknown hosts. Some Ascothoracida have an extensive metamorphosis, but in the most primitive forms there is virtually no morphological change from the a-cypris larva to adult, especially in the males (Grygier 1996a). This proves that the very extensive metamorphic molt that follows settlement of the cirripede cypris is an apomorphy for that taxon (Glenner & Høeg 1993, 1994; Høeg 1985; Walley 1969). We will not deal further with metamorphosis here, although that process could most likely be broken down into a suite of numerous independent apomorphies for the Cirripedia. It is interesting, however, that some Ascothoracida have at least two successive a-cypris instars, while the cirripede molts only once, viz. when it has attached permanently. Again this testifies to the cirripede ontogeny being highly derived.

10   A NOTE ON THE TANTULOCARIDA

In the classification of all recent Crustacea by Martin & Davis (2001), the Thecostraca comprise the Facetotecta, Ascothoracida, and Cirripedia in agreement with our view, but they also discuss the possibility that the parasitic class Tantulocarida should be nested within the Thecostraca. This hypothesis, suggested to them as personal comments, asserts that the facetotectan y-larvae represent some unknown part of the complex tantulocaridan life cycle described by Huys et al. (1993). We agree that the tantulocaridan life cycle may yet hold surprises and that no life cycle can be considered "closed" unless all stages have been cultured in the laboratory (Glenner et al. 2000), but there are strong arguments against a tantulocaridan-facetotectan relation.

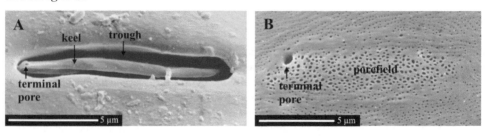

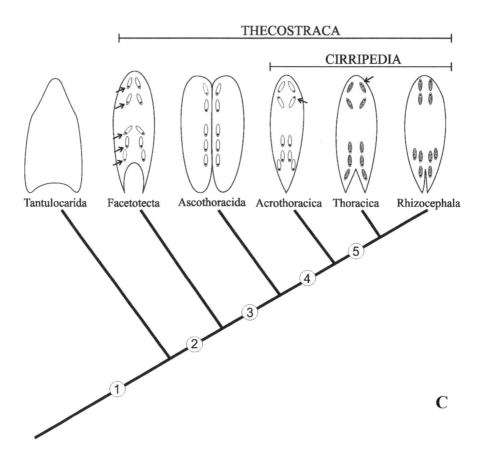

Figure 6. Lattice organs and thecostracan phylogeny. (A) Lattice organ of "crest in a trough" shape (from *Ulophysema oeresundense*, Ascothoracida). (B) Lattice organ of "pore field shape" (from *Scalpellum scalpellum*, Cirripedia). Note the conspicuous terminal pore in both varieties. Despite the very different external appearance, both types are chemosensory organs with very similar TEM level structures. (C) Phylogeny of the Crustacea Thecostraca and the Cirripedia. The cladogram is independently supported by both morphological datasets (Grygier 1987; Høeg & Kolbasov 2002) and 18s rDNA sequence data (Perez-Losada et al. 2002). Apomorphies are principally identified in larval morphology, since adult Facetotecta are unknown, and adult Tantulocarida and Rhizocephala lack almost all relevant characters. Non-larval apomorphies in italics. Shapes of carapace of the cypridi-form larvae, lattice organs and position of their terminal pores illustrated for each taxon. Selected

Like the Thecostraca, the Tantulocarida are sessile organisms (parasites), and it is the role of their tantulus larva to find the substratum. Accepting the said close relation we accordingly consider the tantulus and the cypridiform larva corresponding (homologous) stages. Since the tantulocaridan cycle therefore already incorporates a specialized settlement stage, the tantulus, it would be illogical to expect that another settlement stage, the y-cyprid, could be part of the life cycle. Another strong argument is the total absence of lattice organs in the Tantulocarida, since these structures have proved to be of high value in elucidating thecostracan phylogeny (Høeg & Kolbasov 2002). Within Thecostraca these peculiar chemosensory structures are omnipresent on the carapace of all cypridiform larvae, despite the enormous differences in settlement substrates and mode of attachment. We therefore refute any close relationship between the Tantulocarida and any subgroup of the Thecostraca. Instead we follow the wise decision of Martin & Davis (2001) and consider the Tantulocarida as the sister group of the Thecostraca based primarily on the position of the female gonopore anterior-most on the thorax (Huys et al. 1993).

11 APOMORPHIES IN CYPRIS MORPHOLOGY

In Fig. 6, we enumerate some of the synapomorphies in larval, and mostly cypridiform, morphology that supports the following monophyla: Tantulocarida + Thecostraca (node 1), Thecostraca (node 2), Ascothoracida + Cirripedia (node 3), Cirripedia (node 4), and Rhizo-cephala + Thoracica (node 5). Below we spell out those that concern the morphology of the "true" cypris larva of the Cirripedia compared to the y-cypris of the Facetotecta and the a-cypris larva of the Ascothoracida. Putative apomorphies that have not been convincingly checked in the Facetotecta or Ascothoracida are marked with "(?)". Taken together these characters reflect that all cirripedes posses very similar cypris larvae. This is remarkable considering the vast diversity in substrate and mode of life within the cirripedes and testifies to the enormous success of the cypris morphology in locating and attaching on suit-able substrata and initiating metamorphosis.

1. Terminal pore in second pair of lattice organ situated anteriorly
2. Minute pores in cuticle above lattice organ
3. Frontal horn glands (2 cells per gland)

---

apomorphies in larval morphology: (1) TANTULOCARIDA + THECOSTRACA: Terminal larval instar specialized for permanent attachment, female gonopore on first trunk somite, male gonopore on 7th trunk somite (2) THECOSTRACA: Cypridiform larva with 5 pairs of chemosensory lattice organs on carapace, cypris antennules used for attachment, 7th pair of trunk limbs modified as penis (3) ASCOTHORACIDA + CIRRIPEDIA: Head shield of cypridiform larva drawn ventrolaterally ("bivalved") to from a carapace enclosing the body (due to development of carapace adductor muscle?) (4) CIRRIPEDIA: Nauplii with paired frontolateral horns, true cypris larva attaching with cement (numerous independent apomorphies, see text), pronounced metamorphosis from cypris to next instar, terminal pore in 2nd pair of lattice organs switch from posterior to anterior position, abdomen highly reduced or absent, unique filiform type of spermatozoa, adults are filter feeders using thoracopods as cirri. (5) THORACICA + RHIZOCEPHALA: Terminal pore in 1st pair of lattice organs switches from posterior to anterior position, lattice organs always of "pore field" morphology (Fig. 6B).

4. A pair of multicellular cement glands exiting through a cuticular canal with a muscular sac
5. Numerous unicellular attachment glands in antenna 1
6. Four segmented antennule (detailed pattern different from y-cyprids)
7. First antennulary segment consists of two, movable sclerites
8. Proximal sclerite of first antennulary segment articulates with counterpart in other antennule
9. Third antennulary segment with attachment disc covered by microcuticular projections (hairs)
10. Third antennulary segment with:
    - 8 peripheral setae
    - Axial organ
    - Postaxial seta
    - Postaxial seta 3
11. Attachment disc with:
    - Exit pore for cement gland
    - Numerous unicellular glands
12. Cylindrical fourth segment
13. Fourth segment with:
    - Four, similar, subterminal setae
    - Five, terminal setae of different shapes (which each could qualify for being independent apomorphies)
14. Complex extrinsic and intrinsic antennulary musculature enabling "bipedal" walking on the substratum (could presumably break down into numerous apomorphies)
15. No mouth appendages or mouth
16. (?) Thoracopodal muscle fiber cell of "fast" type enabling fast, discontinuous swimming bouts.
17. (?) Complex system of thoracopodal sclerites
18. Short, serrated grooming setae on thoracopodal exopods
19. Abdomen inconspicuous or non-recognizable
20. Telson small
21. Molts into attached juvenile by means of a complicated metamorphosis

12 CONCLUSION

This chapter has listed a long suite of characters that are present in cypris larvae from all three orders of the Cirripedia, but absent in the homologous cypridiform larvae (a-cyprid, y-cyprid) in the two remaining groups of the Thecostraca. The basic homology of the three types of cypridiform larvae is not in doubt, but the morphological uniqueness of the cirripede cypris supports the monophyly of the Cirripedia, a conclusion also supported by the most recent rDNA sequence data and analysis of *Hox* genes (Harris et al. 2000; Perez-Losada et al. 2002; Deutsch et al. 2003). Many of the characteristics in the cyprid are geared to enable it to carry out a complex exploratory behavior prior to cementation (Lagersson & Høeg 2002). It is therefore likely that the evolution of a "complete cypris" was a prerequisite for enabling the Cirripedia to spread to their extremely diverse types of settling habitats ranging from extremely exposed intertidal rock flats, over a vast range of biotic substrata, to the surfaces of potential host animals.

REFERENCES

Clare, A.S. & Nott, J.A. 1994. Scanning electron microscopy of the fourth antennular segment of *Balanus amphitrite amphitrite. J. Mar. Biol. Ass. UK* 74: 967-970.

Darwin, C. 1852. *A monograph on the subclass Cirripedia. The Lepadidae; or, pedunculated cirripedes.* London: Ray Society.

Darwin, C. 1854. *A monograph on the subclass Cirripedia. The Balanidae, (or sessile cirripedes).* London: Ray Society.

Deutsch, J.S., Mouchel-Vielh, E., Quénnec, È. & Gibert, J.-M. 2003. Genes, segments and tagmata in cirripedes. In: Scholtz, G. (ed), *Crustacean Issue 15, Evolutionary Developmental Biology of Crustacea*: XX-XX. Lisse: Balkema.

Elfimov, A.S. 1995. Comparative morphology of thoracican larvae: studies on the carapace. In: Schram, F.R. & Høeg, J.T. (eds), *New frontiers in barnacle evolution*: 137-152. Rotterdam: A.A. Balkema

Füllen, G., Wägele, J-W. & Giegerich, R. 2001. Minimum conflict: a divide-and-conquer approach to phylogeny estimation. *Bioinformatics* 17: 1168-1178.

Glenner, H. 1999. Functional morphology of the cirripede cypris: A comparative approach. In: Thompson, M.-F. & Nagabhushanam, R. (eds), *Barnacles: The Biofoulers*: 161-187. New Delhi: Regency Publications

Glenner, H. & Høeg, J.T. 1993. Scanning electron microscopy of metamorphosis in four species of barnacles (Cirripedia Thoracica Balanomorpha). *Mar. Biol.* 117: 431-438.

Glenner, H. & Høeg, J.T. 1994. Metamorphosis in the Cirripedia Rhizocephala and the homology of the kentrogon and trichogon. *Zool. Scr.* 23: 161-173.

Glenner, H., Høeg, J.T., Klysner, A. & Larsen, B.B. 1989. Cypris ultrastructure, metamorphosis and sex in seven families of parasitic barnacles (Crustacea: Cirripedia: Rhizocephala). *Acta Zool.* 70: 229-242.

Glenner, H., Høeg, J.T., O'Brian, J.J. & Sherman, T.D. 2000. Invasive vermigon stage in the parasitic barnacles *Loxothylacus texanus* and *L. panopaei* (Sacculinidae): closing of the rhizocephalan life-cycle. *Mar. Biol.* 136: 249-257.

Gould, S.J. & Lewontin, R.C. 1979. The spandrels of San Marco and the Panglossian paradigm: a critique of the adaptationist programme. *Proc. R. Soc. Lond. B* 205: 581-598.

Grygier, M.J. 1987. New records, external and internal anatomy, and systematic position of Hansen's Y-larvae (Crustacea: Maxillopoda: Facetotecta). *Sarsia* 72: 261-278.

Grygier, M.J. 1990. Early planktotrophic nauplii of *Baccalaureus* and *Zibrowia* (Crustacea: Ascothoracida) from Okinawa, Japan. *Galaxea*: 321-337.

Grygier, M.J. 1993. Late planktonic naupliar development of an ascothoracidan crustacean (?Petracidae) in the Red Sea and a comparison to the Cirripedia. *Contrib. Sci. Nat. Hist. Mus. Los Angeles County* 437: 1-14.

Grygier, M.J. 1994. Developmental patterns and updated hypotheses of homology in the antennules of Thecostracan nauplius larvae (Crustacea). *Acta Zool.* 75: 219-234.

Grygier, M.J. 1996a. Ascothoracida. In: Forest, J. (ed) *Traité de Zoologie, 7(2), Crustacés: Géneralités (suite) et systématique (1re partie)*: 433-452. Paris: Masson

Grygier, M.J. 1996b. des Thécostracés (Thecostraca Gruvel, 1905). SousClasse des Facetotecta (Facetotecta Grygier, 1985). In: Forest, J. (ed) *Traité de Zoologie, 7(2), Crustacés: Géneralités (suite) et systématique (1re partie)*: 425-432. Paris: Masson

Grygier, M.J. & Ito, T. 1995. SEM-based morphology and new host and distribution records of Waginella (Ascothoracida). In: Schram, F.R. & Høeg, J.T. (eds), *New frontiers in barnacle evolution*: 209-228: Rotterdam: A.A. Balkema

Harris, J.D., Maxson, L.S., Braithwaite, L.F. & Crandall, K.A. 2000. Phylogeny of the thoracian barnacles based on 18S rDNA sequence. *J. Crust. Biol.* 20: 393-398.

Høeg, J.T. 1985. Cypris settlement, kentrogon formation and host invasion in the parasitic barnacle *Lernaeodiscus porcellanae* (Müller) (Crustacea: Cirripedia: Rhizocephala). *Acta Zool.* 66: 1-45.

Høeg, J.T. 1987. The relation between cypris ultrastructure and metamorphosis in male and female *Sacculina carcini* (Crustacea, Cirripedia). *Zoomorphology* 107: 299-311.

Høeg, J.T. 1992. Rhizocephala. In: Harrison, F.W. & Humes, A.G. (eds), *Microscopic Anatomy of Invertebrates*: 313-345. New York: Wiley-Liss

Høeg, J.T. & Kolbasov, G.A. 2002. Lattice organs in y-cyprids of the Facetotecta and their significance in the phylogeny of the Crustacea Thecostraca. *Acta Zool.* 83: 67-79.

Høeg, J.T., Hosfeld, B. & Jensen, P.G. 1998. TEM studies on the lattice organs of cirripede cypris larvae (Crustacea, Thecostraca, Cirripedia). *Zoomorphology* 118: *195-205*

Huys, R., Boxshall, G.A. & Lincoln, R.J. 1993. The tantulocaridan life cycle: The circle closed. *J. Crust. Biol.* 13: 432-442.

Jensen, P.G., Høeg, J.T., Bower, S. & Rybakov, A.V. 1994. Scanning electron microscopy of lattice organs in cyprids of the Rhizocephala Akentrogonida (Crustacea Cirripedia). *Can. J. Zool.* 72: 1018-1026.

Kolbasov, G.A., Høeg, J.T. 2000. External Morphology of females in the burrowing barnacles *Lithoglyptes mitis* and *L. habei* (Lithoglyptidae) and the phylogenetic position of the Cirripedia Acrothoracica (Crustacea: Thecostraca). *Arthropoda Selecta* 9: 13-27.

Kolbasov, G.A., Høeg, J.T. & Elfimov, A.S. 1999. Scanning electron microscopy of acrothoracican cypris larvae (Crustacea, Thecostraca, Cirripedia, Acrothoracica, Lithoglyptidae). *Contrib. Zool.* 68: 143-160.

Lagersson, N.C. 2002. The Ultrastructure of two types of muscle fiber cells in the cypris *Balanus amphitrite* (Crustacea: Cirripedia). *J. Mar. Biol. Ass. UK* 82: 573-578.

Lagersson, N.C. & Høeg, J.T. 2002. The relation between antennulary biomechanics and settlement behavior in cypris larvae of *Balanus amphitrite* (Crustacea: Thecostraca: Cirripedia). *Mar. Biol.* 141: 513-526.

Lagersson, N.C., Garm, A. & Høeg, J.T. 2003. Notes on the ultrastructure of the setae on the fourth antennulary segment of the *Balanus amphititre* cyprid (Crustacea: Cirripedia: Thoracica). *J. Mar. Biol. Ass. UK* 83: 361-365.

Martin, J.W. & Davis, G.E. 2001. An updated classification of the recent Crustacea. *Science series* 39: 1-124.

Mizrahi, L., Achituv, Y., Katcoff, D. J. & Perl-Treves, R. 1998. Phylogenetic position of *Ibla* (Cirripedia: Thoracica) based on 18S rDNA sequence analysis. *J. Crust. Biol.* 18: 363-368.

Moyse, J., Høeg, J.T., Jensen, P.G. & Al-yahya, H.A.H. 1995. Attachment organs in cypris larvae: using scanning electron microscopy. In: Schram, F.R. (ed), *New frontiers in barnacle evolution*: 153-177. Rotterdam: A. A. Balkema.

Nott, J.A. & Foster, B.A. 1969. On the structure of the antennular attachment organ of the cypris larva of *Balanus balanoides* (L.). *Phil. Trans. R. Soc. Lond. B* 256: 115-134.

Perez-Losada, M., Høeg, J.T., Kolbasov, G.A. & Crandall, G.A. 2002. The phylogenetic position of the Facetotecta and reanalysis of the relationships among the Cirripedia and the Ascothoracida using 18S dDNA sequences (Crustacea: Maxillopoda: Thecostraca). *J. Crust. Biol.* 22: 661-669.

Rybakov, A.V. & Høeg, J.T. 2002. The ultrastructure of retinacula in the Rhizocephala (Crustacea: Cirripedia) and their systematic significance. *Zool. Anz.* 241: 95-104.

Spears, T., Abele, L.G. & Applegate, M.A. 1994. Phylogenetic study of cirripedes and selected relatives (Thecostraca) based on 18s rDNA sequence analysis. *J. Crust. Biol.* 14: 641-656.

Thompson, J.V. 1830. On the cirripedes or barnacles; demonstrating their deceptive character; the extraordinary metamorphosis they undergo, and the class of animals to they indisputably belong. *Zoological researches, and Illustrations; or Natural History of nondescript or imperfectly known animals.* Vol. 1, Part 1, Memoir IV: 69-82; Plates IX & X: 87-88. Cork. (Sherborn Fund Facsimile No. 2, 1968, Soc. Bibliog. Nat. Hist.: London).

Thompson, J.V. 1835. Discovery of the metamorphosis in the second type of Cirripedes, vis the Lepades, completing the natural history of these singular animals, and confirming their affinity with the Crustacea. *Phil. Trans. R. Soc. Lond. 1B*: 355-358.

Thompson, J.V. 1836. Natural history and metamorphosis of an anomalous crustaceous parasite of *Carcinus maenas*, the *Sacculina carcini. Ent. Mag. Lond.* 3: 456.

Walker, G. 1973. Frontal horns and associated gland cells of the nauplii of the barnacles *Balanus hameri, Balanus balanoides* and *Elminius modestus* (Crustacea Cirripedia). *J. Mar. Biol. Ass. UK* 53: 455-463.

Walker, G. 1992. Cirripedia. In: Harrison, F.W. & Humes, A.G. (eds) *Volume 9 Crustacea*: 290-296. New York: Wiley-Liss, inc.

Walley, J. 1969. Studies on the larval structure and metamorphosis of *Balanus balanoides* (L.). *Phil. Trans. R. Soc. Lond.* B 256: 237-280.

Walossek, D. 1993. The upper Cambrian *Rehbachiella* and the phylogeny of the Branchiopoda and Crustacea. *Fossils and Strata* 32: 1-202.

Walossek, D., Høeg, J.T. & Shirley, T.C. 1996. Larval development of the rhizocephalan cirripede *Briarosaccus tenellus* (Maxillopoda: Thecostraca) reared in the laboratory: A scanning electron microscopy study. *Hydrobiologia* 328: 9-47.

On the ontogeny of the Branchiopoda (Crustacea): contribution of development to phylogeny and classification

JØRGEN OLESEN
Zoological Museum, University of Copenhagen, Denmark

1 INTRODUCTION

Branchiopods exhibit a wide range of developmental strategies, extending from a relatively anamorphic, presumably primitive, development with many larval stages in the Anostraca and the Upper Cambrian branchiopod *Rehbachiella kinnekullensis* Walossek & Müller, 1983, to a very abbreviated development without free-living larval stages in most cladocerans. While anamorphic developing branchiopods, such as anostracans, are among the most primitive crustaceans, at least with respect to their gradual development over many stages starting with a nauplius, many of the cladocerans, with their embryonized development, are among the most modified.

The development of the Branchiopoda has received extensive attention and significant contributions are present in the literature dating back almost 250 years (Schaeffer 1756), and various taxa have been studied in almost all possible ways, possibly due to the appealing morphology of many taxa but probably also because they in general are easy to hatch and culture from dried resting eggs. Probably G.O. Sars was the first to explore the possibilities in hatching branchiopods from dried mud from various parts of the world (Sars 1885, 1887), a quality that has been important for the choice of *Artemia salina* Linnaeus, 1758 (Anostraca), as an important model animal used in developmental studies of many types. Much new information about development and phylogeny of the Branchiopoda has been published in recent years. Especially the precise and extensive description by Walossek (1993) of the spectacular *Rehbachiella kinnekullensis,* an Upper Cambrian fossil with many similarities to Recent branchiopods, is of importance. No matter its precise phylogenetic position, *Rehbachiella* has an enormous importance for the understanding of early branchiopod (and crustacean) development and phylogeny, due partly to its great age and partly to its many similarities to Recent branchiopods. Other new information, which is potentially important for considerations of homology and phylogeny, stems from the 'evo-devo' field using different molecular markers. Until now only few studies have been applied to the Branchiopoda (Olesen et al. 2001; Williams et al. 2001; Shiga et al. 2002; Williams 2003). Descriptions of various aspects of the development of many taxa are available in literature; yet only few have attempted broader comparison of major branchiopod taxa, the works of Fryer (1983, 1988) and Walossek (1993) being important recent exceptions. The purpose of the present chapter is to summarize in detail selected aspects of branchiopod development, partly by use of relevant literature but also by including new information, and to explore to what extent developmental data can be used to establish homologies between branchiopod taxa and how they can contribute to the understanding of branchiopod phylogeny and classification. An attempt has been made to integrate information provided by completely different methods (SEM, molecular markers) stemming from recent as well as from fossil taxa. Focus will be on the later part of the ontogeny. At the end of the chapter I provide a phylogeny for the Branchiopoda based mostly on – or at least consistent with – developmental characters or characters relating to external morphology. It is important to note already now that this phylogeny is not the result of a comprehensive and formal phylogenetic analysis of a complete dataset. Instead it focuses on some new characters, identified in connection to this work, and also on some of the more convincing characters from already published works. The presented phylogeny has been translated into a higher-level classification, which will be presented here at the end of the Introduction since many of the higher-level categories will be used throughout the following text.

2 HIGHER-LEVEL CLASSIFICATION OF THE BRANCHIOPODA

Pan-Branchiopoda [1] (= Branchiopoda *s. lat.*)
Rehbachiella kinnekullensis Müller, 1983
Branchiopoda [2] (= Branchiopoda *s. str.* = crown group) Latreille, 1817
    Lepidocaris rhyniensis (Lipostraca Scourfield, 1926)
    Anostraca Sars, 1867
        Phyllopoda Preuss, 1951
            Notostraca [3] Sars, 1867
            Diplostraca Gerstaecker, 1866
                Laevicaudata Linder, 1945
                Spinicaudata [4] Linder, 1945
                Cladoceromorpha Ax, 1999
                      Cyclestheria hislopi (Cyclestherida [5] Sars, 1899)
                      Cladocera Latreille, 1829
                          Ctenopoda Sars, 1865
                          Leptodora kindtii (Haplopoda Sars, 1865)
                          Adantennulata [6] new taxon
                              Anomopoda Sars, 1865
                              Onychopoda Sars, 1865

---

[1] 'Pan-Branchiopoda' (= Branchiopoda *s. lat.*) is used for the Branchiopoda (= Branchiopoda *s. str.*) including its stem lineage (following the terminology suggested by Lauterbach 1989, but also in accordance with the concepts precisely formulated by Ax 1985, 1989).

[2] 'Branchiopoda' (= Branchiopoda *s. str.*) includes the Recent branchiopods (Anostraca, Notostraca, 'Conchostraca' and Cladocera) and *Lepidocaris* no matter the latter's precise position. This is based on a number of shared characters between these taxa (see main text for details). This means that *Rehbachiella kinnekullenis*, based on the phylogenetic position suggested in Fig. 16, is not part of the Branchiopoda in the strict sense.

[3] The extinct, notostracan-like Kazachartra is not considered in this account since no larval stages are known (see McKenzie et al. 1991).

[4] 'Spinicaudata' does not include Cyclestheriidae (*Cyclestheria hislopi*), but only the remaining spinicaudate families, and is therefore used in a more restricted sense than originally suggested by Linder (1945) (similar to the use by Negrea et al. 1999 and Martin & Davis 2001).

[5] 'Cyclestherida' – for which Sars (1899) is here credited since he erected the family Cyclestheriidae – contains the monotypic *Cyclestheria* (similar to the use by Negrea et al. 1999 and Martin & Davis 2001).

[6] 'Adantennulata' means 'antennules (1st antennae) close together' and refers to the special ontogeny of the antennules in the Onychopoda and Anomopoda where the antennular limb buds move closer together during ontogeny (see Olesen et al. 2003 and below for more details).

2.1 *Comment on the classification of the Branchiopoda*

The classification above is a direct translation of the branchiopod phylogeny presented in Fig. 16. It can be discussed whether such a classification is at all necessary since a phylogenetic tree in many ways is sufficient (and better) to express relationships between living organisms. This is well-known and widely accepted in systematic research. However, it is possible to translate a phylogeny directly into a classification, which has been done above. Ideally, such a phylogeny based classification should reflect only sister group relationships, but this has not been possible in all cases since not all sister group relations have been resolved in the phylogeny shown in Fig. 16. The three polytomies included in Fig. 16 (indicated by numbers 1 to 3 in Fig. 16) are illustrated in the classification above by placing the respective taxa at the same level (compare the classification with Fig. 16). These polytomies will eventually be resolved. An important question concerns in what sense to use the term Branchiopoda (= Branchiopoda *s. str.*) if the uppermost polytomy in Fig. 16 (marked as '1') becomes resolved. The question is: should *Lepidocaris rhyniensis* Scourfield, 1926, stay inside the Branchiopoda (Branchiopoda *s. str.*) or should it be placed outside? This, of course, depends on the actual position of *Lepidocaris*. If it turns out to be the sister group to the Anostraca, then, obviously, it should stay inside. If, in contrast, *Lepidocaris* turns out to be in a sister group position to the remaining branchiopods (Anostraca, Notostraca, 'Conchostraca' and Cladocera) then it *could* be argued that it should not be included in the Branchiopoda, but instead should be treated as sister group (or stem group) to this taxon. However, because of the very strong characters supporting *Lepidocaris* as very closely related to the Recent Branchiopoda, I prefer to keep *Lepidocaris* *within* the Branchiopoda (= Branchiopoda *s. str.*) no matter which phylogenetic position – of the those mentioned above - is chosen (see below for the morphological support). Another approach, not preferred here, would be to define the Branchiopoda (Branchiopoda *s. str.*) as including the latest ancestor to all Recent branchiopods and all its descendents. This would have the disadvantage that the content of the Branchiopoda (Branchiopoda *s. str.*) would depend on the phylogenetic position of *Lepidocaris,* which is uncertain at the moment. If *Lepidocaris* were preferred as sister group to the Anostraca, it would be included in the Branchiopoda (= Branchiopoda *s. str.*), whereas, if it were preferred as sister group to all the Recent Branchiopoda, it would not be included.

The classification shown above is in general respects similar to the branchiopod classification suggested by Martin & Davis (2001). The most important differences are 1) that two branchiopod fossils have been included in the present classification, 2) that *Cyclestheria hislopi* (Baird, 1859) (Cyclestherida) and the Cladocera have been combined in the taxon Cladoceromorpha (as suggested by Ax 1999), and 3) that the Anomopoda and Onychopoda have been combined in a new taxon based on similarities in the embryology. Another difference is that no ranks (order, suborder, etc.) have been applied to the higher-level categories used. One reason for this is that, if all the monophyla in Fig. 16 were designated a specific rank in the classification presented above, there would simply not be enough ranking categories. It would be necessary to construct quite a few of these, which would only add to confusion. Another reason is that, if traditional ranks had been used, then some would probably focus on these ranks instead of what is important in a study like this: the monophyly and relationship of the included taxa.

2.2 *Short note on the use non-monophyletic categories and abbreviated taxon names*

Some terms for probable non-monophyletic groups have of convenience been used a various places in the text. These include the 'large' Branchiopoda, which is the traditional way of referring to those Recent branchiopods (Anostraca, Notostraca, and 'Conchostraca') that are not cladocerans. It has been useful in the present account to have a term for these, non-cladoceran, branchiopods since they in many ways are so similar. However, if our current notion of branchiopod phylogeny is correct, this group is paraphyletic with respect to the Cladocera. The 'Conchostraca' (apparently also paraphyletic) is also used sometimes in the text for convenience. When 'Recent Branchiopoda' is used it refers to all branchiopods, excluding *Rehbachiella* and *Lepidocaris*.

In some cases, only the genus name is used when referring to a species. This is done in those cases where only a single species is included in a higher-level taxonomic category, and where no misunderstandings are possible. This is the case for *Rehbachiella, Lepidocaris, Cyclestheria,* and *Leptodora*.

3 DEVELOPMENT OF BRANCHIOPOD TAXA

The larval or embryonic development of all major branchiopod taxa summarized below is based on information from literature and from examination of various taxa by SEM undertaken as part of this study. The literature on the subject is vast, so this paper concentrates on external morphology, mostly focusing on the development of appendages and general body proportions, especially those features important for phylogenetic considerations (given later in the chapter). The function of various appendages is summarized when such information is available. Most considerations of homology and phylogeny are delayed to later sections, but discussion and interpretation of certain morphological details are given here.

Table 1. Collecting data for the material used for various original figures.

Species	Locality	Date	Collector
Eubranchipus (Siphonophanes) grubii (Dybowski, 1860) (Figs. 2, 3)	Denmark	1999-2000	J.O.
Triops cancriformis (Bosch, 1801) (Fig. 5)	Austria	2001	Resting eggs collected by Erich Eder (Vienna) and cultured in lab. by Ole Sten Møller (Copenhagen)
Lynceus brachyurus Müller, 1776 (Figs. 6, 7)	Denmark	1997-2000	J.O.
Eulimnadia braueriana Ishikawa, 1895 (Fig. 9)	Japan	2000-2001	Mark J. Grygier
Diaphanosoma brachyurum (Liévin, 1848) (Fig. 12)	Denmark	1993	J.O.
Simocephalus vetulus (Müller, 1776) (Fig. 13)	Denmark	1993	J.O.
Macrothrix laticornis (Jurine, 1820) (Fig. 13)	Denmark	1985	Ulrik Røen
Polyphemus pediculus (Linnaeus, 1761) (Fig. 14)	Denmark	1993	J.O.

3.1   Rehbachiella kinnekullensis - *an Upper Cambrian fossil with branchiopod affinities*

Rehbachiella kinnekullensis Müller, 1983 (Fig. 1), is an Upper Cambrian (> 500 My) crustacean microfossil with many similarities to Recent Branchiopoda. It belongs to the so-called 'Orsten' fauna of microfossils (< 2 mm) from localities in southern Sweden, characterized by a remarkable preservation of specimens in three-dimensions with plenty of details visible. Due to the detailed reconstruction by Walossek (1993), *Rehbachiella* is now known in considerable detail, which facilitates comparisons with Recent Crustacea. Unless otherwise stated, the following information is from Walossek (1993, 1995).

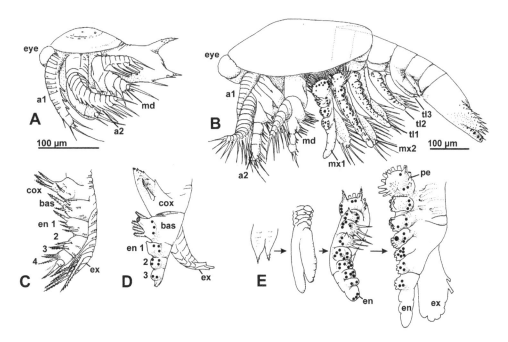

Figure 1. Two larval stages and various limbs of the Upper Cambrian branchiopod 'Orsten' fossil *Rehbachiella kinnekullensis* Müller, 1983. (A) Second naupliar stage. (B) Relatively early post-naupliar stage. (C) Right second antenna of the fourth naupliar stage. (D) Left mandible of the fourth naupliar stage. (E) Development of second maxilla showing degree of development in fourth naupliar stage and three post-naupliar stages. The development of the second maxilla and morphology of largest known stage is basically similar to that of the trunk limbs. Illustrations reproduced from Walossek (1993) with kind permission. Abbreviations: a1 = antennae 1, a2 = antenna 2, bas = basis, cox = coxa, en = endopod, ex = exopod, md = mandible, mx1 = maxilla 1, mx2 = maxilla 2, pe = proximal endite, tl1 = trunk limb 1, tl2 = trunk limb 2, tl3 = trunk limb 3.

According to Walossek (1993), who examined more than 130 specimens, the developmental sequence from the earliest stage to the largest known consists of 30 stages. No stages that with certainty can be considered adult have been described, and no sexual dimorphism has been recognized. The first stage is about 160 μm long, and the largest known stage is about 1.7 mm. The larval series is long and very gradual. Apparently two molts are needed for the addition of one somite and a single pair of limbs, except for a couple of places in the naupliar phase where only one molt is needed for the addition of each somite (Walossek 1993). However, it remains uncertain whether dorsal delineation of segments, seen at the caudal end before the appearance of a pair of limb buds, actually *does* represent a molt or should be interpreted as occurring in the interphase between two molts. If so, it would reduce the number of stages in the development of *Rehbachiella* significantly compared with the interpretation by Walossek (1993), but it would still be among the most anamorphic developing crustaceans.

The lifestyle of *Rehbachiella* must have changed through the lifecycle of the animal. The nauplii were probably not filter feeders but may have swept in particles while swimming. The more posterior appendages were added progressively and became functional gradually, but the naupliar appendages retained their functions and shape over a long period. This means that both the naupliar and thoracopodal apparatus functioned hand in hand for some time in the later stages. In the later stages, *Rehbachiella* was probably a swimmer, propelled by the rhythmic metachronal beat of the trunk limbs and the maxillae, which at the same time was used for filter feeding. In the largest known stage, however, the second antennae still have a considerably size, so these appendages have probably also acted as propulsive organs. Many of these features are strikingly similar to those of anostracans.

The first stage is a nauplius with three pairs of functional limbs of about the same size and a short hind body devoid of limbs (slightly later stage shown in Fig. 1A), shorter than that in larvae of Recent branchiopods. The largest known stage is about 1.7 mm and comprises 13 post-maxillary segments, 12 of which carry limbs developed to varying degrees. The anterior 8-9 pairs of limbs in this largest known stage are well developed; the last 3-4 pairs are less-developed and it is therefore clear that the development has not yet terminated.

The naupliar appendages are more or less completely developed and functional already by the first stage, which is in contrast to various Recent 'large' branchiopods. Presumed eye lobes are present in the early nauplii (Fig. 1A, B), but their fate during late development is unknown. The first antennae of the early nauplii are uniramous, large and robust and composed of a thicker proximal part subdivided into a number of incomplete annuli and three cylindrical podomeres distally (Fig. 1A, B). They increase slowly in size during the early stages, but are never preserved in later stages. The second antennae of the first nauplius are biramous with the corm subdivided into a distinct coxa and basis, an endopod composed of 4 podomeres (in the second nauplius), and a long exopod with 7-8 cylindrical podomeres (Fig. 1C). The coxa, basis and first endopod podomere all have elongate processes that terminate in a rigid spine or spine-like seta directed into the atrium oris and therefore active in food manipulation, as they are in various Recent branchiopods (but they are of a different morphology) (Fig. 1C). In the early stages the second antennae are post-oral and appear to be slightly dominant in food manipulation compared with the mandibles to which they are very similar. After a few stages the second antennae have shifted anteriorly, with long endites directed posteromedially around the corners of the labrum, and start to increase considerably in size. The endopod and the exopod increase in size, and in intermediate post-naupliar stages the latter is comprised of 17 ringlets. This enlargement could be related to a possible function of the second antennae as the major locomotory organ. In later stages, the second antennae seem to undergo reduction related to an increased functionality of the post-mandibular limbs. The spines on the coxa and basis are still directed around the corners of the labrum in the largest known stage. The mandibles are similar in design and size to the second antennae but are smaller than these (Fig. 1D). In the nauplius, the basis is the principal structure, having a large body, which is drawn out medially into a masticatory spine. Later, the coxal endite grows and develops a triturating surface with acute denticles (Fig. 1D). With further growth, the coxal gnathobase enlarges significantly and develops in the latest known stages into an enormous blade-like structure with a concave surface. During growth the mandibles shift slightly anteriorly as do the second antennae. After a few stages the palp (basis and rami) stops increasing in size and undergoes atrophy (as in Recent branchiopods) but is still not completely reduced by the last known stage. The first and second maxillae are both well-developed and trunk limb-like in contrast to those of Recent branchiopods. The first maxillae start their development as a small spine in the second nauplius. In the following stage, an upright bifid lobe has appeared, indicating the future endopod and exopod. In the first post-naupliar stage, the limb has been transformed into an appendage consisting of a wide endite-bearing corm (four endites) continuing into a free, three-segmented inner branch, termed the endopod by Walossek (1993). Laterally, where the oblique corm begins to narrow, is attached a primordial, flattened exopod. These are the basic components of the first maxillae, which become enlarged and more setose during development. Especially the proximal endite becomes very large compared to the more distal endites. The limbs posterior to the first maxillae are virtually identical and deviate only slightly from the first maxillae; most notably, they are larger and have more endites. In contrast to the first maxillae, they are formed as a *bifid* upright limb bud, each bud projecting into a spine (Fig. 1E). The bifid limb bud enlarges and develops a number of spine clusters along the inner margin (Fig. 1E). In an intermediate stage of development, the post-maxillulary limbs have a wide, relatively flattened corm with 6 primordial, bifid endites along the inner margin. Distally the corm narrows and continues, without any profound distinction, into a four-segmented endopod (Fig. 1E). A flattened plate-like exopod with distal setation arises at the oblique lateral margin of the limb where the corm starts narrowing (Fig. 1E). Later in development, the number of endites along the inner margin of the corm increases to nine (?) in the largest known stage. Additionally, the endopod becomes 4-segmented, and the plate-like exopod also enlarges. The setation of each endite typically consists of one or two anterior rows, a semicircular posterior row, and a median group of setae, something that is shared with the early development of setation in various 'large' branchiopods like that of *Cyclestheria hislopi* (Baird, 1859) (compare Walossek 1993, pl. 17:2 with Olesen 1999, Fig. 10). The proximal endites of all post-maxillulary limbs are enlarged, with denser setation, which curve (anteriorly), brush-like, into the food groove. In general, all post-mandibular limbs are similar to the trunk limbs (but with more endites and a segmented endopod) of various Recent 'larger' branchiopods, but the development is different, which will be explored in more detail in a later section.

At least two similarities, the food groove and similar setose endites of the trunk limbs, between *Rehbachiella* and the Branchiopoda, identified by Walossek (1993), qualify as likely synapomorphies, and it appears therefore appropriate to include *Rehbachiella* in the Branchiopoda, as did Walossek (1993), in this chapter treated as Pan-Branchiopoda (or Branchiopoda *s. lat.*).

3.2 *Anostraca*

Among the most important accounts of anostracan development are those of Claus (1873, 1886), Anderson (1967), Fryer (1983) and Schrehardt (1987). Fryer (1983), who provided the most thorough and critical modern treatment, described the larval development of *Branchinecta ferox* (Milne-Edwards, 1840) as strictly anamorphic and noted that approximately 22 instars are passed from hatching as a non-feeding nauplius to the time that an adult way of life is pursued, at which time the animals are about 4.0 mm in length. Fryer also noted that that the species begins to produce eggs at a length of about 17.5 mm and may continue to grow to a length of at least 45 mm. A comparable anamorphic type of development was described for *Artemia salina* in the detailed work of Benesch (1969). Unless otherwise stated, the following is from Fryer (1983).

The first nauplius, which appears from the egg, is a tiny, ellipsoid (< 0.5 mm), non-feeding nauplius with much yolk and almost circular in section. The ellipsoid body is divided into an anterior, appendage bearing half, and a posterior half devoid of limbs. Three pairs of appendages are present, 1st antennae, 2nd antennae and mandibles. The 1st antennae are small tubular limbs, which serve sensory purposes and play no role in either feeding or swimming. The large 2nd antennae are the sole organs of propulsion and long natatory setae of an annulated exopod take care of this function. Posteroventrally each antenna bears a small endopod. The two characteristic masticatory spines posteroventrally at the protopod of each antennae are not yet functional as they lack the setules and spinules that later facilitate food handling. Likewise the rudimentary mandibular gnathobases are not yet functional since they are placed too far apart and a pair of gnathobasic spines with sweeping functions is directed in between them (one spine on each mandible). The mandibular palps (become later segmented into a basis and two probable endopodal segments, see Fig. 3A) are also not functional since their setae are devoid of setules. A large, yolk filled labrum covers the mandibular gnathobases. The stomodeum is in full connection with the mid-gut but the proctodaeum is not, suggesting a non-feeding status. The above-mentioned description can be confirmed and supplemented by SEM observations by Schrehardt (1987) for *Artemia* (species not specified in paper by Schrehardt). Also in *Artemia* the hatching stage is clearly non-feeding since the scanning electron micrographs show that the masticatory spines of the antennae, the mandibular gnathobase and apparently also the setae on the mandibular palp are rudimentary (Schrehardt 1987).

The second nauplius of *Branchinecta ferox* is about 0.5 mm in length. A number of changes between the first and second nauplius are connected to the onset of feeding. These include the development of an armature of spinules or setules on both masticatory setae of the antennae, development of an armature of setae of the mandibular palp, an increase in size of the mandibular gnathobases, and development of a connection between the proctodaeum and the midgut (see naupliar appendages of a comparable larval stage of *Eubranchipus grubii* (Dybowski, 1860) (Fig. 2). Barlow & Sleigh (1980) described the swimming behavior of larvae of *Artemia* but did not deal with feeding. According to Fryer (1983), the natatory setae of the antennae are concerned solely with locomotion, which is in contrast to the belief of Cannon (1928) and Gauld (1959), who thought they were also involved in feeding. Fryer (1983) points out that instead the basipodal masticatory spines of the antennae are specialized for the sieving of particles and conveying them towards the mouth. When the basipodal masticatory spines leave the neighborhood of the labrum and move anteriorly and laterally during the return phase of the antennal cycle, they sweep past the setae of the mandibular palp which are moved in the opposite direction and thereby clean the basipodal masticatory spines of particles (Fryer 1983). The mandibular palps are suited for entrapping particles and for carrying them towards the labral gland secretions and within the ambit of the very mobile coxal masticatory spines of the antennae which pass freely beneath the labrum and which can then convey material to the vicinity of the mandibular gnathobases (Fryer 1983). The described feeding mechanism can be called the 'naupliar feeding apparatus' and persists with few modifications in *B. ferox* very long.

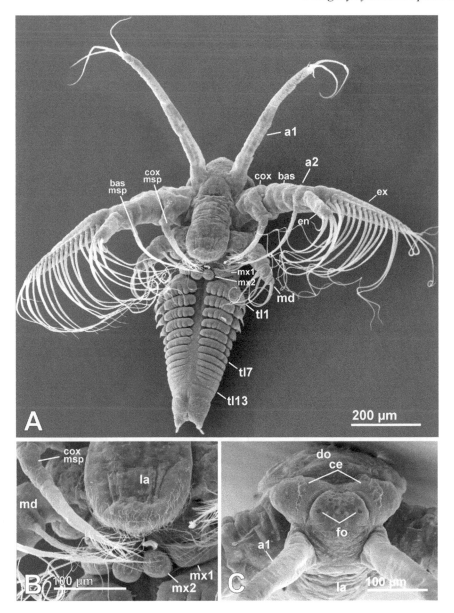

Figure 2. Larval stages of *Eubranchipus (Siphonophanes) grubii* (Dybowski, 1860) (Anostraca). Original. (A) Intermediate larval stage with naupliar appendages. First, second maxillae and trunk limbs still nonfunctional limb buds. (B) Close-up of A showing distal margin of labrum, first and second maxillae and proximal masticatory spine on second antenna. (C) Frontal view of same specimen as in A showing compound eyes, frontal organs, dorsal organ and first antennae. Abbreviations: a1 = antenna 1, a2 = antenna 2, bas = basis, bas msp = basipodal masticatory spine (a2), ce = compound eye, cox = coxa, cox msp = coxal masticatory spine (a2), do = dorsal organ, en = endopod, ex = exopod, fo = frontal organ, la = labrum, md = mandible, mx1 = maxilla 1, mx2 = maxilla 2, tl1 = trunk limb 1, tl7 = trunk limb 7, tl13 = trunk limb 13.

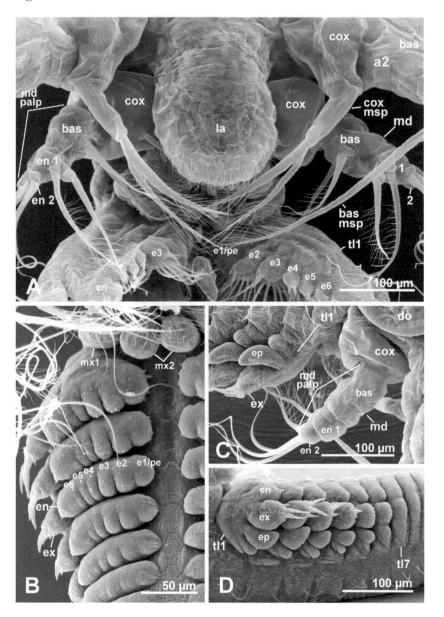

Figure 3. Larval stages of *Eubranchipus (Siphonophanes) grubii* (Dybowski, 1860) (Anostraca). Original. (A) Late larval stage with setose and movable anterior trunk limbs – close-up of labrum area. (B) Buds of right side trunk limbs in intermediate larval stage (same specimen as in 2A). (C) Outer view of right side trunk limbs and mandible of an intermediate stage between those shown in Figs. 2A and 3A. (D) Buds of right side trunk limbs of intermediate stage (same specimen as in Fig. 2A). Abbreviations: a2 = antenna 2, bas = basis, bas msp = basipodal masticatory spine (a2), cox = coxa, cox msp = coxal masticatory spine (a2), e1 to e6 = endite 1 to 6, en = endopod, ep = epipod, ex = exopod, la = labrum, md = mandible, mx1 = maxilla 1, mx2 = maxilla 2, pe = proximal endite, tl1 = trunk limb 1, tl7 = trunk limb 7.

Among these modifications are that - between nauplius 2 and 3 - the molar surfaces of the mandibles enlarge and become functional, the gnathobasic spines, which were earlier directed between the gnathobases, become reduced in size, and the coxal masticatory spines of the antennae become bifurcate, probably to facilitate sweeping of particles towards the mandibular gnathobases (larvae of *E. grubii* with bifurcate coxal masticatory spines of A2 shown in Fig. 3A). Many modifications of the naupliar appendages take place during later development, among which the reduction of the mandibular palp (only the coxa remains) and the reduction/modification of the second antennae are some of the most significant. Some populations of *Polyartemia forcipata* (Fischer, 1851) are partly an exception as a mandibular palp is present also in adults (Ekman 1902).

The post-mandibular appendages develop very gradually in *Branchinecta ferox* (see Fryer 1983) and in *Artemia salina* (see Benesch 1969), as probably is the case in all anostracans, and the trunk limbs become functional usually at the rate of one pair per instar. The trunk limbs do not take over complete responsibility for locomotion and feeding before stage 22 in *B. ferox*, and just before that, the naupliar mechanism operates simultaneously with an almost complete adult mechanism. The following details of the development of the post-mandibular limbs can be mentioned. The *Anlagen* of the maxillules, maxillae, and first five pairs of trunk limbs can be recognized already from the second nauplius stage (see Fig. 2 for a slightly later stage of *Eubranchipus grubii*). The maxillules and the first pair of trunk limbs are the first to become movable, which begin at stage 11, but they are still not functional since the naupliar apparatus takes care of feeding and locomotion. In stage 12 the two first pairs of trunk limbs are functional, the third pair is only movable but not capable of handling particles. With each successive molt an additional pair of trunk limbs becomes active. The contribution of the trunk limbs to feeding and locomotion is gradually increased.

In *Eubranchipus grubii*, the early trunk limbs are formed as pairs of limb buds on the ventrolateral 'corners' of the trunk (Fig. 3B, D), and the same is clearly shown for *Artemia* by Schrehardt (1987) and for *Branchipus schaefferi* Fischer, 1834, by Schlögl (1996). The future proximal part of the adult limb is orientated medially, the future distal part laterally. The limb buds become gradually divided into portions (future endites) beginning proximally, and an exopod and three epipods (Fig. 3). Hence, the full number of portions into which the early limb buds are divided is as many as eleven. There are six endites indicated along the future median margin of the limb buds followed distally by an endopod and a slightly larger exopod and 'dorsally' at the limb buds by three portions representing epipods (Fig. 3). The trunk limbs of adult anostracans have five endites (not six as in larvae) along the inner margin of the trunk limbs followed by an unsegmented endopod, and thereby similar to the trunk limbs of 'conchostracans' and notostracans. This similarity is only superficial since it can be shown that the most proximal, large endite in the adults is formed by a fusion of two small endites present in larval stages (marked e1/pe and e2 in Fig. 3A, B). Fryer (1983) pays some attention to the setal development along the inner margin of the trunk limbs of *Branchinecta ferox* and remarks that the limbs become functional long before they have acquired the full number of spines and setae.

An early stage of *Eubranchipus grubii* has frontally a pair of small depressions (Fig. 2C) that bears some similarity to a pair of 'frontal organs' found in *Lepidocaris* in a similar position (Fig. 4B). The thorax of an adult *Artemia* is comprised of 11 phyllopod-bearing somites and two partly fused genital somites, six limbless abdominal segments and a telson (Schrehardt 1987). This tagmosis is the general for anostracans, but a notable exception is *Polyartemia forcipata*, which has 19 phyllopod-bearing somites (Sars 1896b).

In females the two genital segments give rise to a brood pouch (of different design in different taxa) and to paired penial structures in males. Claus (1886) provided anatomical details on the various types of brood pouches in different taxa, and Schrehardt (1987) gave SEM figures of brood pouches of *Artemia* in different stages of development. Tissue from the first genital segment seems to go into the formation of the anterior side of the brood pouch, while that of the second provides material for the posterior side (Claus 1886).

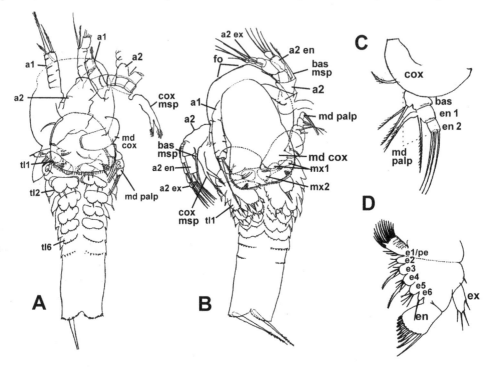

Figure 4. Two larval stages and various limbs of the Devonian branchiopod fossil *Lepidocaris rhyniensis* Scourfield, 1926. (A) Larval stage with six somites. (B) Larval stage with seven somites. (C) Mandible of larva with six somites. (D) First trunk limbs of adult. A-C from Scourfield (1940), D from Scourfield (1926). Abbreviations: a1 = antennae 1, a2 = antenna 2, bas = basis, bas msp = basipodal masticatory spine (a2), cox = coxa, cox msp = coxal masticatory spine (a2), e1 to e6 = endite 1 to 6, en = endopod, ex = exopod, fo = frontal organ, md = mandible, mx1 = maxilla 1, mx2 = maxilla 2, pe = proximal endite, tl1 = trunk limb 1, tl2 = trunk limb 2, tl6 = trunk limb 6.

3.3 Lepidocaris rhyniensis - *a Devonian branchiopod fossil*

Lepidocaris rhyniensis Scourfield, 1926 (Fig. 4), is a very small Middle Devonian (350 My) fossil crustacean from Aberdeenshire, Scotland, with several significant similarities to the Recent Branchiopoda, but also with a number of features that distinguish it very clearly from these. Its extraordinary good preservation in three dimensions, sometimes of nearly complete specimens, has made possible a detailed description by Scourfield (1926, 1940). Unless otherwise stated, the following is from these papers by Scourfield.

The adult female and male and several larval stages are known. The youngest specimen recognized is about 0.3 mm while 3.0 mm is considered the normal length for an adult. Most likely the larval series is not completely known, but enough information is available to get a good impression of the development of most appendages and other external structures. Before summarizing the information, a short note on the counting of somites is necessary. Scourfield referred to both the cephalon and the telson as somites, whereas I will only count segments *between* these two structures as somites (as did Walossek 1993:74), thereby facilitating comparison to other taxa. This yields a total of seventeen somites. I follow the original interpretation of Scourfield (1926, 1940) concerning the reduced status of the second maxillae, but see below on this point.

The earliest stage identified has two somites between the cephalon and the telson (0.3 mm). Stages with three, six, seven and eight somites, as well as a number of later stages, have also been described. Hence, the development can be assumed to have been relatively gradual, and no distinct 'jump' in morphology can be firmly identified in the larval series, but due to the limited number of larval stages described, no definite conclusions can be made. Not all parts of the morphology are known from all stages, but a picture of the ontogeny of various structures can be put together. Information on the larval stages is mostly based on two extraordinarily well-preserved larvae with six and seven somites respectively, described by Scourfield (1940) as an addendum to the original paper of 1926 (see Fig. 4).

The frontal margin of the head appears to have been nearly straight in larvae with seven somites. It is possible that a pair of 'frontal organs' is represented by two shallow depressions on the anterior edge (Fig. 4B) (compare to similar structures in anostracans: Fig. 2C). The labrum is much larger in the larva relative to total body length than in the adults. It is constricted in the middle of its length and has a broad posterior margin fringed with setae. The first antennae in a seven-somite stage consist of a single elongate segment, about three times longer than broad, with several partial rings of minute teeth and with three terminal setae. In the adult, the first antennae are still uniramous but have become three-segmented, something that is unique among branchiopods, with the possible exception of *Rehbachiella*. The morphology of the second antennae is best described from six- and seven-somite stages, but Scourfield was also able to identify largely the same characteristic components from a two-somite stage. The second antennae each have a large protopod of two segments, an endopod of three segments, and an exopod of five. From the base of the protopod (coxa) arises a very large bifid seta (masticatory process/spine), and on its distal segment (basis) there is a strong simple seta. The bifid enditic setae on the protopods (coxa) persist at least until stages of half the adult length, but they have disappeared in the adult. These features are very similar to the larval second antennae of Recent Branchiopoda (compared below). In adults of both sexes the second antennae are very large, with a three-segmented protopod and a long five-segmented exopod with numerous, long, presumed swimming setae. The endopod is two-segmented in females and three-segmented in the presumed males; the tip of the latter is provided with 9 to 12 short conical 'papillae'. The second antennae must have been the most important organs of propulsion in both larvae and adults. A mandible consisting of an enlarged proximal segment (coxa with gnathobase) and a uniramous three-segmented palp (Fig. 4C), of a similar morphology to that of Recent 'large' branchiopods, is already present in the earliest stages examined. It consists of a basis and two probable endopodal segments. The gnathobase is equipped with a gnathobasic seta as is seen in anostracans (see Fryer 1983). The first two segments of the palp (basis and endopod segment 1) are equipped with two setae each, the distal-most with three. The mandibular palp has completely atrophied in the adult, and only an enlarged coxal gnathobase is left.

The first maxillae in stages with seven somites consist of two lobes, the inner of which has the typical branchiopod brush-like morphology. The interpretation of the outer lobe is somewhat unclear. Scourfield suggested that it may represent either a vestigial exopod or an early stage of the male clasper. The first maxillae in the adult females are small semicircular lobes fringed with 8 or 9 forwardly directed plumose setae, very much with the same morphology as the proximal endites of the trunk limbs. In the adult male, the first maxillae are represented by a basal gnathobasic part, as in the females, as well as by a pair of large, inwardly curved, three-segmented claspers (Scourfield 1926, 1940), probably to hold the female during the mating process, as done by modern anostracans but in the latter with the antennal claspers. Eriksson (1934) questioned the interpretation of the claspers as a part of the first maxillae, but the original interpretation of Scourfield seems to be most reasonable. The presence of such claspers, in this position, is unique among branchiopods. The second maxillae in the adults are, according to Scourfield (1926, 1940), supported and illustrated by Cannon & Leak (1933), present as a pair of small rounded structures between the first maxillae and the first pair of trunk limbs. The presence of reduced second maxillae was questioned by Schram (1986), Walossek (1993), and Schram & Koenemann (2001), all of whom interpreted what Scourfield had designated as the first pair of trunk limbs as being the second maxillae, as in the Recent Cephalocarida and a number of 'Orsten' fossils. However, considering how common small and insignificant-looking second maxillae are in the Branchiopoda, and also considering the many detailed similarities between *Lepidocaris* and Recent branchiopods (see below), it appears more likely, as recently remarked by Spears & Abele (2000), that the pair of small lobes in *Lepidocaris* actually does represent the second maxillae. Furthermore, and perhaps the best argument, is that Cannon & Leak (1933) drew a small pore associated with the presumed reduced second maxillae, which is in accordance with the situation in a number of other branchiopods. Other indirect evidence is that, if the original interpretation of Scourfield (1926) is retained, it is the same somites (12 and 13) that give rise to the ventral brood pouches in *Lepidocaris* and the Anostraca.

The development of the trunk limbs can be followed in some detail because a number of the described larval specimens exhibit series of limbs in different degrees of development (Fig. 4A, B). Despite the fact that the three anterior trunk limbs in the adult differ in morphology from the more posterior pairs (see below) all the trunk limbs have a similar morphology, in at least part of the development. Then they consist of a pair of clearly biramous lobes, with the rami (endopod and exopod) directed ventrally, each bearing a robust spine. This is similar to the situation in *Rehbachiella*, not only for the post-maxillary limbs, but also for the first and second maxillae. In *Lepidocaris,* an almost symmetrically biramous condition is maintained in the adults in the posterior-most trunk limbs (endopod and exopod of similar size and morphology). In contrast, the three anterior-most trunk limbs of *Lepidocaris* are asymmetrical with the median lobes (endopod and endites) more developed that the lateral ones (Fig. 4). The morphology of these anterior limbs in *Lepidocaris* is more in accordance with that of trunk limbs in Recent branchiopods than are the posterior limbs. On the anterior limbs, the proximal endite is a large and with the setae arranged brush-like, and along the medial margin are five additional setose endites. The endopod is unsegmented, but it is relatively large compared with the exopod. Distally, the endopod bears a row of strong serrated spines directed somewhat medially. The exopod arises from the oblique lateral margin of the limb. In between the anterior and posterior trunk limbs, the intermediate trunk limbs are basically symmetrically biramous, like the posterior limbs in the series. However, trunk limbs 4-6 have some similarities to the anterior three, as they have an enlarged proximal endite and more pronounced enditic lobes along the inner margin. The morphology of the anterior trunk limbs, with their robust spines distally on the endopod, led Fryer (1985) to suggest that *Lepidocaris* had a scraping lifestyle. In females, posterior to the biramous trunk limbs, is a characteristic egg-pouch with many similarities to egg-pouches described for the Recent Anostraca; even the same limbs appear to be involved. In *Lepidocaris*, the egg-pouch is apparently formed by the two pairs of limbs belonging to somite 12 and 13 (possible also 14), if the second maxillae are considered reduced, as originally suggested, and the somite numbering system introduced above is followed. The limbs on the 12th somite form the egg-pouch cover. The actual egg-pouch is formed by the limbs on the 13th somite, or possibly by the limbs of the 13th and 14th somite together. The biramous limb morphology of the egg-pouch cover (12th somite) can still be clearly recognized. The actual egg-pouch is of more or less spherical shape. Behind it is a pair of rudimentary appendages that probably originated from the 14th somite.

Along the body of adults is a series of lateral scales. Scourfield (1926) discussed the origin of these as either being proximal limb exites or scales originating from the somites, but he did not make strong suggestion on the matter, nor did Borradaile (1926) on the same subject.

The 'caudal segment', in larvae with six or seven somites is relatively long and cylindrical and bifurcated into two broad lobes, which are bordered by small groups of short spines or scales. Each carries a long characteristic seta with setules along the distal half. These setae appear to be articulated with the caudal lobes. The lobes and the setae connected to them were termed the 'primary furca' by Scourfield, and these structures, with a similar morphology, can be found in modern anostracans. Later in development several changes in the morphology and armature of the "caudal segment" take place. A pair of small lateral outgrowths each with a small spine appears just in front of the primary furca. These are termed the 'secondary furca' by Scourfield. They enlarge considerably during development and in the adult the 'secondary furca' are jointed to the telson and are as long as the 'caudal segment' and by far the largest caudal appendages. The 'primary furca' have been pushed into a dorsal position and consist of two small knobs bearing conical setae. An additional pair of appendages is budded off laterally about halfway along the telson. Scourfield suggested that these small appendages and the secondary furca possibly should be regarded as modified limbs. This interpretation was based on their position, which is ventral to some lateral spines that can be followed serially on each segment. These spines are in line with the pleura on more anterior somites, ventral to which the limbs are attached (see Scourfield 1926: fig. 6 and pl. 22:3). All this indicates that the so-called 'caudal segment' as the posterior-most body portion in the adults comprises the telson and at least one true segment.

The detailed similarities between *Lepidocaris* and many Recent branchiopods, some of which can be assigned synapomorphic status (summarized in later section), support a close relationship between these taxa.

3.4 *Notostraca*

The larval development of the Notostraca received pioneering and competent attention as long ago as 1756 by Schaeffer, who provided, for his time, detailed illustrations. Later Zaddach (1841) gave some illustrations of various larval stages, but it appears justified to say that Claus' (1873) account was the first with clear illustrations, of such a quality that they are copied even in recent works. Other contributions of notostracan development are those of Piatakov (1925), who deals only with the earliest embryonic development, and Borgstrøm & Larsson (1974), Fryer (1988), and Williams & Müller (1996). Møller et al. (in press) recently described the five first instars of *Triops cancriformis* by scanning electron microscopy.

The following section focuses on the development of *Triops cancriformis* (Bosch, 1801), which, as convincingly shown by Fryer (1988), retains a number of important primitive features as compared with *Lepidurus arcticus* (Pallas, 1793) (see Borgstrøm & Larsson 1974).

It is not absolutely clear though how many stages *Triops cancriformis* passes through from hatching to adulthood. Claus (1873) and Fryer (1988) followed 5-7 instars in considerable detail and offered only selected comments on further development.

Triops cancriformis hatches as an approximately 600 µm-long larva, richly provided with yolk and with poor swimming capabilities, which causes it to sink frequently to the bottom (Brauer 1872; Claus 1873; Fryer 1988). The first instar is non-feeding, and Fryer (1988) remarked that the animal probably does not feed even in the third instar. In the first instar the first antennae are short, tubular appendages with one seta each. The second antennae are very large and are the only organs of propulsion, as they are until this function is gradually taken over by the trunk limbs in later stages. Both the coxal and basipodal masticatory spines of the antennae, homologous to those of other branchiopod larvae, are present but not yet functional (Claus 1873; Fryer 1988). The tip of the coxal masticatory spine has not yet become bifid. Similarly, the mandibular gnathobases are not yet completely developed, and the mandibular palps have no role in feeding yet (Fryer 1988). In the second instar (up to ca. 750 µm) (see Fig. 5B), the naupliar feeding apparatus is essentially active (Fryer 1988). Food is collected by the three-segmented mandibular palps (basis and two probable endopodal segments) by sweeping relatively coarse bottom particles to the mouth region (Fryer 1988). The mandibular gnathobases apparently are still not functional. However, based on comparison with the anostracan *Branchinecta ferox*, Fryer suggested that the mandibular gnathobasic spine also in *T. cancriformis* pushes forward particles that are swept towards them from behind. The coxal masticatory spines of the second antennae assist, by means of a forward swing, the passage of the food anteriorly. The feeding apparatus of the third instar (up to about 1.25 mm) (see Fig. 5A, C) is basically the same as in the previous stage, but the mandibular gnathobases now lie closer together, and the gnathobasic spines are located relatively more laterally (Fryer 1988). The coxal masticatory spines of the antennae have now become bifid (Fig. 5C).

The trunk limbs are already present in the first instar as 6-7 pairs of ventrolateral limb rows, apparently not yet differentiated into parts (Claus 1873). In the second instar, the limbs are still in a lateral 'flat' position, but various portions (endites, endopod and exopod) can be recognized, at least on the first three pairs of (Fig. 5B and Claus 1873). It is clearly shown on a drawing by Claus (1873) that the first two pairs of trunk limbs in the second instar have developed all the components of the adult limb, which are five endites along the inner margin of the basis, an endopod and an exopod, and possibly also an epipod (see also Fig. 5B and Møller et al. in press). As in other 'large' branchiopods, the trunk limbs develop gradually, and the entire sequence of development can be seen from anterior to posterior in a single specimen (see Fig. 5D). The responsibility for feeding and movement is gradually transferred from the naupliar appendages to the trunk limbs, with a quite long overlap in time where both sets of limbs are active. Fryer (1988) noted that the first three pairs of trunk limbs in instar 3 are capable of feeble movements but are not able to contribute to locomotion. In the fourth and fifth instars, five and seven pairs of limbs are active in locomotion and food collection, respectively, and in each case one or two more pairs are capable of feeble movements (Fryer 1988). Concomitantly, the second antennae's contribution to locomotion diminishes, and Fryer (1988) stated that their contribution to locomotion in stage 8 is probably insignificant. Before adulthood, the second antennae atrophy almost completely.

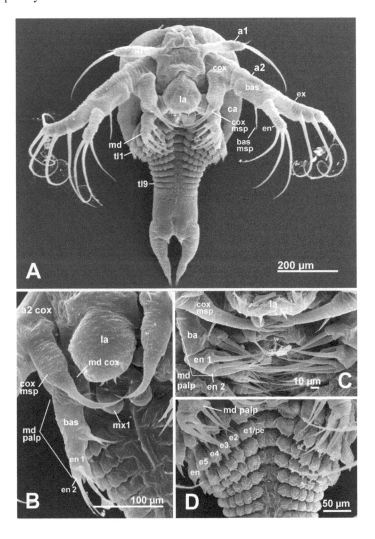

Figure 5. [See Plate 25 in color insert.] Larval stages of *Triops cancriformis* (Bosch, 1801) (Notostraca). (A) Larva of stage 3. (B) Larva of stage 2, labrum area. (C) Larva of stage 4 (or later), close-up of proximal masticatory spine of the second antennae and mandibular palps. (D) Larva of stage 3 (same specimen as in A, close-up of buds of trunk limbs. Staging following that of Claus (1873). Partly from Møller et al. (in press) and partly from unpublished work of Ole Sten Møller (Zoological Institute, University of Copenhagen) with kind permission. Abbreviations: a1 = antenna 1, a2 = antenna 2, bas = basis, bas msp = basipodal masticatory spine (a2), ca = carapace, cox = coxa, cox msp = coxal masticatory spine (a2), e1 to e5 = endites 1 to 5, en = endopod, ex = exopod, la = labrum, md = mandible, mx1 = maxilla 1, pe = proximal endite (endite 1), tl1 = trunk limb 1, tl9 = trunk limb 9.

3.5 Laevicaudata ('Conchostraca')

Grube (1853), Gurney (1926), Botnariuc (1947), and Monakov & Dobrynina (1977) have dealt in great detail with the larval development of the species of *Lynceus*. Others, including Johansen (1925) and Solowiow (1927), have provided briefer observations. Monakov & Dobrynina (1977) identified four larval stages for *Lynceus brachyurus* (Müller, 1776). All instars have basically the same morphology and are, with their peculiar appearance, quite different from any other branchiopod larvae (see Figs. 6, 7).

The dorsal margins of the larvae of *Lynceus brachyurus* are expanded into a large dorsal shield, with a large dorsal organ in the middle. Ventrally, there is a huge, plate-like, immovable labrum. The dorsal shield and ventral labrum together give the larvae a flattened, 'double-layered' appearance (Figs. 6A, 7A). The pre-labral region is drawn out in a conical expansion, and, on each side there is a large lateral horn (Figs. 6A, 7A). The body terminates posteriorly in a pair of caudal lobes. In early descriptions, like that of Grube (1853), the status of the first antennae in the larvae was not clear, but Gurney (1926) pointed out that the small, club-shaped first antennae – as they appear in the first juvenile instar (juveniles = bivalved instars after metamorphosis) – are visible already under the cuticle at the base of the lateral horns in a late larva. In this position, there is a long seta, which is in the same position as an obviously homologous seta attached to the small lobate first antennae in spinicaudatans. The large lateral horns of the *Lynceus* larvae must therefore be interpreted as the modified first antennae, as suggested by Gurney (1926).

The morphology of the post-antennular limbs - hidden between the dorsal shield and the labrum - is more similar to the larval appendages of the branchiopods mentioned previously than are the first antennae. The second antennae, the organs of propulsion, are biramous with, according to the illustrations of Monakov & Dobrynina (1977) and shown in Fig. 7B, C, a two-segmented endopod and four-segmented exopod. Figure 7B, however, shows one, or possibly two, previously unrecognized possible segments between the basis and the first large endopod segment. The protopod has what is, for branchiopods, a typical arrangement of long masticatory spines directed towards the mouth and involved in handling the food. The coxa is equipped with one long spine (coxal masticatory spine), which becomes bifid in a later stage as in other 'large' branchiopods. At the basis is another spine (seta) (the basipodal masticatory spine). The mandibles have a three-segmented palp (basis and two endopodal segments), as in all other 'large' branchiopods. Apparently nothing has previously been known about the armature of the larval mandibular coxa in *Lynceus*, but Fig. 7C shows that its only armature, directed towards the mouth opening, is a masticatory spine (labeled 'md sp'). No information about how larvae of *Lynceus* feed is available.

The trunk limbs develop from the lateral edge of the trunk under the dorsal shield. A few parts of the limbs of an intermediate larval stage are shown in Fig. 6C. However, Monakov & Dobrynina (1977) provided more details for an early juvenile and it is clear that the limbs develop as laterally placed, lobate limb buds, much like that described in detail for *Cyclestheria hislopi* (see Olesen 1999). According to the drawing by Monakov & Dobrynina (1977), the early trunk limb has 4-5 ventral lobes of similar size. Later, the three proximal lobes enlarge and an extra distal lobe buds off.

A striking feature of the larvae of *Lynceus* spp. is the almost circular dorsal shield covering the whole body including the head. Gurney (1926) discussed its relation to the bivalved carapace in the juveniles and adults and found that these two structures are quite different. In late larvae ready to molt, the outline of the future carapace in a juvenile can be seen through the cuticle of the dorsal shield. This observation allowed Gurney (1926) to recognize that the future carapace does not fill the dorsal shield completely. Its anterior margin is immediately behind the large dorsal organ on each side of which it extends forward. Hence, the dorsal shield is more than a precursor to the juvenile/adult bivalved carapace and should probably be interpreted – from its large size – as a specialization of the early larvae of *Lynceus*.

In summary, the many unique features of *Lynceus brachyurus*, including the dorsal shield, the plate-like labrum, and the lateral horns, make it one of the most specialized crustacean larvae. Other aspects of its morphology, including the naupliar feeding apparatus and the development and general morphology of the trunk limbs, are typical for the Branchiopoda.

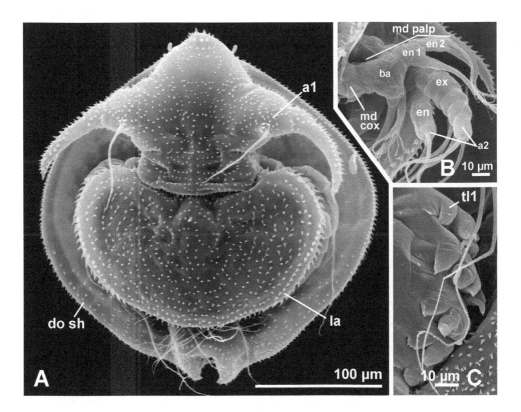

Figure 6. Larval stages of *Lynceus brachyurus* (Müller, 1776) (Laevicaudata). Original. (A) Early larva from ventral. (B) Right mandibular palp of larval stage seen from dorsal. (C) Left side buds of trunk limbs of intermediate stage seen from ventral. Abbreviations: a1 = antenna 1, a2 = antenna 2, bas = basis, cox = coxa, do sh = dorsal shield, en = endopod, ex = exopod, la = labrum, md = mandible, tl1 = trunk limb 1.

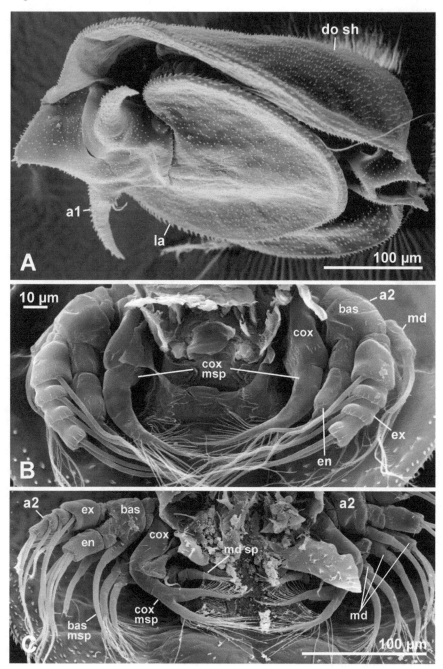

Figure 7. Larval stages of *Lynceus brachyurus* (Müller, 1776) (Laevicaudata). Original. (A) Late larva. (B) Early larva (labrum removed), ventral view of antennae 2. (C) Late larva (with labrum removed), ventral view of antennae 2. Abbreviations: a1 = antenna 1, a2 = antenna 2, bas = basis, bas msp = basipodal masticatory spine (a2), cox = coxa, cox msp = coxal masticatory spine (a2), do sh = dorsal shield, en = endopod, ex = exopod, la = labrum, md = mandible, md sp = mandibular spine.

3.6 *Spinicaudata ('Conchostraca')*

The development of various species of the Spinicaudata has been treated by a number of authors. The development of representatives of the three recognized families of the Spinicaudata (excluding Cyclestheriidae) – Limnadiidae, Cyzicidae, and Leptestheriidae – are treated here.

The development of *Limnadia lenticularis* (Linnaeus, 1758) (Limnadiidae) was competently described by Lereboullet (1866) (as *L. hermanni*, which is a synonym of *L. lenticularis*) and later by Sars (1896b) in even more detail. Zaffagnini (1971) also provided information on this species. Eder (2003) provided a full SEM-based description of *Imnadia yeyetta* Hertzog, 1935, and Anderson (1967) described the development of *Limnadia stanleyana* King, 1855, at both the external and cellular levels. Selected developmental features of other Limnadiidae have been described or briefly mentioned for *Imnadia voitestii* Botnariuc & Orghidan, 1941, by Botnariuc (1947, 1948), for *Eulimnadia stoningtonensis* Berry (1926) by Berry (1926), and for *Eulimnadia texana* (Packard, 1871) by Strenth & Sissom (1975). The following section focuses on *L. lenticularis*, for which the most detailed descriptions appear to be available among the limnadiids. Recently the larval development of *Eulimnadia braueriana* Ishikawa, 1895, has been described by Olesen & Grygier (2003) and photographs of this species are included to illustrate general features of the Limnadiidae (Figs. 8, 9).

Sars (1896b) divided the development of *Limnadia lenticularis* into two convenient periods, the larval (or naupliar) period and the post-larval (or juvenile) period, separated by only one molt, which involves a significant 'jump' in morphological appearance (in my treatment of spinicaudate larval development, the terms 'larvae' and 'nauplii' are used synonymously). In the larval (naupliar) period, which comprises at least seven stages for *L. lenticularis,* the animals have typical larval characteristics, including an elongate labrum with a long, median distal spine, two pairs of antennal spines (or setae) involved in food manipulation, a three-segmented mandibular palp, trunk limbs that are not yet movable, and a carapace which is not yet clearly bivalved. In the first stage of the post-larval (juvenile) period the animal basically looks like a small adult with a laterally flattened appearance, a large bivalved carapace and reduced antennal spines and mandibular palps. This 'jump' in morphology from one stage to another also is seen in other spinicaudatans.

Sars (1896b) described and illustrated seven larval (naupliar) stages for *Limnadia lenticularis* but did not state for certain that this was the complete number. He mentioned, however, that the earliest larva he depicted was the hatching stage, and that the last larva described by him molts into the first post-larval (juvenile) stage. Hence, although it is still possible that Sars missed some stages, at least he described more larvae in *L. lenticularis* than have been described for any other 'conchostracan', most of which are described as having five larval stages, as summarized by Eder (2003).

The hatching stage of *Limnadia lenticularis* already displays all the larval structures, a large pointed labrum, small first antennae, very large biramous second antennae, used for swimming, and a uniramous mandible. All armature of the various limbs is present, including the setae on the limbs and the coxal and basipodal masticatory spines of the second antennae, but these structures appear to be incompletely developed and the hatching stage is probably non-feeding. It also contains considerably more yolk than the following stages, and only weak articulations separate the future segments of the limbs. In the following stages, the yolk is lost, the various setae and the masticatory spines of the second antennae develop more distinct setules, and the limbs become more distinctly articulated.

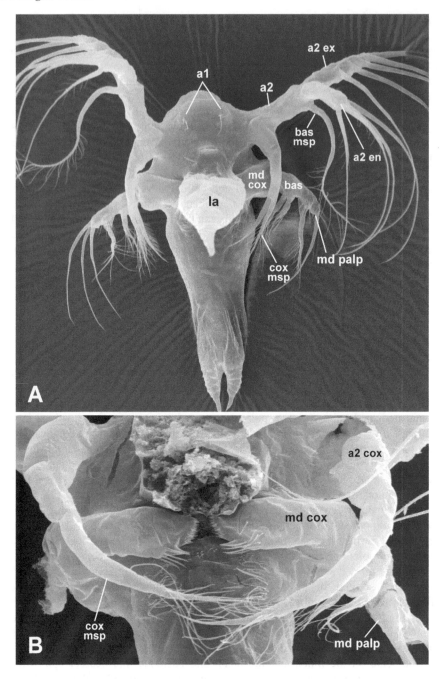

Figure 8. Larval stages of *Eulimnadia braueriana* Ishikawa, 1895 (Spinicaudata). Original. (A) Nauplius 4. (B) Nauplius 2 (labrum removed), mandible (coxa) and antennal coxal masticatory spines. From Olesen & Grygier (2003). Abbreviations: a1 = antenna 1, a2 = antenna 2, bas = basis, bas msp = basipodal masticatory spine (a2), cox = coxa, cox msp = coxal masticatory spine (a2) en = endopod, ex = exopod, md = mandible.

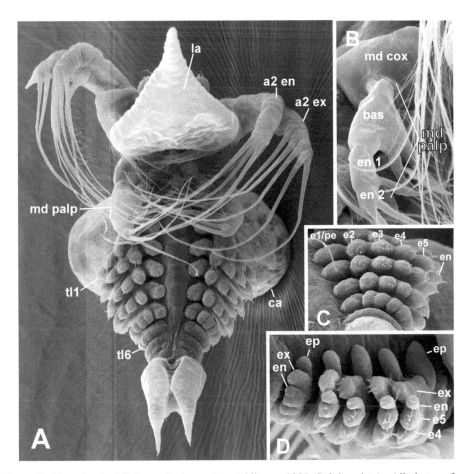

Figure 9. Nauplius 7 of *Eulimnadia braueriana* Ishikawa, 1895 (Spinicaudata). All photos of same specimen. (A) Ventral view. (B) Right mandible. (C) Buds of left trunk limbs seen from postero-ventral. (D) Buds of right trunk limbs seen from lateral. From Olesen & Grygier (2003), except B, which is an original photo. Abbreviations: a2 = antenna 2, bas = basis, ca = carapace, cox = coxa, en = endopod, ep = epipod, ex = exopod, e1 to e5 = endite 1 to 5, la = labrum, md = mandible, pe = proximal endite (= endite 1), tl1 = trunk limb 1, tl6 = trunk limb 6.

Not until the fourth stage depicted by Sars (1896b) the coxal masticatory spines acquire their characteristic bifid tips (Fig. 8 shows comparable stages of *Eulimnadia braueriana*, one with bifid tip and a slightly earlier stage without). The fourth stage depicted by Sars (1896b) is the first to show the *Anlage* of the carapace and the first six pairs of trunk limbs. In the seventh larval (or naupliar) stage depicted by Sars, the future bivalved appearance of the carapace is visible and the limbs, of which there are still only about six immovable pairs, have begun to differentiate. In a comparable stage of *E. braueriana* the trunk limbs are present as two rows of laterally placed, subdivided limb buds in various degrees of development, which have not yet attained the 'vertical' adult orientation (Fig. 9). The early limb bud, as seen in the posterior limbs of a late stage of *E. braueriana* (Fig. 9C, D), is an elongate bud located ventrolaterally on the trunk and subdivided into seven small portions of similar size. By following the development of the limb bud, as illustrated by the serial displacement in the degree of development in a late stage of *E. braueriana* (Fig. 9), it becomes clear that the five proximal portions become the five endites in the adult, while the 6th portion becomes the small unsegmented endopod *and* the most distal (lateral) portion becomes the epipod. Especially the portion of the limb bud representing the exopod in the adult undergoes significant growth and does so in a distal as well as in a proximal direction. From the seventh and latest larval stage described by Sars (1896b) to the first post-larval (or juvenile) stage, a significant change in morphology takes place, a small metamorphosis. The animal now looks considerably more like an adult and the most significant acquisitions of the first post-larval stage are a large bivalved carapace, trunk limbs that are now active, reduced masticatory spines of the second antennae, an almost reduced mandibular palp, and a significant reduction in the size of the long, median acute process of the labrum.

Recently seven stages were described for *Eulimnadia braueriana*, another limnadiid, by Olesen & Grygier (2003). There was no perfect match between the sequences of *E. braueriana* and *Limnadia lenticularis* despite the fact that both species apparently have seven stages (see Olesen & Grygier 2003).

Few accounts are available on the development of other spinicaudatan 'conchostracans', among which the most detailed are those by Sars (1896a) on *Cyzicus packardi* (Brady) (Cyzicidae), Monakov et al. (1980) and Eder (2003) on *Leptestheria dahalacensis* (Rüppel, 1837) (Leptestheriidae), and Ficker (1876) on *Eoleptestheria ticinensis* (Balsamo Crivelli, 1859) (Leptestheriidae). The development of *Leptestheria saetosa* Marinèek and Petrov, 1992 (Leptestheriidae), was described in some detail by Petrov (1992), and single stages of various leptestheriids were dealt with by Botnariuc (1947, 1948). Only selected features are mentioned for these species and compared with those summarized for *Limnadia lenticularis* above.

The development of *Cyzicus packardi*, described by Sars (1896a), is in many ways very similar to that of *Limnadia lenticularis*, despite the fact that they belong to different families. One difference is that while Sars (1896b) described seven stages for *L. lenticularis*, he described only six for *C. packardi* (see comment on this below). The first five stages in the two species are remarkably alike. Similarities include a probably non-feeding first stage containing much yolk, and of the same general shape. Other similarities are that the *Anlagen* of the carapace and trunk limbs appear in the fourth described stage in both species. It is clear that the first five stages in the two species, representing different families, correspond closely in morphology, despite different timing in the appearance of the bifid coxal masticatory spines of the second antennae, which appear in the third stage of *C. packardi,* in the 4th in *L. lenticularis*. Apart from the timing of the appearance of this particular structure, among the few differences between the two species are the shape of the labrum (large and pointed in *L. lenticularis*) and the relative length of the hind body and particularly the paired caudal rami (longer in *L. lenticularis*). The major difference is seen in the number of stages after the fifth stage. For *C. packardi*, Sars (1986a) described only one stage (the sixth) before the first juvenile stage while he described two – practically identical – stages for *L. lenticularis* (sixth and seventh). While Sars (1896a) was very explicit with respect to the number of larval stages in *C. packardi*, he made only tentative statements for those of *L. lenticularis* and did not number the stages precisely as he did for *C. packardi*. *Eulimnadia texana* has been described as having six larval stages (Strenth & Sissom 1975). Other spinicaudatans apparently have only five larval stages, not counting obviously incomplete descriptions where fewer stages have been treated (Botnariuc 1947, 1948; Mattox 1950). Five stages have been described for *Eoleptestheria ticinensis* (Leptestheriidae) by Ficker 1876, for *Leptestheria saetosa* (Leptestheriidae) by Petrov

1992, and for *Leptestheria dahalacensis* (Leptestheriidae) by Monakov *et al.* (1980) and Eder (2003). For all described larvae of these species it is possible to find matching larvae in the larval series described for *C. packardi* by Sars (1896a). When using the development of *C. packardi* as a reference series with its six larval stages, it becomes clear that some larvae are 'missing' in the other spinicaudate species where only five stages have been described. In the sequence of *E. ticinensis*, described by Ficker (1876), the stage corresponding to stage 2 in *C. packardi* is clearly missing. In other cases it is difficult to tell exactly which stage is missing because of the low quality of the illustrations in certain papers, but in *Caenestheria* sp., described by Bratcik (1980), apparently stage 2 (or stage 3?) is missing. In the sequence of *L. dahalacensis* described by Monakov et al. (1980), apparently stage 3 is missing.

The largest number of larval (naupliar) stages documented is six (*Cyzicus packardi* and *Eulimnadia texana*), or seven (*Limnadia lenticularis* and *Eulimnadia braueriana*), while in most species examined in detail only five stages have been reported. It is unclear at the moment whether those shorter sequences are truly shorter, or whether some authors have missed larval stages in some cases.

3.7 Cyclestheria hislopi *(Cyclestherida) ('Conchostraca')*

The development of *Cyclestheria hislopi* (Baird, 1859), monotypic for the Cyclestheriidae, has been dealt with in most detail by Sars (1887), Olesen (1999) and Olesen et al. (2001), and more briefly by Roessler (1995) and Dodds (1926). Its development is direct (with a single known exception, see below) and, in the parthenogenetic part of the life-cycle, it takes place under the carapace, dorsal to the body. The following description is mostly based on Olesen (1999), and photographs of two stages are included as Fig. 10. Throughout development the embryos are attached via long filaments between the forehead of the embryos and the exopods of the mother. The first 3-4 stages are enclosed in a membrane, which is then cast off. The embryos are still attached to the female's exopod after the membrane has disappeared. The first limb buds that can be clearly identified are those of the naupliar appendages, of which those of the second antennae are the largest. In comparison with all other 'large branchiopods' the second antennae have no masticatory spines at any stage of their development. The first antennae are small buds early in development and continue to be so. The mandible also originates as a relatively small, simple limb bud, but it enlarges considerably during development. It grows directly into the adult form, which basically is an enlarged coxa, without the three-segmented palp so typical for larval stages of all other 'large' branchiopods. In this respect, *C. hislopi* is unique among 'large' branchiopods but identical to most cladocerans. The first and second maxillae appear first as small buds and remain small during the entire development, in accordance with their reduced appearance in the adults. In the later part of development the openings from the maxillary glands can be seen in the buds of the second maxillae. The trunk limbs develop in the way typical for branchiopods, as laterally-placed limb buds that become subdivided into a number of portions from proximal to distal before they attain a vertical, adult orientation. Trunk limb development bears, not surprisingly, a particular resemblance to that of the Spinicaudata. The distinction between the endopod, exopod and epipod, the future distal part (but directed laterally), and the distinction between the proximal three endites, appear first. Following this are the more distally-placed endites that gradually become 'budded' off from the future unsegmented endopod.

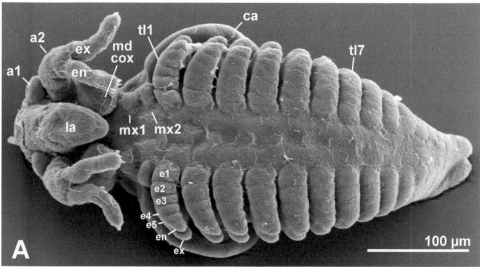

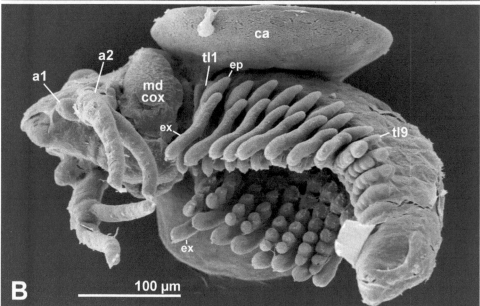

Figure 10. Embryos of *Cyclestheria hislopi* (Baird, 1859) (Cyclestherida). (A) Embryo of sixth stage. (B) Embryo of seventh stage. From Olesen (1999). Abbreviations: a1 = antenna 1, a2 = antennae 2, ca = carapace, cox = coxa, e1 to e 5 = endite 1 to 5, en = endopod, ep = epipod, ex = exopod, la = labrum, md = mandible, mx1 = maxilla 1, mx2 = maxilla 2, tl1 = trunk limb 1, tl7 = trunk limb 7, tl9 = trunk limb 9.

When all limb parts have appeared - endopod, exopod, epipod and five endites - various portions of the limb begin to enlarge so that the whole limb attains a dorsoventral, instead of lateral, orientation. Intermediate embryos have a peculiar, swollen dorsal organ, which is reduced before adulthood. The carapace is already visible in very early stages as a pair of swollen areas immediately posterior to the naupliar region. Later, the posterior margin of the carapace becomes free from the trunk to which it is attached anteriorly in the region of the first and second maxillae. Later the carapace starts to develop anteriorly as well as posteriorly.

According to Roessler (1995), who studied Colombian populations, *Cyclestheria. hislopi* reproduces parthenogenetically for much of the season. Colombia is one of the few places where males have been found (Olesen et al. 1997), and when these appear, resting eggs are produced protected by an ephippium (Roessler 1995). The development of these resting eggs in Colombia was also reported to be direct (Roessler 1995). A puzzling example to the contrary has been mentioned from Cuba by Botnariuc & Viña Bayés (1977) who reported the presence of free-living larvae (heilophore [7] larvae) from a couple of permanent water bodies. These they inferred had hatched from resting eggs as such larvae do in other conchostracans. The morphology of this larva corresponds to the sixth stage of e.g. *Cyzicus packardi* (Cyzicidae) and to other late larvae among the Spinicaudata (see Sars 1896a) and have the arrangement typical for conchostracans (and branchiopods) of the naupliar appendages, which includes antennal protopodal masticatory spines (of which the coxal one is bifid) and a mandibular palp with the typical 'conchostracan' setation. The labrum is pointed in the same way as in certain larvae of the Limnadiidae, such as *Eulimnadia texana* (see Strenth & Sissom 1975), *Limnadia lenticularis* (see Sars 1896b), and *Eulimnadia braueriana* (see Olesen & Grygier 2003). If Roessler (1995) was correct in describing development as direct from resting eggs, and if Botnariuc & Viña Bayés (1977) also were correct when they referred the free-living larval stage to *C. hislopi*, and there are good reasons to assume that both where correct, then we are dealing with a remarkable, population-related difference in developmental strategy *within C. hislopi*. Roessler (1995) explicitly followed and figured the development taking place within the resting eggs, and it is clear from his illustrations that the animal had reached the juvenile stage before hatching and therefore did not hatch as a heilophore larva. Botnariuc & Viña Bayés (1977) are probably also correct when they referred the heilophores they reported to *C. hislopi*, since this species is typical of more permanent water-bodies as the one they studied.

3.8 *Leptodora kindtii* (Haplopoda)

The development of *Leptodora kindtii,* monotypic for the Haplopoda, has been dealt with by a number of workers among which are Müller (1868), Wagner (1868), Sars (1873), Weismann (1876), Samter (1895, 1900), Warren (1991), Andrews (1948), Gerberding (1997), and Olesen et al. (2001, 2003). As in other cladocerans, development in the parthenogenetic part of the life-cycle is direct and takes place in the dorsal brood pouch of the parent, which is in contrast to the development from the resting eggs, from which free-living metanauplii hatch. The following account is based mostly on Olesen et al. (2001, 2003). Photographs of two developmental stages are included as Fig. 11.

---

[7] The term 'heilophore', meaning 'carrier of a lip', was introduced by Botnariuc (1948) as a name for the last larva in the larval (naupliar) phase of the development of the 'Conchostraca'. This larva has, among other characteristics, a small bivalved carapace, only covering a part of the body; antennules as a pair of small setose knobs; strongly developed labrum (see Dumont & Negrea 2002 for more characteristics listed). The late larva depicted in Fig. 9 would be such a 'heilophore' larva.

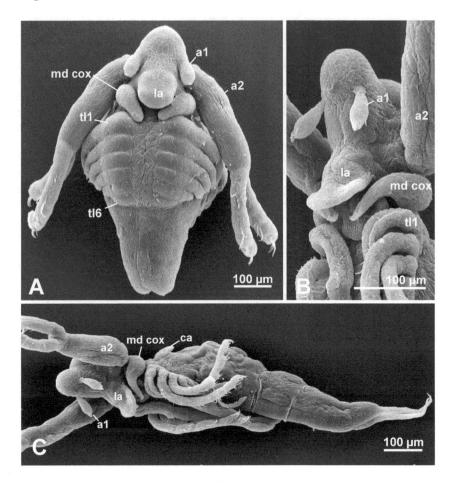

Figure 11. Embryos of *Leptodora kindtii* (Focke, 1844) (Cladocera, Haplopoda). (A) Intermediate embryo. (B, C) Late embryo immediate prior to release from brood pouch. From Olesen et al. (2003). Abbreviations: a1 = antenna 1, a2 = antenna 2, ca = carapace, cox = coxa, la = labrum, md = mandible, tl1 = trunk limb 1, tl6 = trunk limb 6.

As is typical for cladocerans and *Cyclestheria,* all of which have direct development, the first stages in *L. kindtii* are 'egg'-like with the appendages only weakly indicated as limb buds of varying shape. During development, *Leptodora kindtii* undergoes a more dramatic change in morphology than any other cladocerans due to its deviating adult morphology. The homologies of a number of its peculiarities, such as the segmented trunk limbs and the three-lobed 'lower lip' have been explained in a recent study of the species' ontogeny (Olesen et al. 2001, 2003).

The first antennae remain small throughout development. From small, slightly elongate limb buds lateral to the labrum they gradually become more and more elongate (Fig. 11). The second antennae, which are by far the largest appendages in the adult and are used for swimming (as in all other diplostracans), are already large during embryonic development (Fig. 11). In the earliest stages they are attached along their whole length to the sides of the body, but they soon detach, and in later stages are almost as long as the entire body. The mandibles start as a pair of small, triangular limb buds that soon enlarge and become curved with a pointed tip (Fig. 11). There are no signs of a mandibular palp at any stage in the direct developing part of the life-cycle, which is in contrast to the free-living larvae of *Leptodora kindtii* (see below), where mandibular palps are present. The first and second maxillae have often been reported as lacking in *L. kindtii*, which is only partly true. Samter (1900) showed that probably both pairs briefly appear very early in development, and then disappear, but Olesen et al. (2003) showed that the first maxillae actually form the outer lobes of a three-lobed 'lower lip' seen in adults and late juveniles. The median lobe of the 'lower lip' is formed by an elevation of the corresponding region of the sternite. The second maxillae apparently disappear altogether, or the remains of them migrate to a lateral position associated with the openings of the maxillary glands. The trunk limbs begin their development as six pairs of laterally placed limb buds, which become separated into a number of portions (the precursors of the segments in the adults). A study on the limb onto-geny using methods, including expression of the *Distal-less* gene, showed that the two proximal portions of the five embryonic limb buds fuse and form the very long proximal segment in the limbs of the adults (Olesen et al. 2001). During development, the limbs change their position from being attached all along their length to the trunk to a more curved orientation with a characteristic bend in the position of the two proximal embryonic portions (Olesen et al. 2003). Finally they attain an upright, vertical orientation. The adult limbs are stenopodous, with true segments, and lack an exopod and epipod, all of which is in contrast to the morphology of the trunk limbs of most branchiopods. Based on the de-tailed similarity between the early limb buds of *L. kindtii* and *Cyclestheria hislopi* (and other 'large' branchiopods), Olesen et al. (2001) showed that the endites of *C. hislopi*, broadly speaking, are homologous to the limb segments of *L. kindtii*, and that limb segments have evolved from endites and not *vice versa*.

The carapace is paired in the early part of its development. Afterwards, the carapace *Anlagen* from both sides fuse to form a narrow rim, which still shows its paired origin, immediately behind the head and the mandibles. In a late juvenile stage, the carapace is a short, broad flap, attached in the area of the first and second trunk limbs (Fig. 11C). Olesen et al. (2003) found evidence that the displacement of the carapace in a posterior direction, as seen in the adult, takes place by a posterior growth of the carapace during ontogeny and a corresponding fusion of the anterior parts to the dorsal side of the trunk. This corresponds to the observations made by Samter (1895). The actual brood pouch, or carapace, of the adult would then be homologous to only the posterior part of the large bivalved carapaces seen in 'conchostracans'.

From resting eggs hatches a metanauplius that goes through a number of stages before adulthood, as first noted and described by Sars (1873) and later supplemented by Warren (1901) (two stages shown in Fig. 15). By combining the stages depicted by Sars and Warren it can be seen that at least three metanaupliar stages are passed through before a juvenile morphology is attained. As mentioned, *Leptodora kindtii* is the only cladoceran that has free-living larvae (not shown). These free-living larvae (as opposed to the direct developing larvae) have mandibular palps, which is also unique among cladocerans. The palps are different from those of other branchiopod larvae in that they appear to be unseg-mented and are rather long, almost as long as the trunk of the larvae, and have setae only at the tips. Due to this morphology it is difficult to imagine the palps involved in feeding, as mandibular palps are in other branchiopods, and its probably more correct when Sars (1873) suggest them to be stabilizing limbs during swimming.

3.9 *Ctenopoda*

As in most other cladocerans, the development of the Ctenopoda is direct. Morphogenesis in late ontogeny, which is the period covered in this chapter, of species of the Ctenopoda has been dealt with in detail only by a few workers, such as Sudler (1899) on *Penilia schmackeri* (Poppe) and Kotov & Boikova (1998) on *Sida crystallina* (Müller, 1776) and *Diaphanosoma brachyurum* Liévin, 1848. The early embryology of *Holopedium gibberum* Zaddach, 1855, is treated by Agar (1908) and Baldass (1937), and that of *D. brachyurum* by Samassa (1893). The embryonic development of *S. crystallina* and *D. brachyurum* was divided into four well-separated instars, separated by the shedding of egg-membranes, recognized as 'molts' by Kotov & Boikova (1998). Kotov & Boikova (1998) make a convincing case for the recognition of the embryonic stages as 'instars', a term normally reserved for larval development. The following account is focused mostly on the development of *D. brachyurum*, based on the description of Kotov & Boikova (1998) and supplemented by illustrations provided as Fig. 12.

The first instar has a duration of 15 hours and lasts from the entry of the egg mass into the brood pouch to the shedding of the outer membrane. It includes early embryogenesis until a developmental degree where buds of all limbs are present (except trunk limb 6). As typical for cladocerans, at this degree of development, the first antennae are a pair of small buds placed with some distance between, practically at the sides of the embryo. The biramous second antennae are the largest limbs at this stage (and continue to be so); they are directed posteriorly along the sides of the body and reach the first pair of trunk limbs. The mandibles are also small limb buds (larger than the buds of the first antennae). Two pairs of small buds of the first and second maxilla are present lateroventrally. The buds of five trunk limbs are formed as two rows of elongate limb rudiments ventrolaterally on the trunk, apparently yet without subdivision into various portions as seen in adults.

The second instar lasts from approximately 15 hours to approximately 25-26 hours. In this period, many modifications take place, and the early rudiment of the carapace can be recognized. Apart from a general enlargement of all limb buds and the development of incipient setation, the modifications in limb morphology include the following: the first antennae move to a more ventral position (but are not placed closely together as in the Anomopoda and Onychopoda); the mandibles attain a curved shape; the first maxillae become setose, larger than the second maxillae and are placed closer together than the latter; the buds of the second maxillae are smaller and have more distance between them; and the trunk limbs have developed a primordial endopod and exopod and a distinct gnathobasic section. The trunk limbs are still attached along their whole length to the body. In the third instar, after the second molt, the second antennae have become free and are no longer covered by a membrane, whereas the remaining part of the body is still covered. This instar is illustrated by scanning electron microscopy after removal of the covering membrane (Fig. 12). Most features from the previous instar have been retained and are somewhat further developed, with more incipient setation. The first antennae are still a pair of very small, widely separated buds. The second antennae have become very large (about the length of the entire trunk). The mandibles are still relatively small buds. The labrum has elongated and partly covers the tip of the mandibles and first maxillae. The first maxillae have enlarged, and their ventral and setose median edges are close to each other. The second maxillae are still small, widely separated buds, each with a pore opening to the future maxillary glands. The two rows of trunk limbs are separated by a deep rudimentarily food groove. Each trunk limb is now divided distally into a distinct endopod and exopod and a distinct proximal endite. Apart from the proximal endite there is no clear separation into endites along the inner margins. The limbs have still not attained the vertical orientation of the adult. In the fourth instar, the rest of the body becomes free from the membrane, movements of various organs and limbs are now very significant (initiated already in previous stage), and the young animal basically has all the characteristic features of the adults (Kotov & Boikova 1998). The fifth instar is the last in the embryonic period. During this instar, termed the 'neonata,' the animal become released from the brood chamber of the female. Hence, the fifth instar, the 'neonata' (or 'release' stage), has part of its duration inside the brood chamber and part of it outside. The next instar constitutes the first juvenile instar. The whole issue about the status of the 'neonata' instar is summarized and clarified by Kotov (1997).

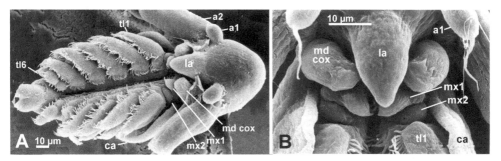

Figure 12. Embryos of *Diaphanosoma brachyurum* Liévin, 1848 (Cladocera, Ctenopoda). (A) Late embryo. From Olesen (1998). (B) Close-up of A. Original. Abbreviations: a1 = antenna 1, a2 = antenna 2, ca = carapace, cox = coxa, la = labrum, md = mandible, mx1 = maxilla 1, mx2 = maxilla 2, tl1 = trunk limb 1, tl6 = trunk limb 6.

3.10 *Anomopoda*

Since the Anomopoda is the most speciose and diverse group within the Branchiopoda, with easily accessible species all over the world, many treatments of various aspects of anomopod development are available in the literature. Among the most important for the present summary, dealing with the later part of the embryology of various species, are those of Grobben (1879), Shan (1969) Shuba & Costa (1972), Murugan & Sivaramakrishnan (1973), Murugan & Venkataraman (1977), Kotov (1996), and Kotov & Boikova (2001). Early development of various species was treated by Wotzel (1937) and Baldass (1941) among others.

It is not possible to treat development of the Anomopoda in a short summary due to the large diversity of the taxon. I will focus on the development of *Daphnia hyalina* (Leydig, 1860), for which a detailed account exists (Kotov & Boikova 2001) even though there are no reasons to assume that development of this species is typical of, or ancestral to, the Anomopoda. Kotov & Boikova (2001) identified four embryonic instars, as in the Ctenopoda, but the third instar is not completely free in *D. hyalina* (Anomopoda) due to an only partial molt of one of the covering membranes. The first appendages to become visible are those of the first and second antennae, which appear as transverse furrows in the anterior and middle portion of the embryo. Slightly later the rudiments of the labrum and the mandibles develop, and the first signs of the biramous structure of the second antennae appear. The three anterior trunk limbs appear as bars laterally on the trunk, and the two pairs of maxillae appear as small buds. Further modifications include, most notably, a significant enlargement of the second antennae, movement of the small first antennae buds to a ventral position so that they eventually fuse, movement of the mandibles to a more ventral position, reduction in size of the second maxillae relative to the first maxillae, and movement of these to a more lateral position. Five trunk limbs become developed, and the serial heterogeneity present in the adults, begins to develop.

The limb development of *Moina rectirostris* (Leydig, 1860) as described by Grobben (1879) is largely similar to that of *Daphnia hyalina*. A significant difference is that the first antennae of *M. rectirostris* enlarge during development and remain separated from each other, which is in contrast to *D. hyalina* where they remain small and fuse to each other.

The diversity of the Anomopoda is enormous, and the type of development summarized above is just one example. A preliminary SEM investigation of various taxa (Fig. 13, showing an example for each of the families Macrothricidae, Chydoridae and Daphniidae) and published information, reveals however, that the same basic pattern is present. All taxa pass through intermediate stages with basically identical morphology, which includes: (1) small first antennae limb buds that migrate ventrally and in most cases fuse; (2) relatively large biramous second antennae; (3) small mandibular limbs buds; (4) a labrum in the center between the 'naupliar' appendages; (5) first and second maxillae as small buds, mostly serially aligned in position with the proximal endites of the trunk limbs, but in later stages the second maxillae become displaced laterally. Normally five pairs of elongate trunk limb buds are present divided into various parts reflecting the morphology in the adults. In accordance with the serial heterogeneity of the morphology of trunk limbs in the adults, the embryonic trunk limbs 1-5 at this stage differ already from each other.

Continued comparative studies on the embryogenesis of especially trunk limbs of the Anomopoda, with strong attention to homologies, will provide important new information for the phylogeny of this taxon and also for its relationship to other cladoceran taxa. Interesting similarities in the embryology of the trunk limbs and first antennae of *Polyphemus pediculus* Linnaeus, 1761 (Onychopoda) and certain anomopods are already available (see below, and Olesen et al. 2003), and suggest a close relationship between the Anomopoda and Onychopoda.

3.11 *Onychopoda*

Little information is available in the literature on morphogenetic aspects of development of the Onychopoda. Kühn (1913), for example, treats only the very early development of *Polyphemus pediculus* . Sufficient information is presented here for *P. pediculus* (Fig. 14) to provide a brief summary of late ontogeny of the limbs and general body proportions. Three stages of development are shown. The earliest (Fig. 14A) has an extremely large head section, which reflects the enormous size of the eye, and a slightly larger trunk section. The head section is bent forward, the trunk section backward, giving the embryo an s-shaped appearance. Ventrally, between the head and the trunk, is a small globular labrum. On each side of the labrum are the small, undifferentiated buds of the mandibles, and slightly lateroanterior to these are the buds of the first antennae. More laterally arise the large, biramous second antennae. They are posteriorly directed and still connected to the trunk along their whole length. Neither first nor second maxillae are indicated externally at this stage. A carapace fold has appeared posterolaterally approximately at the level of the buds of the mandibles. Four pairs of trunk limbs are indicated, but they are not yet divided into clear portions.

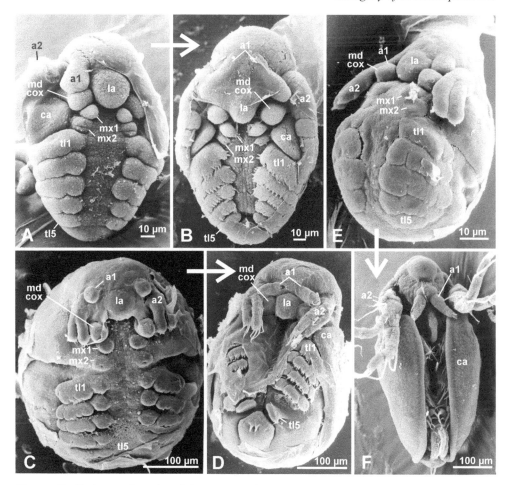

Figure 13. Embryos of species of the Anomopoda (Cladocera) representing three different families. (A, B) Early and intermediate embryos of *Macrothrix laticornis* Jurine, 1820 (Macrothricidae). Originals. (C) Early embryo of *Eurycercus glacialis* Lilljeborg, 1887. From Olesen (1998). (D) Intermediate embryo of *Eurycercus glacialis*. From Olesen et al. (2003). (E, F) Early and late embryo of *Simocephalus vetulus* O. F. Müller, 1776. Originals. Abbreviations: a1 = antenna 1, a2 = antenna 2, ca = carapace, cox = coxa, la = labrum, md = mandible, mx1 = maxilla 1, mx2 = maxilla 2, tl1 = trunk limb 1, tl5 = trunk limb 5.

The next stage shown (Fig. 14B) has the same shape and proportions as the previous stage. The most significant change is that the first antennae have become slightly more elongate, have moved to a ventral position, and are almost fused at their bases and have together a characteristic v-shaped appearance (similar to that of certain anomopods). Other modifications include the completely free second antennae, further differentiation of the mandibles, and division of the buds of trunk limbs into a number of portions.

In the third and latest stage shown (Fig. 14C), the morphology of the adult has largely been attained. Important modifications from the previous stage are that the first antennae are now placed very close together and are directed away from the body. The trunk limbs have become free and have attained an 'upright' orientation, with a distinct exopod and a long inner branch not yet showing any segmentation. Another modification is that the long caudal appendage with a pair of setae at the tip has developed.

Adult trunk limbs of *Polyphemus pediculus* (and other onychopods) are segmented, in contrast to most other branchiopods, *Leptodora kindtii* being the single exception, with a large corm and a three-segmented inner branch and an unsegmented outer branch. Here I provide a short comment on the possible evolutionary origin of these segmented appendages.

The morphology of the early limb buds in the second stage of *Polyphemus pediculus* (Fig. 14B) corresponds very closely to what is found in an early stage of the anomopod *Macrothrix laticornis* Jurine, 1820 (Fig. 13B), with respect to their size relative to each other, the early setation, and the relative proportions of the various limb portions (explored in more detail in Olesen et al. 2003). The difference is that while these limb buds develop into segmented limbs in *P. pediculus* they develop into phyllopodous unsegmented limbs in *M. laticornis*. This could indicate that *P. pediculus*, and thereby other onychopods, had their origin among anomopod cladocerans and that the segmented limbs are of secondary origin. This is similar to the mechanism shown for the origin of the segmented trunk limbs in *Leptodora kindtii* (Olesen et al. 2001). This interpretation implies that the three-segmented inner branch of the limbs in *P. pediculus*, and in other onychopods, is not precisely homologous to the unsegmented endopod found in the larger branchiopods.

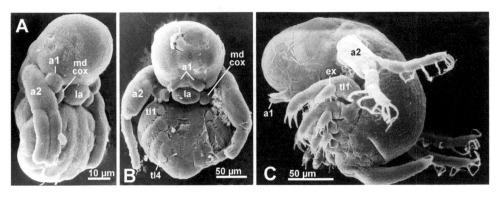

Figure 14. *Polyphemus pediculus* Linnaeus, 1761 (Cladocera, Onychopoda). (A) Early embryo. From Olesen et al. (2003). (B) Intermediate embryo. From Olesen (1998). (C) Late embryo. Original. Abbreviations: a1 = antenna 1, a2 = antenna 2, cox = coxa, la = labrum, md = mandible, tl1 = trunk limb 1, tl4 = trunk limb 4.

4 USEFULNESS OF DEVELOPMENTAL CHARACTERS FOR RECONSTRUCTION OF BRANCHIOPOD PHYLOGENY

In the following will be explored to what extent developmental information can be used for phylogenetic considerations within the Branchiopoda. As demonstrated above, branchiopods exhibit large differences in developmental types. Nearly all possible variants, with either long or more abbreviated larval series, or with direct development, are represented. In some cases this complicates comparison between taxa since it can be difficult to identify homologous stages of development.

The following takes the form of an extended 'character discussion' but should *not* be considered as a complete treatment of all characters relevant for branchiopod phylogeny. Focus is on characters relating to the ontogeny of the Branchiopoda. Characters related to the naupliar feeding system and to the early development of trunk limbs are treated in most detail because of a presumed importance of these characters in elucidating early branchiopod phylogeny. Another important aspect of branchiopod evolution, free-living larvae *versus* direct development, is also treated in some detail because of its assumed major role in the radiation of the evolutionarily successful cladocerans.

5 THE BRANCHIOPOD NAUPLIAR FEEDING APPARATUS

A very characteristic feature of the Branchiopoda *s. str.* is the morphology of the naupliar feeding and swimming apparatus. It is found, with some variations, in the Anostraca, Notostraca, Spinicaudata, Laevicaudata, Cyclestherida (in free-living larva), and *Lepidocaris* (Lipostraca), and constitutes a possible synapormorphy for these taxa (and which is presumably lost in the Cladocera). It consists of the following characteristic morphological features:

1. Natatory, biramous second antennae - most often very large - each with a pair of characteristic, often masticatory spines (endites) originating from the coxa and basis, respectively.
2. A three-segmented, uniramous mandibular palp in free-living larvae with almost identical setation in all taxa. The palp consists of a basis and two endopodal segments (often not clearly segmented in very early stages).

5.1 *Second antennae of larvae large with coxal and basipodal antennal masticatory spines (endites) of special morphology*

The second antennae are strikingly similar in free-living larvae of most Recent 'large' branchiopods, both at the general and more detailed level. This similarity is one of many characters supporting a monophyletic Branchiopoda *s. str.* First, they are very large relative to the size of the body (especially the protopod), compared to most other crustacean larvae, with the exception of larvae of the Laevicaudata, which do not have very large second antennae. The exact size of the second antennae of the larvae in *Lepidocaris* is slightly uncertain due to the preservation stage of this fossil, but they appear to be not as large as the antennae of anostracans, notostracans and spinicaudatans. The second antennae are of a smaller type in *Rehbachiella*. The large size of second antennae of branchiopod larvae was pointed out by Sanders (1963) as setting the branchiopod nauplii apart from other crustacean nauplii, but the similarities go much further than this.

All larvae of 'large' Branchiopoda have, on the protopod of each second antenna, a pair of masticatory 'spines' or setae, one on the coxa and one on the basis, termed the 'proximal masticatory spine' and the 'distal masticatory' spine, respectively, by Fryer (1983, 1988) for the Anostraca and Notostraca. These mostly very long spines are quite similar in morphology in major taxa of the 'large' branchiopods and have a comparable function at least in the Anostraca and Notostraca (see Fryer 1983, 1988), where they are involved in the collecting and further handling of food items.

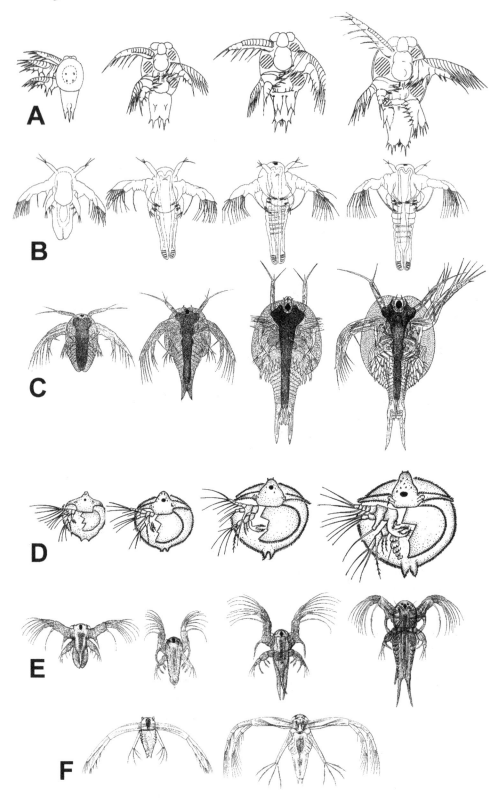

Based on their similar morphology, these structures are likely to have a similar function in spinicaudatan and laevicaudatan 'conchostracans' and in the Devonian fossil *Lepidocaris*. The similarities further include an almost exact correspondence in the timing of the change in morphology of the coxal masticatory spines from having an undivided tip to having a bifid tip, in the Anostraca, Notostraca and Spinicaudata. This change takes place between the molt from the second to the third instar in *Branchinecta ferox* (see Fryer 1983) and *Artemia salina* (see Anderson 1967) (Anostraca), in *Triops cancriformis* (Notostraca) (see Fryer 1988; Claus 1873) and in various spinicaudatan 'conchostracans', such as *Imnadia yeyetta* (Limnadiidae) and *Leptestheria dahalacensis* (Leptestheriidae) (see Eder 2003), *Cyzicus packardi* (see Sars 1896a), *Limnadia stanleyana* (see Anderson 1967), and *Eulimnadia texana* (Limnadiidae) (see Strenth & Sissom 1975). Among the spinicaudatans, the well-examined *Limnadia lenticularis* (Limnadiidae) seems to be an exception to this pattern, as three instars are passed before the coxal masticatory spines become bifid in the fourth instar (Sars 1896b), and the same was recently found in *Eulimnadia braueriana* (see Olesen & Grygier 2002). In the Devonian *Lepidocaris rhyniensis*, the few larvae known also have enlarged bifid coxal masticatory spines, which are very similar to those of intermediate and late larvae of various Recent branchiopods (see Scourfield 1940). No shift in morphology from an unbranched appearance to a bifid appearance of these particular spines was documented by Scourfield (1940) (so in theory they could be bifid from hatching), but this could be due to the fact that no very early and short bodied larva has been described in detail; only relatively advanced larvae with numerous trunk limb buds are known in necessary detail.

The second antennae of the larvae of *Rehbachiella kinnekullensis* clearly have masticatory spines on the basis and coxa of the second antennae, but apparently these are not of the same morphology as described above for larvae of the Anostraca, Notostraca, Spinicaudata, Laevicaudata, and *Lepidocaris*, i.e., there are no very elongate coxal, masticatory spines with a tip that becomes bifid after the second (third) instar. It should be noted, however, that spines and setae tend to be broken in *Rehbachiella*, and therefore the masticatory spines of the antennal protopod in the living animal may have been longer than the figures in Walossek (1993) indicate (Waloszek pers. comm.).

In all taxa with direct developing embryos, i.e., in *Cyclestheria hislopi* and the Cladocera, no sign of the masticatory endites/spines of the second antennae are present at any stage, which is not surprising as no feeding takes place in these embryos. Not even in the free-living larvae of *Leptodora kindtii* (the single exception to direct development within the Cladocera) is there any sign of the masticatory spines (Sars 1873; Warren 1901). Hence, the free-living larvae of *L. kindtii* are probably non-feeding, which is also supported by the presence of large amounts of yolk (Warren 1901). The free-living larvae of *L .kindtii* are also in many other respects quite different from other branchiopod larvae (treated in more detail below).

Figure 15. Overview of early larval sequences of branchiopods with free-living larvae. Except for *Leptodora kindtii* (F), the depicted stages represent the four earliest known stages for each species. (A) *Rehbachiella kinnekullensis* Müller, 1983. From Walossek (1993). (B) *Artemia salina* (Anostraca). From Anderson (1967). (C) *Triops cancriformis* (Notostraca). From Claus (1873). (D) *Lynceus brachyurus* (Laevicaudata). From Monakov & Dobrynina (1977). (E) *Limnadia lenticularis* (Spinicaudata). From Sars (1896b). (F) Two free-living larval stages of *Leptodora kindtii* (Cladocera, Haplopoda). From Sars (1873).

Other non-malacostracan crustaceans, such as the mystacocarids, copepods, cirripedes and cephalocarids, also have elongate spine-bearing endites on the second antennae in early larvae, of different morphologies, but in no case is the same characteristic pattern found as outlined above for the Branchiopoda *s. str.* (see Sanders 1963; Dahms 1987, 1991, 2000; Dahms & Hicks 1996; Olesen 2001).

Phylogenetic interpretations of these features are presented in a later section.

5.2 Three-segmented mandibular palp in larvae

The mandibles in the larvae of all Recent 'large' branchiopods and in the extinct *Lepidocaris* are composed of a proximal enlarged segment (coxa) and a three-segmented 'palp'. The first segment of the mandibular palp is the 'basis', followed by what is probably a two-segmented endopod. Anderson (1967) interpreted the mandibular palp of the branchiopods he examined as mainly consisting of a retained exopod. However, based on a brief comparison of the general morphology and setation of the palp of the larvae of Recent 'large' branchiopods (and *Lepidocaris*) with that of *Rehbachiella*, is appears more likely that the palp in the Recent branchiopods consists of the basis and two *endopodal* segments. The exopod of the mandible in *Rehbachiella* is multi-segmented (flagelliform), with one outwardly directed seta on each segment. The mandibular palp of the Recent branchiopods (and *Lepidocaris*) does not have such a morphology, but is more similar to the short mandibular endopod in *Rehbachiella*. The similarities include that most often each segment has more than one seta, which are inwardly directed and involved in food manipulation.

As mentioned by Scourfield (1940) and Fryer (1988), not only are the segmental patterns the same among such different taxa as the Anostraca, Notostraca, *Lepidocaris,* and the 'Conchostraca' but also the setal arrangement is strikingly similar in these taxa, some of which split off from each other at least 350 Mya since *Lepidocaris* is from the Devonian. The setation in all taxa includes two setae on the proximal palp segment (basis), one ('Conchostraca') or two (Anostraca, Notostraca, and *Lepidocaris*) on the middle segment (endopod segment 1), and three on the distal (endopod segment 2). The free-living larvae of *Leptodora kindtii* (Cladocera) also has a mandibular palp, but, apart from the fact that it has only one branch, there is no specific similarity to that of the free-living larvae of other branchiopods. In contrast to the three-segmented palp with setae spread along the whole length, as seen in the 'large' Branchiopoda and *Lepidocaris*, that of *L. kindtii* is unsegmented and has setae only on the tip. This is probably another indication of the non-feeding nature of its free-living larvae.

The larvae of the 'large' Branchiopoda (Anostraca, Notostraca and 'Conchostraca') and *Lepidocaris* are unique in having a larval mandibular palp with the above-mentioned morphology. I suggest these similarities to be a synapomorphy for the mentioned taxa (lost or modified in the Cladocera) supporting the monophyly of the Branchiopoda *s. str.*

In contrast to these taxa, larval stages of *Rehbachiella kinnekullensis* (out-group to the Branchiopoda *s. str.*) have a biramous mandible with a basis carrying a three-segmented endopod and a multi-segmented exopod, in general terms more similar to mandibles found in larvae in taxa outside the Branchiopoda such as the Copepoda, Thecostraca (see Rainbow & Walker 1976) and Mystacocarida (see Olesen 2001). The free-living nauplii in the Euphausiacea and the Dendrobranchiata (Decapoda) also have biramous mandibular palps (Sars 1898; Scholtz 2000).

6 DEVELOPMENT OF POST-MAXILLARY LIMBS (TRUNK LIMBS)

The trunk limbs of all Recent Branchiopoda pass through a stage as elongate limb buds with a bifid tip (endopod/exopod) directed *laterally*, and ventrally they are subdivided into a number of portions *before* the limbs begin to attain a vertical orientation (Figs. 2, 3, 5, 6, 9-14). These portions will in most cases become the median endites in the adults, with, however, at least one exception, which is in the predatory cladoceran *Leptodora kindtii* where instead they become true segments of the adult limbs (see Olesen et al. 2001). This type of trunk limb development has long been known for various branchiopods, e.g., Claus (1873). It was also treated by Williams & Müller (1996). It was suggested as an autapomorphy for the Recent Branchiopoda in Olesen (1999), despite some obvious variation, especially within the Cladocera. This type of early trunk limb development is different from that seen in larvae of the two branchiopod fossils, *Rehbachiella kinnekullensis* and *Lepidocaris rhyniensis*, where the limbs start their development with the bifid tip (endopod/ exopod) directed *ventrally,* and only *after* this do the median endites and various limb portions develop (Figs. 1, 4). The latter type of limb development is also found in a number of other Crustacea, such as the Euphausiacea (Sars 1898) and at least one other representative of the 'Orsten' fauna, *Bredocaris admirabilis* (see Müller & Walossek 1988). It must therefore be considered plesiomorphic for the Branchiopoda. The presence of elongate, subdivided limb buds in early larvae or embryos must be considered a potential autapomorphy for all the Recent Branchiopoda. Schram & Koenemann (2001) recently identified an '*Artemia* model of limb development' in Recent branchiopods as a supporting character for these, not found in *Rehbachiella* and *Lepidocaris*. This 'model' is essentially the same as the type of limb development summarized at the beginning of this paragraph for the Recent Branchiopoda, and was mentioned by Olesen (1999) as a character possibly placing *Rehbachiella* outside the Recent branchiopods. Although *Lepidocaris* also lacks this type of trunk limb development, in contrast to Schram & Koenemann (2001), I yet hesitate to place it outside the group comprised of the Recent Branchiopoda with certainty since at least one important character, the possession of a ventral brood pouch, suggests a close relationship to the Anostraca, as suggested by Walossek (1993, 1995). The brood pouches in these two taxa are even generated from the same two somites (12 and 13), if the original somite numbering of Scourfield (1926) is followed (and thereby accepting that the second maxillae *are* present but reduced as in other branchiopods). The fact that these two characters (mode of trunk limb development and the presence of a ventral brood pouch) conflict phylogenetically, is in the present treatment reflected as an unresolved *Lepidocaris*/Anostraca/Phyllopoda node. The phylogenetic implications of these and various other characters will be treated in more detail in the phylogenetic section below.

The specific number of endites along the inner margin in the adults of various branchiopod taxa is another aspect of possible phylogenetic interest. The trunk limbs of the most advanced stages of *Rehbachiella* have about 8 endites along the inner margin, followed by a 3-4 segmented endopod. Based on the detailed similarity in general morphology and setation between the endites of the limbs of *Rehbachiella* and those of such branchiopods as *Cyclestheria hislopi*, there are good reasons to consider them homologous. Adult trunk limbs in the Anostraca appear at first glance to have five endites followed by an unsegmented endopod. The first endite is large and is followed by a smaller endite and three even smaller ones. However, a study of early larvae of *Eubranchipus grubii* has shown that the proximal large endite in the adult is actually a fusion of two smaller endites that are present in the larvae (Fig. 3). Already Borradaile (1926) discussed the possibility that two proximal endites had become merged into one in the Anostraca, but he could not reach any firm conclusion. Fryer (1983) was aware that the large proximal endite of the trunk limbs of the Anostraca is a fusion product, and the issue was also discussed by Walossek (1993). This brings the number of endites in the Anostraca up to the same number as in *Lepidocaris*, which is six (at least in the most anterior limbs) and also here followed by an unsegmented endopod. In the Notostraca and all 'conchostracans' (Laevicaudata, Spinicaudata and *Cyclestheria*), five endites are present followed by an unsegmented endopod. The number of endites (or segments in *Leptodora* and the Onychopoda) in the trunk limbs in the Cladocera is always fewer than for the taxa mentioned hitherto, but, because of the enormous differences within the taxon, the subject is not treated here. Phylogenetic interpretation of the significance of the number of endites is not simple. However, the identical number of endites (six) in the Anostraca and *Lepidocaris* should probably be considered homologous, as should the same number of endites (five) in the Notostraca and 'Conchostraca'. Whether six endites (or fewer) can be considered an autapomorphy for the Branchiopoda *s. str.* and five endites (or fewer) an autapomorphy for the Phyllopoda is uncertain. In contrast, it is clear that the unsegmented endopod present in most Branchiopoda *s. str.* should be treated as a synapomorphy for this group. Onychopods fall somewhat aside here as they have a three-segmented 'endopod'. However, if the Onychopoda obtained segmented trunk limbs from enditic lobes in a similar way to that recently shown for *Leptodora* (see section on 'Haplopoda' above and Olesen et al. 2001), then this three-segmented structure is, against all expectations, not homologous to the unsegmented endopod of the 'large' branchiopods (see above in Onychopoda section). Other authors have also considered a connection between endites of phyllopodous limbs and segments of stenopodous limbs. Borradaile (1926) envisioned segmented crustacean limbs as derived from phyllopodous limbs, somewhat similar to the view of Fryer (1992), while Snodgrass (1956) held the opposite view.

Another trunk limb feature of possible phylogenetic importance is the presence of easily recognizable, naked, balloon-like epipods in all Recent 'large' branchiopods. The function of the epipods in the Branchiopoda appears to be a complicated issue, but Martin (1992) discusses that they appear to be osmoregulatory rather than respiratory. Balloon-like epipods occur also in a number of cladocerans, at least on some trunk limbs, but are completely absent in the raptorial taxa, the Haplopoda and Onychopoda. An epipod is apparently lacking in the two branchiopod fossils *Rehbachiella* and *Lepidocaris,* which could indicate that its presence should be treated as a possible synapomorphy of the Recent Branchiopoda. However, if the branchiopod balloon-like epipod is considered homologous to the gills in the Malacostraca, and there actually are certain similarities to the development and position of the epipods in embryos of *Nebalia* sp. (Olesen & Walossek 2000), then the polarization of this character would be reversed.

7 LARVAL SEQUENCES WITHIN THE BRANCHIOPODA AND THEIR USE FOR PHYLOGENY

The 500 My old fossil *Rehbachiella kinnekullensis* and some of the Recent anostracans have by far the longest larval sequences with the most gradual development, not only for the Branchiopoda but also for the Crustacea as a whole (Walossek 1993). If it can be assumed that this mode of development is primitive for the Crustacea, then the Branchiopoda is primitive in this respect. Indeed, the development of *Rehbachiella* is extremely gradual since it seemingly needs two molts for the addition of one trunk segment; at least this was inferred by Walossek (1993) from his detailed studies. The earliest recognized stage of *Rehbachiella* is a very short-bodied larva with only the typical three pairs of naupliar appendages, to which more segments and appendages are gradually added (Walossek 1993). Because of its presumed phylogenetic position as a putative sister group to the Branchiopoda *s. str.*, or at least as closely related to this taxon, and based on the great age of this fossil, it appears justified to assume that this type of anamorphic, very gradual development is ancestral to the Branchiopoda *s. str.*, from which other developmental strategies within the Branchiopoda, or even the developmental sequences for other Crustacea, have been derived, as summarized in detail by Walossek (1993).

Within the Branchiopoda *s. str.* the Anostraca exhibits the most anamorphic development, with as many as 20 stages described for *Artemia salina* (by Benesch 1969) and *Branchinecta ferox* (by Fryer 1983). Only the three naupliar appendages are developed in the hatching stage of these taxa, as in *Rehbachiella*, but it is in contrast non-feeding and lecitotrophic. In general, one segment and one pair of appendages are added for each molt in the continued development, which makes the development of these species slightly less anamorphic than that of *Rehbachiella*. However, they are still the most anamorphic of Recent Crustacea even when compared to the, in many ways, primitive Cephalocarida, as pointed out by Fryer (1983) and Walossek (1993). The development of *Lepidocaris* is too poorly known to mention in this context.

Notostracan development is in general slightly more abbreviated. The first stage of *Triops cancriformis* has some of the characteristics mentioned for the hatching stage of certain anostracans, including having only three pairs of developed naupliar limbs; it is non-feeding and lecitotrophic (Claus 1873; Fryer 1983). From the first to the second stage, 6-7 trunk limbs become visible, but not yet functional, and already in the third stage about 10 trunk limbs have been developed to a varying extent, and the general appearance is somewhat adult-like. *Artemia salina* (Anostraca) would need as many as 13 molts to achieve a similar degree of development to that reached by *T. cancriformis* in only two molts. The development of *Lepidurus* is even more abbreviated (Borgstrøm & Larsson 1974; Fryer 1988).

The development of 'conchostracans' with free-living larvae is more abbreviated than that of *Rehbachiella* and the Anostraca. Different from any of the taxa mentioned, the development of the 'conchostracans' with free-living larvae can be divided in two distinctly different phases, a *naupliar* (larval) phase and a *juvenile* (post-larval) phase, separated by a metamorphosis-like jump in morphology from one instar to another. The naupliar phase lasts 5-7 instars depending on the species, and is characterized by the presence of (1) an active naupliar feeding apparatus, (2) a very large and often spine-bearing labrum, (3) undeveloped trunk limbs, and (4) a small dorsal carapace (not yet bivalved and not yet embracing the whole trunk). The remarkable larvae of *Lynceus* have some specialities, such as two very large horn-like first antennae. Basically all these features are lost during one molt from the last naupliar to the first juvenile instar. While the nauplii have a dorso-ventrally flattened appearance, juveniles have a laterally flattened appearance, basically looking like small adults enclosed in their large bivalved carapaces.

There appears to be no simple way to extract information of phylogenetic value from the larval sequences themselves, such as the number of stages. If one attempts to do so, it should probably be in cases where there is an *exact* correspondence in, for example, the number of larval stages, but this alone is not enough, since it should also be documented that the sequence is homologous by identifying similarities between taxa stage by stage. The only possible example within the Branchiopoda that satisfies such strict criteria is that of the larval sequences for some spinicaudatan 'conchostracans' (*Cyclestheria* not included). Eder (2003) identified five larval stages of a similar morphology for the two species he examined, and found a similar pattern for various species mentioned in literature. When these five stages have been passed, a small metamorphosis to the juvenile phase takes place and adult morphology is approached. This similarity in the developmental sequence could be treated as a possible autapomorphy for the Spinicaudata. However, it should be mentioned that the number of larval stages for the spinicaudatans is not identical for all species, as Sars (1896a, b) Strenth & Sissom (1976) and Olesen et al. (2003) have described species with six and seven stages. Certain elements of the direct development of *Cyclestheria hislopi* also fit into this pattern, such as the approximate same number of stages and the fact that *Cyclestheria* changes abruptly, metamorphosis-like, in morphology between two stages, as do spinicaudatans (see Olesen 1999). The embryonized larval stages in the dorsal brood pouch of the female in *Cyclestheria* correspond more or less one by one to the free-living larval stages of the Spinicaudata. Exactly how the larval development of the Laevicaudata applies here is uncertain, since this group is not yet studied in sufficient detail.

8 DIRECT DEVELOPMENT WITHIN THE BRANCHIOPODA

Direct development is found in the Cladocera and in *Cyclestheria hislopi* (and in some cases in *Artemia salina*, see Viña Bayes & Botnariuc 1977) and was considered by Olesen (1999) and Taylor *et al.* (1999) to be homologous in the two taxa. However, if direct development was actually already present in the common ancestor to *Cyclestheria* and the Cladocera, what is then the status of the free-living larvae of the cladoceran *Leptodora kindtii* and in the Cuban population of *Cyclestheria hislopi?* With regard to *Leptodora* two possibilities exist when accepting the direct development in the parthenogenetic part of the life-cycle in the brood chamber as homologous to that of *Cyclestheria*. Either the free-living larvae are retained from earlier in evolution, which means that the embryonization of the development in the resting eggs has been delayed compared to that of the brood pouch, or the larvae have become secondarily free-living. The question is analogous to a case recently discussed for malacostracan Crustacea, where the free-living nauplii of the Euphausiacea and Dendrobranchiata were argued to have been evolved secondarily within the Malacostraca based on a parsimony argument and on certain embryonic characteristics of the nauplii that could indicate an 'egg-naupliar' origin (Scholtz 2000). Similar arguments could be applied to the free-living larvae of *Leptodora kindtii*. No matter which phylogeny is preferred, it will be more 'parsimonious' (less number of evolutionary steps involved) to view these larvae as secondary, having been regained from direct developing embryos. Similarly it can be noted that the larvae of *Leptodora* is quite different from those of other branchiopods with free-living larvae, which, in contrast share a number of characteristics related to their feeding, as addressed in detail above. The larvae of *Leptodora* lack appropriate structures to collect food and contain a high amount of yolk. However, these features are common in early free-living stages of crustaceans, even within the Branchiopoda, and these larvae are not argued to have evolved secondarily. The only argument for a secondary origin of the free-living larvae of *Leptodora* is the parsimony argument. We are unlikely, unfortunately, to come to the answer to such a question much closer. The conclusion will always depend on which kind of evidence one is focusing, e.g., whether one prefers parsimony arguments (favors secondary status of larva) or complicated structures (like free-living larvae) not to be lost and regained (favors larva of *Leptodora* as plesiomorphic).

All the above-mentioned considerations rest on the assumption that the direct development in *Cyclestheria* and the Cladocera is homologous. If this turns out not to be the case, then the parsimony argument of a secondary origin of the free-living larvae of *Leptodora* would vanish. However, there appear to be good reasons for homology between the direct development in the two taxa. First of all, there are now convincing arguments that *Cyclestheria* is actually the sister group to the Cladocera both from a morphological and molecular point of view (various types of evidence cited in Olesen 2000, but see Fryer 1999, 2001, 2002 for an alternative view). If this is correct, then it is unparsimonious to accept convergent evolution of these characteristics in such closely related taxa. Furthermore, a number of similarities can be mentioned between the embryos of *Cyclestheria* and those of the Cladocera, like the lack of mandibular palp, lack of masticatory spines on second antennae, and a overall similarity in the relative size of the first antennae, second antennae, mandibles, which would support further the homologous status of the direct development in *Cyclestheria* and the Cladocera. However, since one can argue that most of these characters are related to the embryonic status of the stages, and since also many *differences* between *Cyclestheria* and the Cladocera as well as in between the cladoceran orders can be mentioned, the similarities may not be very convincing. It appears plausible that a convergent evolution of direct development in various branchiopods would end up with embryos having these characteristics.

An interesting question is how to homologies the free-living larvae of various 'large' branchiopods with the embryonic stages of *Cyclestheria hislopi* and various cladocerans. As mentioned in the previous section, the embryonic stages of *Cyclestheria* correspond more or less one-by-one to the free-living larvae of certain Spinicaudata, to which *Cyclestheria* obviously bears many homologous similarities (symplesiomorphies). Considering the comparable number of stages and many other detailed similarities in limb development and the mentioned similar 'jump' in morphology separating the 'naupliar' phase from the 'juvenile' phase, it is not difficult to see that the embryos of *Cyclestheria* are embryonized free-living spinicaudate larvae. Often the term 'embryo' is applied to the developmental stages of *Cyclestheria* even though these are actually homologous to the free-living larvae of the Spinicaudata. In a strict sense, this is unprecise from the point of view of homology, which is why the terms 'larvae' and 'post-larvae' were applied for the developmental stages of *Cyclestheria* by Olesen (1999). This is rather obvious when *Cyclestheria* is considered because of the mentioned one-to-one correspondence to larvae of the Spinicaudata. Considering the similarities between the embryos of *Cyclestheria* and the Cladocera and their likely close relationship, it appears likely that the embryos of the Cladocera are also modified free-living larvae, just more embryonized than those of *Cyclestheria*. Kotov & Boikova (2001), based on a summary of the number of egg membranes in various branchiopod taxa, reach a similar conclusion. Both parthenogenetic and gametogenetic eggs of the Anomopoda and Ctenopoda are covered by two egg-membranes which is similar to the eggs of the Anostraca, and Kotov & Boikova (2001) suggest, based on Hoshi (1951), that the 'moment' of casting off of the inner (second) membrane in the mentioned cladoceran taxa is homologous to hatching in 'large' branchiopods. From this it follows that the period in the development of the Ctenopoda and Anomopoda, which takes place in the brood pouch, is homologous to the phase with free-living larvae in the larger branchiopods (Kotov & Boikova 2001).

The Cladocera has often been said to have a neotenic origin from a conchostracan-like ancestor due to the similarity of certain adult cladocerans (mainly ctenopods) to the larvae of certain conchostracans (Claus 1876; Grobben 1892; Margalef 1949; Schminke 1981). However, as mentioned by Olesen (1999), this idea conflicts with the hypothesis of

Cyclestheria as sister group to the Cladocera. If this is the case, then the common ancestor to these taxa most likely had direct development with embryonized larvae from which no *neotenic* processes could have given rise to the cladocerans.

9 PHYLOGENY OF THE BRANCHIOPODA ESTABLISHED BY USE OF DEVELOPMENTAL CHARACTERS

The phylogeny presented in Fig. 16 is partly a consensus tree based on Walossek (1993), Olesen et al. (1997), Olesen (1998) and Spears & Abele (2000), and is partly based on ontogenetic characters, some of which are presented for the first time in this study. It is not the result of a comprehensive analysis considering all relevant characters. Instead, focus has been on characters relating to ontogeny.

Since many convincing characters can be mentioned in support for the Recent Branchiopoda and *Lepidocaris,* this taxon must be considered monophyletic with a high level of confidence, in this account referred to as Branchiopoda *s. str.*

Supporting characters (some already discussed) for the Branchiopoda *s. str.* include:

- A similar naupliar feeding apparatus, including 1) very large second antennae, each with a characteristic pair of masticatory spines/setae, the proximal (coxal) of which become bifid approximately in the same molt in all taxa, and 2) a uniramous, three-segmented palp of the mandibles (consists of basis and two endopodal segments). Secondary loss or modifications of these structures must be assumed in the embryos of taxa with direct development and in the free-living larva of Leptodora.
- An unsegmented endopod on the trunk limbs of the adult. The status of the three-segmented inner branch of the Onychopoda is uncertain, but see above. The trunk limbs of the Cladocera are in general extremely diverse in morphology.
- Second maxillae are small. In *Rehbachiella,* maxillae 2 are well developed and basically look like trunk limbs.

These characters exclude *Rehbachiella* from the Branchiopoda *s. str.* Furthermore, the characters previously suggested by Walossek (1993) for a sister group relationship between *Rehbachiella* and the Recent Anostraca (plus *Lepidocaris*) are not convincing, with the possible exception of the early development of the compound eyes, which could then be considered a symplesiomorphy. However, I accept the characters suggested by Walossek (1993) as indicating a close relationship between *Rehbachiella* and the Branchiopoda and therefore include it in this account in a taxon referred to as Pan-Branchiopoda or Branchiopoda *s. lat.* The characters suggested by Walossek (1993) as shared between *Rehbachiella* and the remaining branchiopods are (1) a ventral sternitic food groove and (2) posteriorly concave trunk limbs corms with a characteristic set of three rows of setae on the endites (see Walossek 1993 for more details). However, while I consider the Branchiopoda *s. str.* to be very well supported, I consider the Pan-Branchiopoda as less well supported, and the possibility that certain other crustacean taxa are more closely related to the Branchiopoda *s. str.* than is *Rehbachiella* should not be excluded.

Within the Branchiopoda *s. str.,* the position of *Lepidocaris* as closely related to the Recent Anostraca, suggested tentatively by Scourfield (1926) and supported by Walossek

(1993, 1995), is likely, but another plausible possibility most be taken into account (see below). The general similarity between *Lepidocaris* and the Anostraca in body proportions and tagmosis is striking, but this could be symplesiomorphic. The following specific characters can be identified as potential synapomorphies for *Lepidocaris* and the Anostraca:

Supporting characters for the Anostraca and *Lepidocaris*:

- With a ventral brood pouch, which is apparently generated from the same two pairs of limbs in the two taxa.
- Early limb buds of the second maxillae are placed closer together (see Figs. 3B, 4B) than in other branchiopods.
- Trunk limbs have six endites (two proximal endites fused to one large in adults of the Anostraca), while more are present in *Rehbachiella* and fewer in all other branchiopods.

The most convincing of these characters by far is the presence of the ventral brood pouch, which is difficult to dismiss as convergence. An important similarity between the ventral brood pouches in *Lepidocaris* and the Recent Anostraca is that they appear to be formed from appendages belonging to the same trunk segments. Other similarities concern the general morphology of the brood pouches. While the brood pouch in a number of species of the Recent Anostraca have a quite different morphology from that of *Lepidocaris*, it is interesting that the brood pouch in certain developing stages of *Artemia salina* (as shown in SEM by Schrehardt 1987) looks strikingly similar to that of the adult of *Lepidocaris* with the anterior part developing from the anterior genital segment and the posterior part from the posterior genital segment.

On the other hand some characters are found only in the Recent Branchiopoda, not in *Lepidocaris* (or *Rehbachiella*), thereby supporting a monophyletic taxon consisting of the Anostraca, Notostraca, 'Conchostraca', and the Cladocera, with *Lepidocaris* as the sister group (see also Schram & Koenemann 2001).

Supporting characters for the Recent Branchiopoda (conflicting with characters supporting the Anostraca and *Lepidocaris*):

- The trunk limbs develop as elongate, laterally placed limb buds, which become subdivided into a number of enditic portions before they attain an 'upright' vertical orientation.
- The trunk limbs have epipods. This implies that the epipods found in other Recent Crustacea, like the Malacostraca, are not homologous with those of the Branchiopoda.
- Larval second antennae, especially the protopods, are very long relative to the body size. The large size of the second antennae is striking when comparing representatives of the Notostraca, Anostraca and Spinicaudata (see Figs. 2A, 5A, 8A) - significantly longer than those of *Rehbachiella* and *Lepidocaris*. Also, the free-living larvae of *Leptodora* have very large second antennae. The smaller-sized second antennae of the larvae of the Laevicaudata and the direct developing embryos of *Cyclestheria* must then be assumed secondarily reduced.
- First antennae short and unsegmented in the adults (a few taxa like *Lynceus* spp. and *Ilyocryptus* spp. have two-segmented first antennae). In *Lepidocaris*,

the first antennae are three-segmented in the adults, and in *Rehbachiella* they are long, annulated, and with three segments distally.

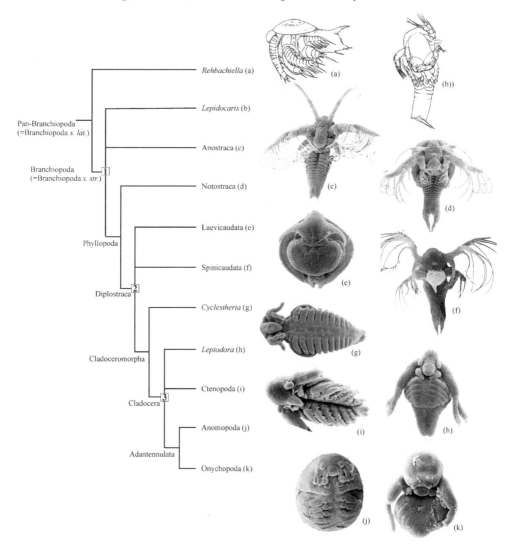

Figure 16. Phylogeny of the Branchiopoda based partly on developmental characters. The nodes numbered 1 to 3 are polytomies.

Because of these conflicting sets of characters the relationships among *Lepidocaris*, the Anostraca, and the Phyllopoda are left unresolved, and shown as a polytomy in Figure 16. It can with some justification be argued that the mentioned potential synapomorphies for the Recent Branchiopoda, excluding *Lepidocaris*, are more likely to have appeared convergently than the ventral brood pouch in *Lepidocaris* and the Recent Anostraca. An alternative explanation would be to interpret the ventral brood pouch as a symplesiomorphy.

The Phyllopoda, consisting of the Notostraca and Diplostraca, is included as a mono-phyletic group in Fig. 16, but this is mostly based on earlier published morphological and molecular information (Walossek 1993, 1995; Negrea 1999, Taylor et al. 1999; Spears & Abele 2000). It is striking, however, to note that all members of this taxon in addition share two reduction characters, which are (1) 'five or fewer endites on the trunk limbs' and (2) the number of segments of the exopod of the second antennae is reduced to about four in the larvae, while it is considerably higher in larvae of the Recent Anostraca and in *Reh-bachiella*. In *Lepidocaris*, however, this number is also low, which, based on this character alone, would suggest *Lepidocaris* as the sister group to the Phyllopoda. The latter possibi-lity may sound less likely but should not be excluded altogether since the usable characters supporting the Phyllopoda are so few that it does not take much to change the phylogenetic branching. In this connection it is interesting that *Lepidocaris* has no external eyes. It has often been stated to be blind, but another possibility would be that it had internalized eyes like the Notostraca, 'Conchostraca' and the Cladocera (Phyllopoda). However, this is pure speculation.

A final decision of the character distribution of these and other characters must await a more complete phylogenetic analysis of a more complete set of characters.

The Diplostraca, consisting of the former 'Conchostraca' and the Cladocera, is included as a monophyletic group, but this is based on earlier published information mostly relating to the morphology of the male claspers and an apparently similar development of the carapace, suggested by Walossek (1993, 1995) (see Olesen 1998, 2000, 2002), but see Fryer (1987, 1996, 1999, 2001, 2002) for a differing view on the status of these characters and of the Diplostraca.

The position of *Cyclestheria hislopi* as a sister group to a monophyletic Cladocera is partly based on the characters listed below, some of which are developmental characters.

Supporting characters for *Cyclestheria* and the Cladocera (Cladoceromorpha):
- Direct development - embryos are similar with respect to relative size of limb buds (see above).
- Ephippium. Must then be assumed lost in the Ctenopoda, Haplopoda and Onychopoda 1-3 times, depending on the intrinsic phylogeny of the Cladocera.
- Production of two types of eggs: parthenogenetic eggs and resting eggs after mating.
- Fused compound eyes.
- Molecular data including sequence data and stem-loop structures for the 18S rRNA.

The Cladocera is here accepted as a monophyletic group, following Martin & Cash-Clark (1995), Olesen et al. (1997), Negrea et al. (1999), Olesen et al. (2003) and Olesen (1998, 2000, 2002), but see Fryer (1987, 1999, 2001, 2002) for a different view.

The phylogeny of the Cladocera has traditionally been difficult to elucidate. At least three, partly conflicting, possibilities have now been put forward. Apart from numerous suggestions in older literature, Martin & Cash-Clark (1995) and Olesen (1998) suggested that the Haplopoda and Onychopoda were sister groups, while Wingstrand (1978) and Negrea et al. (1999) suggested the Haplopoda as sister group to the remaining cladocerans. The latter possibility is also implied by the scheme put forward by Eriksson (1934), who suggested the term 'Eucladocera' for the remaining cladocerans. Olesen et al. (2003)

introduced the possibility of the Anomopoda and Onychopoda as sister groups based on a similar development of the first antennae and trunk limbs. In the present account, I have introduced a new taxonomic category 'Adantennulata,' which means 'antennules (= 1st antennae) close together,' to accommodate these two taxa. The taxon name refers to the fact that the 1st antennae in these two taxa during development, in a strikingly similar way, move considerably closer together than in other cladocerans and in other branchiopods (treated and illustrated in more detail in Olesen et al. 2003). This suggested grouping of higher cladoceran taxa (Onychopoda and Anomopoda) conflicts with recent molecular evidence put forward by Taylor et al. (1999) and Braband et al. (2002), which supports the Onychopoda and Haplopoda as sister taxa.

ACKNOWLEDGMENTS

Ole Sten Møller (Zoological Institute, University of Copenhagen) is gratefully acknowledged for allowing me to include photographs of *Triops canriformis* from his unpublished work. I am also grateful to Mark J. Grygier (Lake Biwa Museum, Japan) for letting me use photographs of *Eulimnadia braueriana* from our work on the larval development of Japanese 'conchostracans'. Danny Eibye-Jacobsen (Zoological Museum, Copenhagen) helped translating parts of certain papers. Joel W. Martin (Los Angeles, U.S.A.), Geoffrey Fryer (Cumbria, England), Dieter Waloszek (Ulm, Germany), Stefan Richter (Berlin, Germany), and Gerhard Scholtz (Berlin, Germany) all kindly spent valuable time going through different versions of the manuscript and offered many useful suggestions to both content and writing. This work was directly supported by a postdoctoral grant from the Danish Research Council. Gerhard Scholtz (Berlin, Germany), editor of the present book, is thanked for the invitation to write this chapter.

REFERENCES

Agar, W. E. 1908. Note on the early development of a cladoceran (*Holopedium gibberum*). *Zool. Anz.* 33: 420-427.
Anderson, D. T. 1967. Larval development and segment formation in the branchiopod crustaceans *Limnadia stanleyana* King (Conchostraca) and *Artemia salina* (L.) (Anostraca). *Austr. J. Zool.* 15: 47-91.
Andrews, T. F. 1948. The parthenogenetic reproductive cycle of the cladoceran *Leptodora kindtii*. *Trans. Amer. Microscop. Soc.* 67: 54-60.
Ax, P. 1985. Stem species and the stem lineage concept. *Cladistics* 1: 279-287.
Ax, P. 1989. The integration of fossils in the phylogenetic system of organisms. *Abhandl. Naturwiss. Ver. Hamburg (NF)* 28: 27-43.
Ax, P. 1999. *Das System der Metazoa II. Ein Lehrbuch der phylogenetischen Systematik.* Stuttgart: Gustav Fischer Verlag.
Baldass, F. 1937. Entwicklung von *Holopedium gibberum. Zool. Jahrb. Anat.* 63: 399-454.
Baldass, F. 1941. Entwicklung von *Daphnia pulex. Zool. Jahrb. Anat.* 67: 1-60.
Barlow, D. I. & Sleigh M. A. 1980. The propulsion and use of water currents for swimming and feeding in larval and adult *Artemia*. In: Persoone, G., Sorgeloos, P., Roels, O. & Jaspers, E. (eds.): *The brine shrimp* Artemia, *1, morphology, genetics, radiobiology, toxicology*: 61-73. Wetteren: Universa Press.
Benesch, R. 1969. Zur Ontogenie und Morphologie von *Artemia salina* L. *Zool. Jahrb. Anat.* 86: 307-458.

Berry, W. 1926. Description and notes on the life history of a new species of *Eulimnadia*. *American J. Sci.* 65: 429-433.

Borgstrøm, R. & Larsson, P. 1974. The first three instars of *Lepidurus arcticus* (Pallas), (Crustacea: Notostraca). *Norw. J. Zool.* 22: 45-52.

Borradaile, L. A. 1926. Notes upon crustacean limbs. *Ann. Mag. Nat. Hist., Series 9*, 17: 193-213, 4 pls.

Borradaile, L. A. 1926. On the primitive phyllopodium. *Ann. Mag. Nat. Hist., Series 9*, 18: 16-18.

Botnariuc, N. 1947. Contributions à la connaissance des phyllopodes Conchostracés de Roumanie. *Notationes Biologicae* 5: 68-159.

Botnariuc, N. 1948. Contribution a la connaissance du développement des phyllopodes Conchostracés. *Bull. Biol. France Belgique* 82: 31-36.

Botnariuc, N. & Viña Bayés, N. 1977. Contribution à la connaissance de la biologie de *Cyclestheria hislopi* (Baird), (Conchostraca: Crustacea) de Cuba. In: Orghidan, T. (ed), *Résultats des expéditions biospéléogiques cubano-roumaines à Cuba*, 2: 257-262. Bucuresti: Editura Academiei Republicii Socialiste Romania.

Braband, A., Richter, S., Hiesel, R. & Scholtz, G. 2002. Phylogenetic relationships within the Phyllopoda (Crustacea, Branchiopoda) based on mitochondrial and nuclear markers. *Mol. Phyl. Evol.* 25: 229-244.

Bratcik, R. J. 1980. Morfologiceskie osobenosti postembrionalnogo razvitia *Caenestheria* sp. (Conchostraca, Cyzicidae). *Tr. Inst. Biol. Vnutr. Vod. AN SSSR* 44/47: 66-71.

Brauer, F. 1872. Beiträge zur Kenntniss der Phyllopoden. *Sitzungsberichte der Akademie der Wissenschaften in Wien Mathematisch-naturwissenschaftliche Klasse* 65: 279-291.

Cannon, H. G. 1928. On the feeding mechanism of a fairy shrimp, *Chirocephalus diaphanus* Prévost. *Trans. R. Soc. Edin.* 55: 807-822.

Cannon, H. G. & Leak, F. M. C. 1933. On the feeding mechanism of the Branchiopoda. Appendix on the mouth parts of the Branchiopoda. *Phil. Trans. R. Soc. Lond. B* 222: 340-352.

Claus, C. 1873. Zur Kenntniss des Baues und der Entwicklung von *Branchipus stagnalis* und *Apus cancriformis*. *Abhandlungen der Königlichen Gesellschaft der Wissenschaften in Göttingen* 18: 93-140.

Claus, C. 1876. *Untersuchungen zur Erforschung der genealogischen Grundlage des Crustaceen-Systems*. Wien: Carl Gerold's Sohn.

Claus, C. 1886. Untersuchungen über die Organisation und Entwicklung von *Branchipus* und *Artemia* nebst vergleichenden Bemerkungen über andere Phyllopoden. *Arbeiten aus dem Wiener Zoologischen Institut* 6: 1-104, pl. I-XII.

Dahms, H.-U. 1987. Die Nauplius-Stadien von *Bryocamptus pygmaeus* (Sars, 1862) (Copepoda, Harpacticoida, Canthocamptidae). *Drosera* 87: 47-58.

Dahms, H.-U. 1991. Usefulness of postembryonic characters for phylogenetic reconstruction in Harpacticoida (Crustacea, Copepoda). *Proceedings of the Fifth International Conference on Copepoda; Bulletin of the Plankton Society of Japan, Special Volume:* 87-104.

Dahms, H.-U. 2000. Phylogenetic implications of Crustacean nauplius. *Hydrobiologia* 417: 91-99.

Dahms, H.-U. & Hicks, G. R. F. 1996. Naupliar development of *Parastenhelia megarostrum* (Copepoda: Harpacticoida), and its bearing on phylogenetic relationships. *J. Nat. Hist.* 30: 11-22.

Dodds, G. S. 1926. Entomostraca from the Panama Canal zone with description of one new species. *Occasional papers of the Museum of Zoology. University of Michigan* 174: 1-27.

Eder, E. 2003. SEM investigations of the larval development of *Imnadia yeyetta* and *Leptestheria dahalacensis* (Branchiopoda: Conchostraca). *Hydrobiologia* 486: 39-47.

Dumont, H. J. & Negrea, S. V. 2002. Introduction to the class Branchiopoda. In: Dumont H. J. F. (ed) *Guides to the Identification of the Microinvertebrates of the Continental Waters of the World*. Leiden: Backhuys Publishers.

Ekman, S. 1902. Beiträge zur Kentniss der Phyllopodenfamilie Polyartemiidae. *Bihang till K. Svenska vetenskapsakademiens handlinger* 29.

Eriksson, S. 1934. Studien über die Fangapparate der Branchiopoden nebst einigen phylogenetischen Bemerkungen. *Zoologiske Bidrag från Uppsala* 15: 23-287.

Ficker, G. 1876. Zur Kenntniss der Entwicklung von *Estheria ticinensis* Bals. Criv. *Arbeiten aus dem zoologisch-vergleichend-anatomischen Institue der Wiener Universität* 74: 407-421.

Fryer, G. 1983. Functional ontogenetic changes in *Branchinecta ferox* (Milne-Edwards) (Crustacea, Anostraca). *Phil. Trans. R. Soc. Lond. B* 303: 229-343.

Fryer, G. 1985. Structure and habits of living branchiopod crustaceans and their bearing on the interpretation of fossil forms. *Trans. R. Soc. Edinb. Earth Sciences* 76: 103-113.

Fryer, G. 1987. A new classification of the branchiopod Crustacea. *Zool. J. Linn. Soc.* 91: 357-383.

Fryer, G. 1988. Studies on the functional morphology and biology of the Notostraca (Crustacea: Branchiopoda). *Phil. Trans. R. Soc. Lond. B* 321: 27-124.

Fryer, G. 1992. The origin of the Crustacea. *Acta Zool.* 73: 273-286.

Fryer, G. 1996. The carapace of the branchiopod Crustacea. *Phil. Trans. R. Soc. Lond. B* 351: 1703-1712.

Fryer, G. 1999. A comment on a recent phylogenetic analysis of certain orders of the branchiopod Crustacea. *Crustaceana* 72: 1039-1050.

Fryer, G. 2001. The elucidation of branchiopod phylogeny. *Crustaceana* 74: 105-114.

Fryer, G. 2002. Branchiopod phylogeny: facing the facts. *Crustaceana* 75: 85-88.

Gauld, D. T. 1959. Swimming and feeding in crustacean larvae: the nauplius larva. *Proc. Zool. Soc. Lond.* 132: 31-50.

Gerberding, M. 1997. Germ band formation and early neurogenesis of *Leptodora kindti* (Cladocera): first evidence for neuroblasts in entomostracan crustaceans. *Invert. Repr. Dev.* 32: 63-73.

Grobben, C. 1879. Die Entwicklungsgeschichte der *Moina rectirostris*. Zugleich ein Beitrag zur Kentniss der Anatomie der Phyllopoden. *Arbeiten aus dem Zoologischen Institut der Universität Wien und der Zoologischen Station in Triest* 2: 203-268.

Grobben, K. 1892. Zur Kenntniss des Stammbaumes und des Systems der Crustaceen. *Sitzungsberichte Kaiserlichen Akademie Wissenschaften Wien* 101: 237-274.

Grube, A. E. 1853. Bemerkungen über die Phyllopoden, nebst einer Uebersicht ihrer Gattungen und Arten. *Arch. Naturgesch.* 19: 71-172.

Gurney, R. 1926. The Nauplius larva of *Limnetis gouldi. Rev. ges. Hydrobiol. Hydrogeo.* 16: 114-117.

Hoshi, T. 1951. Studies on the physiology and ecology of plankton. VI. Glycogen in embryonic life of *Simocephalus vetulus* with some notes on the energy source of development. *Biol. Repts Tohoku Univ. 4th ser. (Biology)* 19: 123-133.

Johansen, F. 1925. Further observations on Canadian Euphyllopoda. *Canad. Field-Natural.* 39: 105-108.

Kotov, A. 1996. Fate of the second maxilla during embryogenesis in some Anomopoda Crustacea (Branchiopoda). *Zool. J. Linn. Soc.* 116: 393-405.

Kotov, A. A. 1997. A special moult after the release of the embryo from the brood pouch of Anomopoda (Branchiopoda, Crustacea): a return to an old question. *Hydrobiologia* 354: 83-87.

Kotov, A. A. & Boikova, O. S. 1998. Comparative analysis of the late embryogenesis of *Sida crystallina* (O.F. Müller, 1776) and *Diaphanosoma brachyurum* (Lievin, 1848) (Crustacea: Branchiopoda: Ctenopoda). *Hydrobiologia* 380: 103-125.

Kotov, A. A. & Boikova, O. S. 2001. Study of the late embryogenesis of *Daphnia* (Anomopoda, 'Cladocera', Branchiopoda) and a comparison of development in Anomopoda and Ctenopoda. *Hydrobiologia* 442: 127-143.

Kühn, A. 1913. Die Sonderung der Keimebezirke in der Entwicklung der Sommereier von *Polyphemus pediculus* de Geer. *Zool. Jahrb. Anat.* 35: 243-340.

Lauterbach, K. E. 1989. Das Pan-Monophylum - Ein Hilfsmittel für die Praxis der Phylogenetischen Systematik. *Zool. Anz.* 223: 139-156.

Lereboullet, M. 1866. Observation sur la génération et le développement de la Limnadie de Hermann (*Limnadia hermannni* Ad. Brogn.). *Ann. Scie. Nat. Zool. ser* 5 (5): 283-308.

Margalef, R. 1949. Importancia de la neotenia en la evolutión de los Crustáceos de agua dulce. *Publicaciones del Instituto Biologia Aplicada* 6: 41-51.

Martin, J. W. 1992. Branchiopoda. In: Harrison F. W. (ed), *Microscopic Anatomy of Invertebrates, Vol. 9:* 25-224. New York: Wiley-Liss.

Martin, J. W. & Cash-Clark, C. 1995. The external morphology of the onychopod "cladoceran" genus *Bythotrephes* (Crustacea, Branchiopoda, Onychopoda, Cercopagididae), with notes on the morphology and phylogeny of the order Onychopoda. *Zool. Scripta* 24: 61-90.

Martin, J. W. & Davis, G. E. 2001. An updated classification of the Recent Crustacea. *Natural History Museum of Los Angeles County, Science Series* 39: 1-164.

Mattox, N. T. 1950. Notes on the life history and description of a new species of conchostracan phyllopod *Caenestheriella gynecia*. *Trans. Amer. Micros. Soc.* 69: 50-53.

McKenzie, K. G., Chen P.-J. & Majoran, S. 1991. *Almatium gusevi* (Chernyshev 1940): redescription, shield-shapes; and speculations on the reproductive mode (Branchiopoda, Kazacharthra). *Paläont. Zeitschr.* 65: 305-317.

Møller, O. S, Olesen, J. & Høeg, J. T. 2003. SEM studies on the early development of *Triops cancriformis* (Bosch) (Crustacea, Branchiopoda, Notostraca). *Acta Zool.* 84: in press.

Monakov, A. V. & Dobrynina, T. I. 1977. Postembryonic development of *Lynceus brachyurus* O. F. Müller (Conchostraca). *Zool. Zhurnal.* 56: 1877-1880.

Monakov, A. V., Paveljeva, E. B. & Bratcik, R. J. 1980. Postembrionalnoe razvitie i rost *Leptestheria dahalacensis* (Rüppel) (Branchiopoda, Conchostraca). *Tr. Inst. Biol. Vnutr. Vod. AN SSSR* 41/44: 53-58.

Müller, K. J. & Walossek, D. 1988. External morphology of the Upper Cambrian maxillopod *Bredocaris admirabilis*. *Fossils Strata* 23: 1-70.

Murugan, N. & Sivaramakrishnan, K. G. 1973. The biology of *Simocephalus acutirostratus* King (Cladocera: Daphnidae) - laboratory studies of life span, instar duration, egg production, growth and stages in embryonic development. *Freshw. Biol.* 3: 77-83.

Murugan, N. & Venkataraman, K. 1977. Study of the in vitro development of the parthenogenetic egg of *Daphnia carinata* King (Cladocera: Daphniidae). *Hydrobiologia* 52: 129-134.

Negrea, S., Botnariuc, N. & Dumont, H. J. 1999. Phylogeny, evolution and classification of the Branchiopoda. *Hydrobiologia* 412: 191-212.

Olesen, J. 1998. A phylogenetic analysis of the Conchostraca and Cladocera (Crustacea, Branchiopoda, Diplostraca). *Zool. J. Linn. Soc.* 122: 491-536.

Olesen, J. 1999. Larval and post-larval development of the branchiopod clam shrimp *Cyclestheria hislopi* (Baird, 1859) (Crustacea, Branchiopoda, Conchostraca, Spinicaudata). *Acta Zool.* 80: 163-184.

Olesen, J. 2000. An updated phylogeny of the Conchostraca-Cladocera clade (Branchiopoda, Diplostraca). *Crustaceana* 73: 869-886.

Olesen, J. 2001. External morphology and larval development of *Derocheilocaris remanei* Delamare-Deboutteville & Chappuis, 1951 (Crustacea, Mystacocarida), with a comparison of crustacean segmentation and tagmosis patterns. *Kong. Danske Vidensk. Selskab. Biol. Skrift.* 53: 1-59.

Olesen, J. 2002. Branchiopod phylogeny - continued support for higher taxa like the Diplostraca and Cladocera and paraphyly of 'Conchostraca' and 'Spinicaudata'. *Crustaceana* 75: 77-84.

Olesen, J. & Grygier, M. J. 2003. Larval development of Japanese 'conchostracans' : part 1, larval development of *Eulimnadia braueriana* (Crustacea, Branchiopoda, Spinicaudata, Limnadiidae) compared to that of other limnadiids. *Acta Zool.* 84: 41-61.

Olesen, J. & Walossek, D. 2000. Limb ontogeny and trunk segmentation in *Nebalia* species. *Zoomorphology* 120: 47-64.

Olesen, J., Martin, J. W. & Roessler, E. W. 1997. External morphology of the male of *Cyclestheria hislopi* (Baird, 1859) (Crustacea, Branchiopoda, Spinicaudata), with comparison of male claspers among the Conchostraca and Cladocera and its bearing on phylogeny of the 'bivalved' Branchiopoda. *Zool. Scripta* 25: 291-316.

Olesen, J., Richter, S. & Scholtz, G. 2001. The evolutionary transformation of phyllopodous to stenopodous limbs in the Branchiopoda (Crustacea) - Is there a common mechanism for early limb development in arthropods? *Int. J. Dev. Biol.* 45: 869-876.

Olesen, J., Richter, S. & Scholtz, G. 2003. On the ontogeny of *Leptodora kindtii* (Crustacea, Branchiopoda, Cladocera), with notes on the phylogeny of the Cladocera. *J. Morph.* 256: 235-259.

Petrov, B. 1992. Larval development of *Leptestheria saetosa* Marinek & Petrov, 1992 (Leptesheriidae, Conchostraca, Crustacea). *Arc. Biol. Sci. Belgrade* 44: 229-241.

Piatakov, M. L. 1925. Zur Embryonalentwicklung von *Lepidurus apus* und *Triops cancriformis. Zool. Anz.* 62: 234-236.

Rainbow, P. S. & Walker, G. 1976. The feeding apparatus of the barnacle nauplius larva: a scanning electron microscope study. *J. Mar. Biol. Ass. UK* 56: 321-326.

Roessler, E. W. 1995. Review of Colombian Conchostraca (Crustacea) - ecological aspects and life cycles - family Cyclestheriidae. *Hydrobiologia* 298: 113-124.

Samassa, P. 1893. Die Keimblätterbildung bei den Cladoceren. II. *Daphnelle* und *Daphnia. Arch. Mikrosk. Anat.* 41: 650-688.

Samter, M. 1895. Die Veränderung der Form und Lage der Schale von *Leptodora hyalina* Lillj. während der Entwicklung. *Zool. Anz.* 483; 484: 334-338; 341-344.

Samter, M. 1900. Studien zur Entwicklungsgeschichte der *Leptodora hyalina* Lillj. *Zeitschr. wissensch. Zool.* 68: 169-260.

Sanders, H. L. 1963. The Cephalocarida. Functional Morphology, Larval Development, Comparative External Anatomy. *Memoirs of the Connecticut Academy of Arts and Sciences* 15:1-80.

Sars, G. O. 1873. Om en dimorph udvikling samt generationsvexel hos *Leptodora. Forhandlinger i Videnskabsselskabet i Kristiania*: 1-15.

Sars, G. O. 1885. On some Australian Cladocera, raised from dried mud. *Forhandlinger i Videnskabsselskabet i Kristiania* 8: 1-46.

Sars, G. O. 1887. On *Cyclestheria hislopi* (Baird), a new generic type of bivalve Phyllopoda; raised from dried Australian mud. *Forhandlinger i Videnskabsselskabet i Kristiania* 1: 223-239.

Sars, G. O. 1896a. Development of *Estheria packardi* as shown by artificial hatching from dried mud. *Archiv for Mathematik og Naturvidenskab B* 18: 1-27.

Sars, G. O. 1896b. *Fauna Norvegiae. I. Descriptions of the Norwegian species at present known belonging to the suborders Phyllocarida and Phyllopoda* Vol. 1. Aschehoug, Christania: Joint-Stock Printing Company.

Sars, G. O. 1898. On the propagation and early development of Euphausiidæ. *Archiv for Mathematik og Naturvidenskab* 20: 1-41, pls. 1-4.

Schaeffer, J. C. 1756. *Der krebsartige Kiefenfuss mit der kurzen und langen Schwanzklappe.* Regensburg.

Schlögl, T. 1996. Die postembryonale Entwicklung der Männchen des Feenkrebses *Branchipus schaefferi. Stapfia* 42: 137-148.

Schminke, H. K. 1981. Adaptation of Bathynellacea (Crustacea, Syncarida) to life in the Interstitial ("Zoea Theory"). *Intern. Revue gesamt. Hydrobiol.* 66: 575-637.

Scholtz, G. 2000. Evolution of the nauplius stage in malacostracan crustaceans. *J. Zool. Syst. Evol. Res.* 38: 175-187.

Schram, F. R. 1986. *Crustacea.* New York: Oxford University Press.

Schram, F. R. & Koenemann, S. 2001. Developmental genetics and arthropod evolution: part 1, on legs. *Evol. Dev.* 3: 343-354.

Schrehardt, A. 1987. A scanning electron-microscope study of the post-embryonic development of *Artemia.* In: Sorgeloos, P., Bengtson, D.A., Declair, W. & Jasper, E. (eds), Artemia *Research and its applications. 1. Morphology, Genetics, Strain Characterization, Toxicology*: 5-32. Wetteren: Universa Press.

Scourfield, D. J. 1926. On a new type of crustacean from the Old Red Sandstone (Rhynie Chert Bed, Aberdeenshire) - *Lepidocaris rhyniensis* gen. et sp. nov. *Phil. Trans. R. Soc. B* 214: 153-187.

Scourfield, D. J. 1940. Two new and nearly complete specimens of young stages of the Devonian fossil crustacean *Lepidocaris rhyniensis. Zool. J. Linn. Soc.* 152: 290-298.

Shan, R. K. 1969. Life cycle of a Chydorid Cladoceran, *Pleuroxus denticulatus* Birge. *Hydrobiologia* 34: 513-523.

Shiga, Y., Yasumoto, R., Yamagata, H. & Hayashi, S. 2002. Evolving role of Antennapedia protein in arthropod limb patterning. *Development* 129: 3555-3561.

Shuba, T. & Costa, R. R. 1972. Development and growth of *Ceriodaphnia reticulata* embryos. *Trans. American Microsc. Soc.* 91: 429-435.

Snodgrass, R. E. 1956. Crustacean Metamorphoses. *Smiths. Misc. Coll.* 131: 1-78.

Solowiow, P. 1927. Zur Biologie von *Limnetis brachyura* Müll. (Euphyllopoda, Crustacea). *Zool. Anz.* 74: 151-157.

Spears, T. & Abele, L. G. 2000. Branchiopod monophyly and interordinal phylogeny inferred from 18S ribosomal DNA. *J. Crust. Biol.* 20: 1-24.

Strenth, N. E. & Sissom, S. L. 1975. A morphological study of the post-embryonic stages of *Eulimnadia texana* Packard (Conchostraca, Crustacea). *Texas Journal of Science.* 26: 137-154.

Sudler, M. T. 1899. The development of *Penilia schmaeckeri* Richard. *Proc. Boston Soc. Nat. Hist.* 29: 107-133.

Taylor, D. J., Crease, T. J. & Brown, W. M. 1999. Phylogenetic evidence for a single long-lived clade of crustacean cyclic parthenogens and its implications for the evolution of sex. *Proc. R. Soc. Lond. B* 266: 791-797.

Wagner, N. 1868. *Hyalosoma dux,* a new form of the group Daphnida (Crustacea Cladocera). Arbeiten der erste Sitzung russ. Naturforscher in St. Petersburg 1868. *Trudy I.S.R.E. Otd Zoologii*: 218-239.

Walossek, D. 1993. The Upper Cambrian *Rehbachiella* and the phylogeny of Branchiopoda and Crustacea. *Fossils Strata* 32: 1-202.

Walossek, D. 1995. The Upper Cambrian *Rehbachiella,* its larval development, morphology and significance for the phylogeny of Branchiopoda and Crustacea. *Hydrobiologia* 298: 1-13.

Warren, E. 1901. A preliminary account of the development of the free-living nauplius of *Leptodora hyalina* (Lillj). *Proc. R. Soc. London B* 68: 210-218.

Weismann, A. 1876. Zur Naturgeschichte der Daphniden. I. Über die Bildung von Wintereiern bei *Leptodora hyalina. Zeitschr. Wissensch. Zool.* 27: 51-112.

Williams, T.A. 2003. The evolution and development of crustacean limbs: an analysis of limb homologies. In: Scholtz, G. (ed), *Crustacean Issues 15, Evolutionary Developmental Biology of Crustacea*: 169-193. Lisse: Balkema.

Williams, T. A. & Müller, G. B. 1996. Limb development in a primitive crustacean, *Triops longicaudatus*: subdivision of the early limb bud gives rise to multibranched limbs. *Dev. Genes Evol.* 206: 161-168.

Williams, T., Nulsen, C. & Nagy, L. M. 2001. A complex role for Distal-less in Crustacean appendage development. *Dev. Biol.* 241: 302-312.

Wingstrand, K. G. 1978. Comparative Spermatology of the Crustacea Entomostraca. 1. Subclass Branchiopoda. *Kong. Danske Vidensk. Selskab. Biol. Skrift.* 22: 1-67.

Wotzel, R. 1937. Zur Entwicklung des Sommereies von *Daphnia pulex. Zool. Jahrb. Anat.* 63: 455-470.

Zaddach, E. G. 1841. *De apodis cancriformis Schaeff. anatome et historia evolutionis. Dissertatio inauguralis zootomica.* Bonnae.

Zaffagnini, F. 1971. Alcune precisazioni sullo sviluppo post-embrionale del concostraco *Limnadia lenticularis* (L.). *Mem. dell'Istit. Ital. Idrobiol.* 27: 45-60.

Contributors

Abzhanov, Arhat: Harvard Medical School, Department of Genetics, 200 Longwood Avenue, Boston MA 02115, USA; Email: aabzhano@rascal.med.harvard.edu

Deutsch, Jean S.: Université P. et M. Curie, 9 quai St-Bernard, case 241, 75252 Paris Cedex 5, France ; Email: jean.deutsch@snv.jussieu.fr

Dohle, Wolfgang: Freie Universität Berlin, Institut für Biologie, Zoologie, Königin-Luise-Str. 1-3, 14195 Berlin, Germany; Email: wdohle@mail.zedat.fu-berlin.de

Gerberding, Matthias: University of California Berkeley, 142 Life Sciences Addition #3200, Berkeley CA 94720-3200, USA; Email: mgerberd@uclink.berkeley.edu

Gibert, Jean-Michel: Department of Zoology, Downing Street, Cambridge, CB2 3EJ, United Kingdom; Email: jmg59@cam.ac.uk

Glenner, Henrik: Zoological Institute, University of Copenhagen, Universitetsparken 15, DK-2100, Copenhagen, Denmark; Email: hglenner@zi.ku.dk

Hejnol, Andreas: Technische Universität Braunschweig, Institut für Genetik, Spielmannstr. 7, 38106 Braunschweig, Germany; Email: a.hejnol@tu-bs.de

Høeg, Jens T.: Zoological Institute, University of Copenhagen, Universitetsparken 15, DK-2100, Copenhagen, Denmark; Email: jthoeg@zi.ku.dk

Kaufman, Thomas C.: Howard Hughes Medical Institute, Department of Biology, Indiana University, Bloomington IN 47405, USA; Email: kaufman@bio.indiana.edu

Koenemann, Stefan: Institute for Biodiversity and Ecosystem Dynamics, University of Amsterdam, Postbox 94766, NL-1090 GT Amsterdam, The Netherlands; Email: koenemann@science.uva.nl

Lagersson, Niklas C.: Zoological Institute, University of Copenhagen, Universitetsparken 15, DK-2100, Copenhagen, Denmark; Email: nclagersson@zi.ku.dk

Mouchel-Vielh, Emmanuèle: Université P. et M. Curie, 9 quai St-Bernard, case 241, 75252 Paris Cedex 5, France; Email: emmanuele.mouchel-vielh@snv.jussieu.fr

Olesen, Jørgen: Zoological Museum, University of Copenhagen, Universitetsparken 15, DK-2100 Copenhagen OE, Denmark; Email: jolesen@zmuc.ku.dk

Quéinnec, Éric: Université P. et M. Curie, 9 quai St-Bernard, case 241, 75252 Paris Cedex 5, France ; Email: eric.queinnec@snv.jussieu.fr

Scholtz, Gerhard: Humboldt-Universität zu Berlin, Institut für Biologie/Vergleichende Zoologie, Philippstr. 13, 10115 Berlin, Germany; Email: gerhard.scholtz@rz.hu-berlin.de

Schram, Frederick R.: Institute for Biodiversity and Ecosystem Dynamics, University of Amsterdam, Postbox 94766, NL-1090 GT Amsterdam, The Netherlands; Email: schram@science.uva.nl

Whitington, Paul M.: Department of Anatomy and Cell Biology, University of Melbourne, Victoria, Australia 3010; Email: P.Whitington@Anatomy.unimelb.edu.au

Williams, Terri A.: Department of Ecology and Evolutionary Biology, Yale University, P. O. Box 208106, 165 Prospect St., New Haven CT 06520, USA; Email: terri.williams@yale.edu

Index

P

Pacifastacus leniusculus · 177, 180
pagurid · 99
paguroids · 8
Pagurus
    bernhardus · 148
    kennerlyi · 188
pair rule · 119
    segmentation gene · 9
Palaemonetes · 171
    argentinus · 115, 138, 143, 144, 155, 156, 158, 159, 160
palp · 54, 57, 179, 223, 224, 227, 229, 234, 241, 245, 251, 254, 260
Pan-Arthropoda · 35
Pan-Branchiopoda · 218, 223, 260
Pancarida · 99, 105
Pan-Crustacea · 86, 89
Panulirus argus · 148, 157
paragnaths · 55
para-Hox genes · 34, 35, 37
paralogs · 112, 119
parasegment · 32, 36, 88, 118, 123
parasite · 21-23, 27, 198, 203, 209, 211
Parastacoidea · 99
parsimony · 6, 7, 258, 259
parthenogenetic eggs · 263
pathway · 95, 99, 154, 155, 156
pattern
    cell division · 8, 10, 11, 95, 115, 116, 119, 123, 126, 142
    expression · 3, 7, 9, 27, 31, 32, 34-37, 45, 47, 51, 52-54, 56, 58, 60, 62-69, 75, 79, 80, 83, 85, 89, 96, 119, 120, 122-124, 126, 136
    grid-like · 105, 107, 108, 146
    of neuroblasts · 114, 142, 143, 145
    tagmatic · 46
    underlying · 88

patterning · 45, 51, 56, 57, 67-69, 119, 123, 175, 178-181, 184, 185, 188
Pax-6 · 10, 12
pCC neuron · 122, 154, 155
PCR · 35, 68
pedipalp · 52-54, 77
peduncle · 20, 26
pedunculate
    barnacles · 20, 26, 28, 30
    thoracicans · 26, 28
pelagic instar · 197
penaeid decapod · 124
penial structures · 228
Penilia schmackeri · 246
penis · 20, 27, 28, 209, 211
pentastomids · 89
Peracarida · 67, 79, 101, 103, 104, 109, 116, 138, 142, 149, 153, 154
pereion · 44, 47, 49, 50, 59, 61, 63, 64, 66, 69
pereionic
    leg · 63, 64, 67
    limb · 49, 63, 66
peripheral
    nerve · 154-156
    nervous system · 181
    sensory structures · 182
phenotype · 11
photoreceptors · 161
Phyllocarida · 138, 149
phyllopod · 170, 171, 187, 227
Phyllopoda · 218, 255, 256, 262
phyllopodous limbs · 9, 126, 173, 178, 182, 188, 256
phyllopods · 171, 172, 174, 187, 188
phylogenetic
    analyses · 7, 9, 10, 38, 44, 218, 263
    systematics · 5, 7, 96
phylogeny · 7-9, 23, 24, 32, 43-45, 61, 76, 96, 102, 169, 172, 183, 198, 210, 211, 217, 219, 220, 248, 250, 251, 256, 258, 260, 263
phylotypic stage · 5, 12
phylum · 3-7, 86, 198
pioneer neurons · 8
pioneering axon growth · 154
Pleocyemata · 67, 99, 138

pleon · 8, 37, 44, 47, 49, 50, 59, 61-67, 69, 79, 83, 84, 86, 88, 137, 144
pleonic appendage · 62, 66
pleopod · 51, 66, 68, 82, 125, 126, 170-174, 183, 184, 186, 187
plesiomorphic character · 7, 113
pleurobranch · 173, 174
podocopid · 86
podomeres · 56, 65, 66, 222
podomerous (pdm) · 119, 177, 180
Pollicipes pollicipes · 149
Polyartemia · 86
    forcipata · 227
Polyartemiella · 86
Polyphemus pediculus · 220, 248, 250
polyphyletic origin · 99
polytomy · 219, 262
Porcellio · 47, 50, 54, 57, 59, 62-64, 108, 111, 113, 120, 122, 181
    laevis · 145, 149-151, 153
    scaber · 34, 37, 38, 44, 46, 49, 51-53, 55, 56, 61, 65-67, 69, 76, 83, 107, 119, 125, 150, 153-155, 177
positional homology · 75, 80, 84
postaxial seta · 206, 207
post-cephalic segments · 60, 199
posterior
    commissure · 137, 146, 149, 151, 153, 155
    head · 53, 65
    Hox genes · 59, 79, 89
post-esophageal commissure · 153
post-gnathal trunk · 66
post-larval stage · 157, 159, 240
post-naupliar
    region · 126, 136, 138
    segments · 110, 113, 120, 147
    stage · 221, 222
preantennal segment · 77, 123
precursor cells · 96, 99, 103, 113, 139, 148
precypris · 26

CRUSTACEAN ISSUES
ISSN 0168-6356

1 Crustacean phylogeny
 Schram, Frederick R. (ed)
  1983 ISBN 90 6191 231 8 Sold out

2 Crustacean growth: Larval growth
 Wenner, Adrian (ed)
  1985 ISBN 90 6191 294 6

3 Crustacean growth: Factors in adult growth
 Wenner, Adrian (ed)
  1985 ISBN 90 6191 535 X

4 Crustacean biogeography
 Gore, Robert H. / Heck, Kenneth L. (eds)
  1986 ISBN 90 6191 593 7

5 Barnacle biology
 Southward, Alan J. (ed)
  1987 ISBN 90 6191 628 3

6 Functional morphology of feeding and grooming in Crustacea
 Felgenhauer, B.E. / Thistle, A.B. / Watling, L. (eds)
  1989 ISBN 90 6191 777 8

7 Crustacean egg production
 Wenner, Adrian / Kuris, Armand (eds)
  1991 ISBN 90 6191 098 6

8 History of carcinology
 Truesdale, F.M. (ed)
  1993 ISBN 90 5410 137 7

9 Terrestrial Isopod Biology
 Alikhan, A.M.
  1995 ISBN 90 5410 193 8

10 New Frontiers in Barnacle Evolution
 Schram, Frederick R. / Hoeg, J.T. (eds)
  1995 ISBN 90 5410 626 3

11 Crayfish in Europe as alien species - How to make the best of a bad situation?
 Gherardi, Francesca / Holdich, David M. (eds)
  1999 ISBN 90 5410 469 4

12   The biodiversity crisis and crustacea - Proceedings of the fourth international
        crustacean congress, Amsterdam, Netherlands
        Vaupel Klein, J. Carel von / Schram, Frederick R. (eds)
            2000   ISBN 90 5410 478 3

13   Isopod systematics and evolution
        Kensley, Brian / Brusca, Richard C. (eds)
            2001   ISBN 90 5809 327 1

14   The Biology of Decapod Crustacean Larvae
        Anger, K.
            2001   ISBN 90 2651 828 5

15   Evolutionary Developmental Biology of Crustacea
        Gerhard Scholtz (ed)
            2004   ISBN 90 5809 637 8

T - #0039 - 101024 - C6 - 254/178/17 [19] - CB - 9789058096371 - Gloss Lamination